XXXI Conference on Electrical and Electronic Engineering

XXXI Conference on Electrical and Electronic Engineering

Editors

Soraya Lucia Sinche Maita
Fernando Carrera Suarez
Pablo Lupera
Jackeline Abad
Jaime Cepeda

MDPI

Basel • Beijing • Wuhan • Barcelona • Belgrade • Novi Sad • Cluj • Manchester

Editors

Soraya Lucia Sinche Maita
Escuela Politécnica Nacional
Quito
Ecuador

Fernando Carrera Suarez
Escuela Politécnica Nacional
Quito
Ecuador

Pablo Lupera
National Polytechnic School
Quito
Ecuador

Jackeline Abad
Escuela Politécnica Nacional
Quito
Ecuador

Jaime Cepeda
Escuela Politécnica Nacional
Quito
Ecuador

Editorial Office
MDPI
St. Alban-Anlage 66
4052 Basel, Switzerland

This is a reprint of articles from the Proceedings published online in the open access journal *Engineering Proceedings* (ISSN 2673-4591) (available at: https://www.mdpi.com/2673-4591/47/1).

For citation purposes, cite each article independently as indicated on the article page online and as indicated below:

Lastname, A.A.; Lastname, B.B. Article Title. *Journal Name* **Year**, *Volume Number*, Page Range.

ISBN 978-3-7258-0371-2 (Hbk)
ISBN 978-3-7258-0372-9 (PDF)
doi.org/10.3390/books978-3-7258-0372-9

Contents

About the Editors

Soraya Lucia Sinche Maita

Soraya Lucia Sinche Maita received the Engineer degree of Electronic and Telecommunications (1999) and the master's degree of Connectivity in Telecommunications Networks (2016) from the Escuela Politécnica Nacional, Ecuador; and the master degree of Wireless Systems and Relative Technologies from Polytechnic of Turin, Italy, in 2004. She received his PhD in In-formation Science and Technology in January 2020 from the University of Coimbra, Portugal. Her current research interests include Internet of Things, wireless communications, wireless sensor networks and mobile phone sensing. Since 2000, she has been Lecture at the Department of Electronic, Telecommunications and Networks of Escuela Politécnica Nacional, Quito, Ecuador. She is a Member of IEEE since 2004 and Senior Member of IEEE since 2015.

Fernando Carrera Suarez

Fernando Carrera Suarez received the B.S. degree from Escuela Politécnica Nacional (2001), Ecuador, the M.S. degree in Technologies, Systems and Communications Networks from Universitat Politècnica de València (2010), and Ph.D. degree in Electrical Engineering from Universitat Politècnica de València (2015). He was recipient of grants from Ecuadorian Government CEREPS 2002 and SENESCYT 20010 for postgraduate studies. Dr. Carrera is currently a lecturer with the Departamento de Electrónica, Telecomunicaciones y redes de Información, Facultad de Ingeniería Eléctrica y Electrónica, Escuela Politécnica Nacional, Quito. His current research interests include RF passive and active circuit design, antenna design, multibeam antennas, reconfigurable antennas at millimiter frequencies.

Pablo Lupera

Pablo Lupera received his B.S and Ph.D. degrees in electronic and telecommunications from The National Polytechnic School, Quito, Ecuador and Bonch-Bruevich Saint Petersburg State University of Telecommunications, Saint Petersburg, Russia, in 2002 and 2009, respectively. Dr. Lupera is now a lecturer at National Polytechnic School of Quito, Ecuador. His current research interests include physical layer signal processing in wireless networks.

Jackeline Abad

Jackeline Abad received the B.S. degree from Escuela Politécnica Nacional (2008), Ecuador, and the M.S. and Ph.D. degree in Electrical Engineering from Washington State University (2012, 2014). She held a Fulbright grant in 2010. She is currently an Associate Professor with the Departamento de Automatización y Control Industrial, Facultad de Ingeniería Eléctrica y Electrónica, Escuela Politécnica Nacional, Quito. Her current research interests include structural analysis and controller design of dynamical networks with applications to sensor/vehicle networking, epidemic control, and power systems network control.

Jaime Cepeda

Jaime Cepeda received the Electrical Engineering degree from Escuela Politécnica Nacional in 2005, the PhD degree in Electrical Engineering from National University of San Juan, Argentina in 2013 and the Master's degree in Big Data from the European University Miguel de Cervantes, Spain in 2021. In 2013, he served as a WAMS specialist at CENACE. From 2014 to 2021 he was the Head of Research and Development and the General Manager of Technical Development at the

Ecuadorian Independent System Operator ISO CENACE, while from 2021 to 2022 he held the position of Executive Director of the Agency for the Regulation and Control of Energy and Non-Renewable Natural Resources. Currently, he works as a University Professor of Master's and PhD Programs at Escuela Politécnica Nacional and other Universities, and as a consultant on issues related to smart grids, power system operations, power system security and vulnerability assessment, power system modeling and control and application of data science to power systems.

engineering proceedings

Editorial

Preface of the XXXI Conference on Electrical and Electronic Engineering †

Soraya Sinche [1,*], Fernando Carrera [1], Pablo Lupera [1], Jackeline Abad [2] and Jaime Cepeda [3]

1 Departamento de Electrónica Telecomunicaciones y Redes de Información, Escuela Politécnica Nacional, Quito 170525, Ecuador; fernando.carrera@epn.edu.ec (F.C.); pablo.lupera@epn.edu.ec (P.L.)
2 Departamento de Automatización y Control Industrial, Escuela Politécnica Nacional, Quito 170525, Ecuador; jackeline.abad@epn.edu.ec
3 Departamento de Energía Eléctrica, Escuela Politécnica Nacional, Quito 170525, Ecuador; jaime.cepeda@epn.edu.ec
* Correspondence: soraya.sinche@epn.edu.ec
† All proceeding papers published in the volume are presented at the XXXI Conference on Electrical and Electronic Engineering, Quito, Ecuador, 29 November–1 December 2023.

Citation: Sinche, S.; Carrera, F.; Lupera, P.; Abad, J.; Cepeda, J. Preface of the XXXI Conference on Electrical and Electronic Engineering. *Eng. Proc.* **2023**, *47*, 25. https://doi.org/10.3390/engproc2023047025

Published: 5 February 2024

The XXXI Electrical and Electronic Engineering Conference (XXXI JIEE-2023) is an annual event organized by the Faculty of Electrical and Electronic Engineering at the Escuela Politécnica Nacional, Quito, Ecuador. This conference brings together students, professors, researchers, and professionals with the aim of disseminating their graduation projects, curricular integration initiatives, research project advancements, and engagement activities, as well as presenting academic papers.

The JIEE serves as an opportunity to promote the exchange of knowledge and experience in various fields of expertise, establish professional connections, stay updated on the latest technologies, and acquire skills to tackle the challenges presented by industry.

With more than 180 participants, the XXXI JIEE 2023 included 6 international keynote speakers, 24 oral session talks on technical papers, and 21 oral session talks on student works.

The following six tutorials were provided at the XXXI JIEE 2023:

- Mariano Baquero Escudero (Universidad Politécnica de Cataluña, Spain)—"Reconfigurable antennas for broadband communications at millimeter frequencies".
- Luigi Vanfretti (Rensselaer Polytechnic Institute (RPI), USA)—"Using Probing Input Signals for Enhanced Power Grid Monitoring and Control".
- Jorge Sá Silva (University of Coimbra, Portugal)—"Why do we need Humans in the Future Internet?"
- Marcelo Molina (University of San Juan (UNSJ) and researcher on the National Scientific and Technical Research Council (CONICET), Argentina)—"Energy storage technologies and their decisive role in accelerating the Energy Transition".
- Sean Charles Warnick (Brigham Young University, USA)—"IDeA Labs: An Educational Model Interweaving Computational Thinking Throughout the University".
- Boris Bellalta (Universitat Pompeu Fabra, Spain)—"Next-generation Wi-Fi networks: From Extremely High Throughput to Ultra High Reliability".

The XXXI JIEE-2023 was a space for collaboration between the academic community and the energy, telecommunications, and automation sectors in which the EPN showed the technological advances and challenges that the community tackled in several research and final degree projects relating to telecommunications, electronics, information technologies and electrical engineering. Among the main topics addressed in the conference were (i) smart grids, (ii) the modeling and simulation of power systems, (iii) the deployment of novel technologies for planning and operating power systems, (iv) renewable energy and storage, (v) automation, (vi) robotics, (vii) power electronics, (viii) control, (ix) radiofrequency material characterization, (x) signal processing in radar applications and (xi) 5G network performance evaluations.

The XXXI JIEE-2023 aimed to contribute to the sustainable development of the Ecuadorian energy, telecommunications, and production sectors, including the electricity sector in its generation, transmission, and distribution stages and the control, automation, and security of essential infrastructure in the country's production. In this sense, the XXXI JIEE-2023 provided space to disseminate national and international research and development aimed at guaranteeing the future availability of energy and telecommunications services and improving the production chain in the country while promoting a lower impact on the environment.

In submitting conference proceedings to *Engineering Proceedings*, the volume editors of the proceedings certified to the publisher that all papers published in this volume have been subjected to a peer review administered by the volume editors. Reviews were conducted by expert referees to the professional and scientific standards expected of a proceedings journal.

- Types of peer review: single-blind; double-blind; triple-blind; open; and other (please describe): single-blind.
- Conference submission management system: SuSy (https://susy.mdpi.com/user/myprofile).
- Number of submissions sent for review: 29.
- Number of submissions accepted: 24.
- Acceptance rate (number of submissions accepted/number of submissions received): 82.75%.
- Average number of reviews per paper: 1.
- Total number of reviewers involved: 34.
- Any additional information on the review process is provided below.

Peer-review process and regulations: The review process was conducted in a single-blind manner, with the entire procedure administered through the MDPI platform. Evaluations were performed by reviewers from various universities and countries. Based on their feedback, revisions were made to each article, encompassing thorough examinations of the structure, content and writing.

Author Contributions: Conceptualization, S.S. and J.C.; validation, F.C., P.L., J.A. and J.C.; writing—original draft preparation, S.S.; writing—review and editing, J.A, J.C., F.C. and P.L. All authors have read and agreed to the published version of the manuscript.

Conflicts of Interest: The authors declare no conflict of interest.

engineering proceedings

MDPI

Proceeding Paper

Design and Implementation of a Robotic Arm Prototype for a Streamlined Small Chocolate Packaging Process †

Esteban Andrade, Gustavo Cerecerez, Melyssa Garzòn and Angèlica Quito *

Mechatronic School, Technical Science Faculty, International University of Ecuador UIDE, Quito 170411, Ecuador; esandradeco@uide.edu.ec (E.A.); gucerecerezji@uide.edu.ec (G.C.); kagarzonzu@uide.edu.ec (M.G.)

* Correspondence: anquitoca@uide.edu.ec; Tel.: +593-0999187809

† Presented at the XXXI Conference on Electrical and Electronic Engineering, Quito, Ecuador, 29 November–1 December 2023.

Abstract: This report presents the development of a robotic arm for the efficient packaging of chocolates in a small-scale process. The robotic arm has a PRR (prismatic–rotational–rotational) configuration and uses a universal vacuum grip (UVG) to manipulate the chocolates with precision. Various materials and components were used, such as 3D-printed PLA parts for the mechanical elements and flexible TPU bands for optimal movement control. In terms of electronics, NEMA motors, an Arduino board, and a CNC interface were used for the precise control of the motors. The programming was conducted in Python and a graphical user interface (GUI) was created to manage the sequence of movements of the robotic arm. Key parameters, such as the accuracy in the placement of the chocolates, the grip strength, and the shape recovery ability were evaluated. The results demonstrate the successful integration of the robotic arm in the packaging process, achieving an efficient and consistent placement of the chocolates in the plastic trays. This project highlights the importance and potential of automation in the food packaging industry, by improving productivity, reducing human error, and ensuring packaging quality. The knowledge and results obtained in this project contribute to the field of robotics.

Keywords: PRR robot; UVG; chocolate packaging; process automation

Citation: Andrade, E.; Cerecerez, G.; Garzòn, M.; Quito, A. Design and Implementation of a Robotic Arm Prototype for a Streamlined Small Chocolate Packaging Process. *Eng. Proc.* **2023**, *47*, 1. https://doi.org/10.3390/engproc2023047001

Academic Editor: Jackeline Abad

Published: 26 September 2023

1. Introduction

This research focuses on developing and applying a precision robotic arm (PRR), specifically designed to improve the chocolate packaging process by reducing human errors and accelerating cycle times. The main objective is to increase productivity by freeing up human resources for higher-value tasks.

The uniqueness of this study lies in its ability to transform the way chocolate packaging is carried out, resulting in more efficient production and a more uniform distribution of chocolates in plastic trays, thus improving the quality of the final product.

However, it is important to note that this project faces significant challenges that require meticulous attention to both control and maintenance by specialized personnel. The creation of the PRR robotic arm involves considerable demands in terms of financial investment and technical resources, as well as in-depth engineering knowledge. Furthermore, adapting the PRR to the specific needs of different chocolate production lines can be a complex process. Finally, the ongoing costs associated with the maintenance and operation of the PRR, which require qualified personnel, are key aspects to consider in this engineering project.

2. Estate of Art

Automatic food packaging plays a fundamental role in the food industry, as it contributes significantly to improving efficiency and precision, and reducing possible errors.

This is achieved by eliminating repetitive tasks typically performed by human workers, which, in the long term, can slow down the process and lead to errors. As delineated in the study by Smith and Jones (2018) [1], the integration of automation into food packaging processes has demonstrated its efficacy in boosting productivity, curtailing cycle times, and safeguarding the final product's quality. This is facilitated through the adept deployment of robotic systems that are proficient in swift and precise grasping and placing tasks.

The landscape of robotic arm technologies has witnessed manifold strides and innovations, each tailored to adapt these capabilities to the realm of food packaging. H. Lee's work (2017) [2], for instance, introduces a robotic arm with a PRR configuration, offering heightened dexterity and accuracy in food manipulation.

This project builds upon previous efforts by addressing the inherent challenges in automated packaging and leveraging a PRR robotic arm. However, its distinctiveness emerges from the incorporation of a universal vacuum gripper (UVG) bolstered by a mini-pump and balloon, as evidenced in the investigation conducted by Tong Chen (2020) [3]. This pioneering approach facilitates the secure and efficient manipulation of chocolates during the packaging process, adeptly adapting to diverse product shapes and dimensions.

The selection of NEMA 17 motors for this project is rooted in their extensive applicability and well-established advantages within the domain of robotics [4]. NEMA 17 motors, renowned for their compact form factor, superior precision, and suitable torque output for effective robotic arm motion, bear a nominal current capacity of up to 2.5 A, furnishing the requisite force for seamless and accurate joint and gripper movements. Additionally, these motors boast a standard resolution of 200 steps per revolution, which ensures an acceptable degree of positional accuracy when controlling robotic arm components [5,6].

3. Methodology

The decision to opt for the UVG was underpinned by its capacity to ensure the secure handling of chocolates, thus averting any potential damage during the manipulation process [7]. Moreover, its straightforward design and user-friendly nature facilitated a seamless integration into our robotic system. This choice harmonizes seamlessly with our overarching aim of attaining impeccable, high-caliber packaging, thereby safeguarding the chocolates' integrity and fostering a gratifying consumer experience [8].

To avoid deformations in the PLA parts that make up a large part of the robot, it is essential to keep it at a temperature below 50 °C [9], especially if it is exposed to the sun and high temperatures for long periods [10].

The programming defines how the robot operates, including its ability to pick and place chocolates accurately and reliably. The code ensures smooth task execution with the help of the user interface, which provides a user-friendly means to operate the robot manually or set up automated tasks, making it accessible to a wide range of users. To carry out the pick & place task for the chocolates, we selected the trajectory that required the fewest possible movements and accomplished the task in the shortest amount of time [11].

The primary parameter taken into consideration during programming was the activation and deactivation time of the vacuum gripper when picking up and releasing the chocolate. After several tests, it was determined that the optimal time for ensuring good synchronization between movements was 5 s. A time shorter than this did not guarantee a proper grip of the chocolate by the gripper.

3.1. Mechanical Design

The connection system entailed steel rods and an endless screw for the prismatic link (Link 1). This configuration facilitated seamless linear movement, which is pivotal for the precise functioning of the robotic arm during grasping and displacement.

Using the CAD design software SolidWorks, a 3D model of the assembly of the robotic arm was created. The CAE simulation software Ansys Workbench was used for obtaining a generative design. The components of the robot that can respond to the movements required by the robot were obtained. Next, we present a summary regarding the design of the gears that allowed the force transmission and movement execution relative to the load weight. This information is crucial when selecting the appropriate motors for the robotic arm design, as can see in the Table 1.

Table 1. Mechanical Data

Pulleys	Gear Module	Primitive Diameter	Internal Diameter	Circular Pitch
and	$M = Ed/N + 2$	$Pd = M * N$	$Id = Pd - (2 * M)$	$P = \pi * M$
Gears	0.625	38.75 mm	37.5 mm	1.96 mm

3.2. Gripper

The chocolate used in the project had a spherical, irregular shape. Its approximate diameter varied between 3 and 4 cm, which gave it a compact and round appearance.

The determination of the grip force was also a significant challenge. A balance had to be found between the pressure exerted by the gripper's fingers to hold the chocolate without damaging it. Excessive force could have deformed or broken the chocolate, and insufficient force did not ensure a secure and reliable grip.

Finally, the inclusion of force sensors to measure the pressure applied to the chocolate added complexity to the claw design. These sensors had to be precisely integrated into the gripper's fingers to obtain accurate grip force measurements. The addition of sensors also required additional considerations in terms of wiring and calibration.

The vacuum grip system consisted of using the vacuum principle to achieve a safe and stable grip on the chocolate. The clamp used a bell with a line of silicone or other material that conformed to the chocolate, as can be seen in the blue highlighted area in Figure 1, creating a tight seal around it. By generating a vacuum in the space between the bell and the chocolate, a gripping force was achieved that held the chocolate in place without the need to apply excessive pressure. The goal was to achieve a gripping system that was more efficient in terms of time and design compared to the three-fingered claw as can be seen in Table 2.

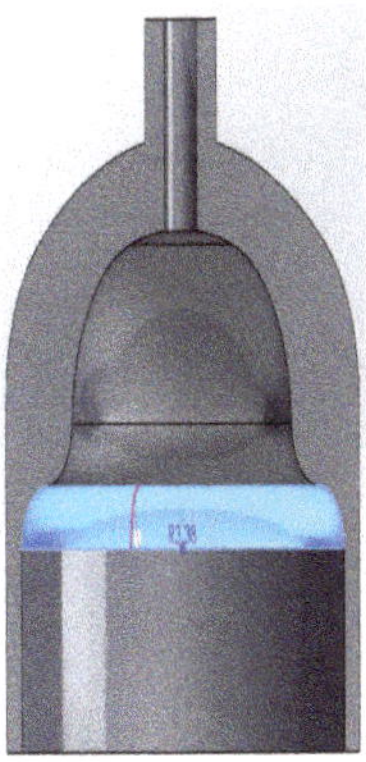

Figure 1. Design of the bell.

This system allowed faster handling, and by not applying direct mechanical force to the chocolate, the risk of damaging it during the gripping process was minimized, as can be seen in Figure 2.

The UVG (universal vacuum gripper) was identified as a superior method to the second gripper, as it was highly effective in gripping delicate objects. It was specifically adapted and optimized for Ferrero Rocher chocolates. The system consists of three main components: a 3D-printed bell, a globe, and fine-grained material. Table 2 summarizes the different types of grippers that were considered throughout the project.

Figure 2. Grip concept of the vacuum gripper.

Table 2. Comparison of grippers.

Category	3-Finger G.	Vacuum Gr.	UVG
Object adaptation	4	5	5
Construction difficulty	4	2	3
Durability	4	5	5
Cost	3	2	2
Control difficulty	3	1	1
Maintenance	3	4	3
Power requirements	3	3	3

3.3. *Denavit–Hartenberg*

This section presents the Denavit–Hartenberg parameter table, which is a key tool in the analysis and kinematic modeling of the robotic arm. These parameters, such as link lengths, offset angles, and center distances, allow one to accurately describe the geometry and relationship of motion between the links of the arm. Accurately determining these parameters is essential to achieve precise and efficient control of the gripper positions and orientations during packaging tasks. In Figure 3, you can see how the frames are placed for DHP calculation.

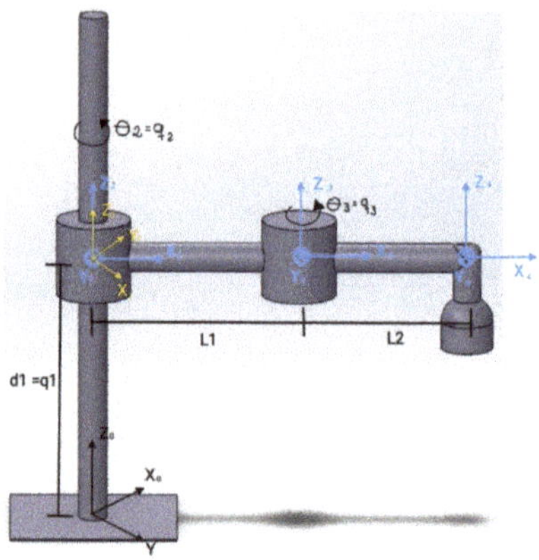

Figure 3. Frame positions.

The Denavit–Hartenberg parameter (DHP) for our PRR robot is presented in Table 3, along with the homogeneous transformation matrix obtained in MATLAB 2020.

Table 3. Table of DHP.

i	Alpha i − 1	ai − 1	di	Theta i
1	0	0	0	q1
2	0	0	q2	0
3	0	L1	q3	0
4	0	L2	0	0

From the table obtained in MATLAB 2020, the equations were obtained based on the parameters to be analyzed in the movements of the robotic arm.

$$\begin{pmatrix} cos(\theta 1+\theta 2) & -sin(\theta 1+\theta 2) & 0 & 13cos(\theta 1+\theta 2)+14cos(\theta 1) \\ sin(\theta 1+\theta 2) & cos(\theta 1+\theta 2) & 0 & 13sin(\theta 1+\theta 2)+14cos(\theta 1) \\ 0 & 0 & 1 & d3 \\ 0 & 0 & 0 & 1 \end{pmatrix}$$

3.4. *Jacobian Matrix*

Using a 2 × 2 Jacobian matrix, we target crucial aspects of packaging. As we prioritize precise chocolate placement over angular orientation, streamlining the Jacobian matrix accelerates the velocity calculations. Through MATLAB programming, we ensure effective robotic control, ensuring seamless and accurate handling of chocolate during packaging. This concept is illustrated in the following matrix:

$$\begin{pmatrix} -13sin(\theta 1+\theta 2)-14sin(\theta 1) & -13sin(\theta 1+\theta 2) \\ 13cos(\theta 1+\theta 2)+14cos(\theta 1) & 13cos(\theta 1+\theta 2) \end{pmatrix}$$

3.5. *Electronic Connections*

The NEMA motors used in our project have a maximum current rating of 2.5 A, but they typically operate within a range of 1 to 1.5 A. By employing the Shield V3 and meticulous fine-tuning of the current limits, we ensure the effective and precise operation of our robotic arm.

4. Results

This text discusses the use of coffee as a fine-grained solid and the advantages and disadvantages of using a mini pump DC5 V/12 V −55 kPa vacuum pump air compressor for its application.

Table 4 summarizes the test results, evaluating them from 1 to 5 (with 5 being the best property).

Table 4. The results table of the tests with different fine-grained materials in the UVG [12].

Fine-Grained Material	Compression	Adaptability
Coffee	4	2
Sugar	1	1
Coffee + Sugar	3	2
Sea Salt	4	4

Table 5 shows the type of grain of each material.

Table 5. Grain size and moisture absorption of different fine-grained materials.

Fine-Grained Material	Grain Size	Hygroscopicity
Coffee	200 µm–1000 µm	7–8% RH
Sugar	0.5 mm–1 mm	84–85.9% RH
Coffee + Sugar	Mix	Mix
Sea Salt	0.2 mm–0.5 mm	75% RH

To ensure consistent and accurate positioning, a trajectory training mechanism was devised. This mechanism empowered the robotic arm to accurately follow predefined paths, facilitating the precise placement of chocolates within the blister. The paths can be seen in Figure 4. This training process leveraged machine learning techniques to bolster the arm's accuracy and efficiency in adhering to the designated trajectories.

	Chocolate 5							
Z-Axis	-30	-80	-80	-30	-30	-70	-70	-30
Link 1	-55	-55	-55	-55	0	0	0	0
Link 2	0	0	0	0	-34	-34	-34	-34
Gripper	OFF	OFF	ON	ON	ON	ON	OFF	OFF
	Chocolate 6							
Z-Axis	-30	-80	-80	-30	-30	-70	-70	-30
Link 1	-41	-41	-41	-41	17	17	17	17
Link 2	-63	-63	-63	-63	-81	-81	-81	-81
Gripper	OFF	OFF	ON	ON	ON	ON	OFF	OFF
	Chocolate 7							
Z-Axis	-30	-80	-80	-30	-30	-70	-70	-30
Link 1	-55	-55	-55	-55	19	19	19	19
Link 2	0	0	0	0	-49	-49	-49	-49
Gripper	OFF	OFF	ON	ON	ON	ON	OFF	OFF
	Chocolate 8							
Z-Axis	-30	-80	-80	-30	-30	-70	-70	-30
Link 1	-41	-41	-41	-41	37	37	37	37
Link 2	-63	-63	-63	-63	-89	-89	-89	-89
Gripper	OFF	OFF	ON	ON	ON	ON	OFF	OFF

Figure 4. Configurations for the last 4 chocolate sets.

Within the framework of this project's development, an exploration of various robotic arm configurations was conducted. This included devising a methodology to instruct the robotic arm along predetermined trajectories, as well as creating a mechanism for executing an automated cycle to position six chocolates within a blister packaging, as can be seen in the Figure 5. Notably, there is potential to refine the system through the implementation of algorithms aimed at automatically detecting the chocolate placements. Nonetheless, it is important to recognize that, in the context of an online production chain, the spatial arrangements of products consistently adhere to standardized layouts. Thus, this present endeavor is a valuable contribution to contemporary chocolate enterprises, aligning seamlessly with the established sequence of operations within their production processes. The final prototype can see in the Figure 5.

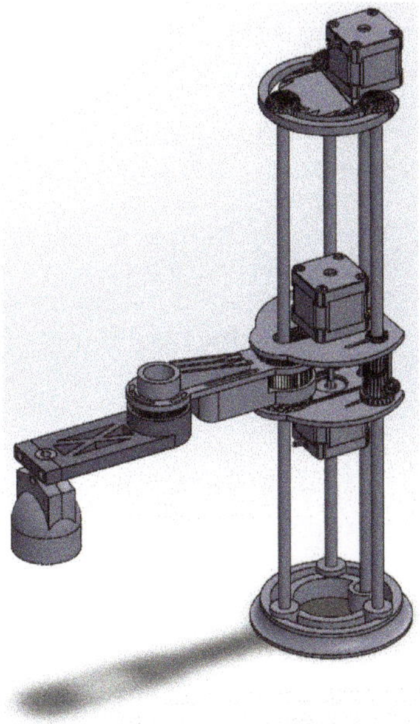

Figure 5. Assembly of the robotic arm on SolidWorks.

The project's development centered on the creation of a robotic arm system. This involved initially identifying and establishing ideal arm positions to ensure the precise placement of chocolates in blister packaging. These positions were determined through a comprehensive approach that integrated mathematical calculations, simulations, and physical experimentations, all aimed at achieving the exact alignment of the chocolates through the interface made in python 3.11 shown in Figure 6.

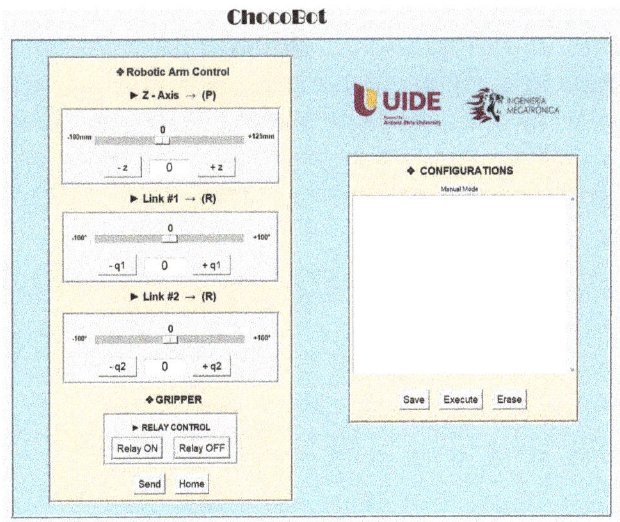

Figure 6. User interface.

5. Conclusions

- The chocolate packaging process,using the final prototype of our PRR robotic arm with the universal vacuum gripper (UVG) shown in Figure 7, has proven to be successful and efficient. Throughout the tests and measurements, an average time of 5 min and 20 s was achieved to complete the packaging process.
- A total of 64 different configurations were employed to achieve precise positioning of the chocolates on the plastic trays. These settings were determined by programming and controlling the robotic arm, ensuring that each chocolate was placed correctly.
- We are confident that this approach has the potential to improve efficiency and quality in the food packaging industry. The conclusions obtained in this report support the implementation of robotic technologies in the automation of similar processes, suggesting the potential for even greater speed and precision in the future.

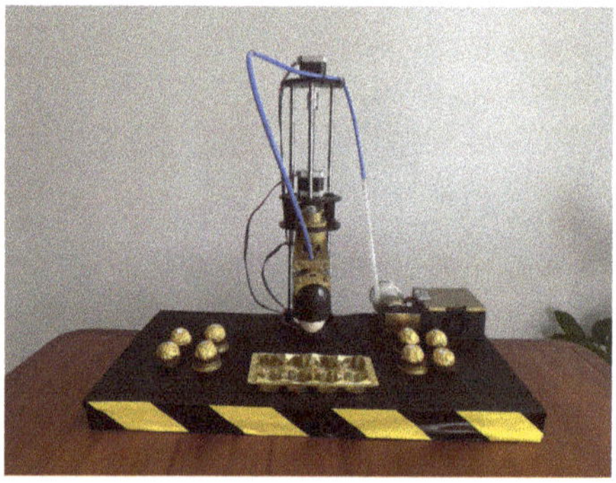

Figure 7. Final prototype.

Author Contributions: This work presents a low-cost alternative to conventional grippers for the manipulation of delicate objects. For the development, the contribution was: Investigation and Methodology: E.A.; Software and validation: G.C.; Data curation, resources and visualization: M.G. and for writting—review and editing: A.Q. All authors have read and agreed to the published version of the manuscript.

Funding: This work did not have any type of financing.

Institutional Review Board Statement: Not applicable.

Informed Consent Statement: Not applicable.

Data Availability Statement: Data is contained within the article.

Conflicts of Interest: The authors declare no conflict of interests.

References

1. Smith, C.; Jones, D. Automation in food packaging: Benefits and impacts. *Food Autom. J.* **2018**, *12*, 45–58.
2. Lee, H.; Kim, J.; Park, S. Development of a PRR robotic arm for food handling. In *Robotics and Automation in Food Industry Conference Proceedings*; Frontiers: Lausanne, Switzerland, 2021; pp. 112–117.
3. Chen, T. Universal Vacuum Gripper (UVG) for efficient food packaging. *IEEE Trans. Robot.* **2020**, *25*, 78–85.
4. Brown, A.; Johnson, M. Performance Analysis of NEMA 17 Stepper Motors in Robotic Arm Systems. In Proceedings of the International Conference on Robotics and Automation Proceedings, Virtually, 1 June 2020; pp. 112–117.
5. Smith, C.; Jones, D. A Comparative Study of NEMA 17 Motors for Robotic Applications. *J. Robot. Autom.* **2019**, *25*, 45–58.
6. Chen, L.; Wang, J. Design and Control of a Robotic Arm Using NEMA 17 Motors. *Robot. Autom. Mag.* **2018**, *15*, 78–85.
7. Kumar, B. Vacuum Cup Grippers for Material Handling in Industry Researchgate. ReserchGate, Jun-2017. Available online: https://www.researchgate.net/publication/317823110_Vacuum_Cup_Grippers_for_Material_Handling_In_Industry (accessed on 15 February 2017).
8. Universal Robots. SMC Vacuum Gripper Unit. Available online: https://www.smcworld.com/newproducts/ (accessed on 13 May 2023).

9. IMPRESORAS3D.COM. Guía de Uso: Filamento PLA para Impresora 3D. Available online: https://www.impresoras3d.com/filamento-pla-consejos-caracteristicas-y-mucho-mas/ (accessed on 15 May 2023).
10. Tandfonline.com. Measuring the Deliquescence Point of Crystalline Sucrose as a Function of Temperature Using a New Automatic Isotherm Generator. Available online: https://www.tandfonline.com/doi/full/10.1080/10942910903474393 (accessed on 15 May 2023).
11. Human Focus. How to Prevent Human Error in Workplace. Available online: https://humanfocus.co.uk/blog/human-error-and-workplace-safety/ (accessed on 15 May 2023).
12. Sciencedirect.com. Instant Coffee. Available online: https://www.sciencedirect.com/topics/agricultural-and-biological-sciences/instant-coffee (accessed on 15 May 2023).

engineering proceedings

MDPI

Proceeding Paper

Brief Survey: Machine Learning in Handover Cellular Network †

Viviana Párraga-Villamar [1,*], Pablo Lupera-Morillo [1], Felipe Grijalva [2] and Henry Carvajal [3]

1 Departamento de Electrónica, Telecomunicaciones y Redes de Información (DETRI), Escuela Politécnica Nacional, Quito 170525, Ecuador; pablo.lupera@epn.edu.ec
2 Colegio de Ciencias e Ingenierías "El Politécnico", Universidad San Francisco de Quito USFQ, Quito 170901, Ecuador; fgrijalva@usfq.edu.ec
3 Faculty of Engineering and Applied Sciences (FICA), Telecommunications Engineering, Universidad de Las Américas (UDLA), Quito 170503, Ecuador; henry.carvajal@udla.edu.ec
* Correspondence: viviana.parragav@epn.edu.ec
† Presented at the XXXI Conference on Electrical and Electronic Engineering, Quito, Ecuador, 29 November–1 December 2023.

Abstract: The proposed work offers a concise review of the application of machine learning (ML) to cellular network handovers (HO) via the Systematic Mapping Study (SMS) methodology, emphasizing the problem areas and requirements. The key points include the paramount role of high-quality data, with meticulous data acquisition and preprocessing as vital steps in ML dataset construction. The article identifies prevalent parameters for HO enhancement and underscores the diversity of ML algorithms, aligning them with specific data input and tasks. This study establishes a robust basis for forthcoming research in applying machine learning to cellular network HOs.

Keywords: machine learning; handover; cellular network; methodology

Citation: Párraga-Villamar, V.; Lupera-Morillo, P.; Grijalva, F.; Carvajal, H. Brief Survey: Machine Learning in Handover Cellular Network. *Eng. Proc.* **2023**, *47*, 2. https://doi.org/10.3390/engproc2023047002

Academic Editor: Soraya Lucia Sinche Maita

Published: 26 September 2023

1. Introduction

Cellular wireless technology is rapidly advancing due to the growing demand for faster and more efficient communication systems. Within these systems, effective mobility management is crucial, particularly in the context of HO mechanisms. HO, defined as the process of transferring an ongoing connection from one base station to another to prevent connection loss [1], plays a pivotal role in ensuring seamless communication.

ML is an automated learning approach that extracts knowledge from data by uncovering hidden patterns. These acquired patterns are leveraged to predict, optimize, or automate tasks, offering solutions to diverse problems across various domains [2]. Numerous researchers have embraced ML techniques and algorithms to enhance HO mechanisms within cellular networks.

The application of ML to HO mechanisms in cellular networks stems from the realization that mobile cellular networks have evolved into massive generators and carriers of data. ML can analyze these data to address issues such as failed HOs. Current research in this field is geared toward enhancing the quality of service provided to subscribers [3]. Consequently, many studies have gathered data from cellular networks for analysis. For instance, telecom companies utilize data on voice and SMS usage patterns to predict user churn [4], while others have examined signal measurements, network performance logs, and heterogeneous data to develop interference management algorithms and predict HOs [5,6].

However, processing such extensive datasets requires suitable methods and tools. Some researchers employ statistical methods to select the most significant data features [4], but the use of ML techniques and algorithms for large-scale data analysis has gained prominence in this domain. The ML process hinges on selecting the appropriate algorithm, including supervised, unsupervised, or reinforcement learning. In supervised learning, a labeled analysis variable is essential, such as distinguishing between soft HOs (ensuring no interruption during the process) and hard HOs (characterized by an actual interruption of the connection during the transition between base stations) [7]. Unsupervised learning, in contrast, clusters data without relying on labels, as seen in the co-clustering algorithm based on the Latent Block Model (LBM) for grouping similar Long-Term Evolution (LTE) cells according to Key Performance Indicators (KPIs) for congestion prediction [8]. Reinforcement learning is also applicable, where rewards and penalties guide decision-making based on input data. For instance, a Q-Table modeled algorithm may evaluate the received link beam power using Reference Signal Received Power (RSRP), cell ID, and reward values during HO [1].

Furthermore, within the ML process, defining the specific task the algorithm aims to solve is crucial. In the context of ML in cellular network HOs, tasks may include self-organizing, which automates the HO process through self-organization of connections [9]; the management of HO control parameters (HCP), where parameters are automatically configured to enhance efficiency [10]; and self-optimizing networks, which achieve automatic system optimization [6].

In summary, this article provides a concise review of the state-of-the-art application of machine learning to cellular network HOs. Emphasis is placed on the prerequisites for applying machine learning to cellular network data to enhance HO performance.

2. Methodology

The literature review was conducted using the Systematic Mapping Study (SMS) methodology, which delineates the steps for identifying available academic evidence and areas necessitating further research on a given topic. This methodology involves the analysis of an extensive collection of primary studies, encompassing articles and publications, to carry out an initial examination of them [11]. Subsequently, the studies enabling this review were selected and organized in accordance with the framework illustrated in Figure 1.

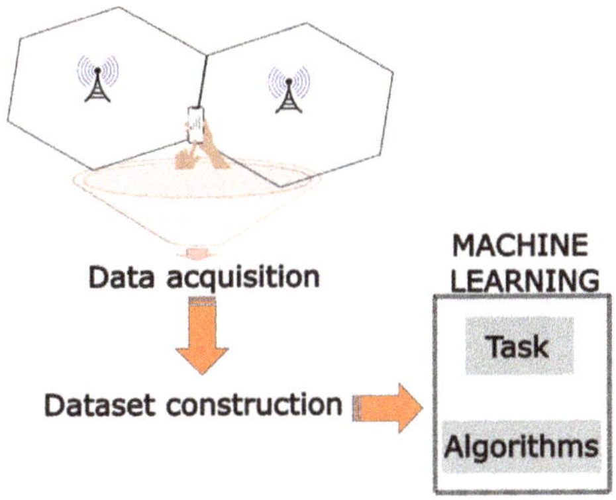

Figure 1. Paper structure.

Articles pertaining to ML in HO cellular networks have underscored that the principal resource for an ML process is the input data. Within this context, two pivotal processes can be distinguished. Firstly, the data acquisition process, which concerns the manner in which HO data are obtained within cellular networks for subsequent analysis. Secondly, the construction of the dataset that will be incorporated into the modeling process, wherein the data are subject to cleansing and new variables are generated in accordance with the study's requirements. Following the delineation of the data input, the ML process advances to encompass the algorithms employed, contingent upon the chosen learning approach and the modeling objectives referred to as the task of the ML process.

3. Results

The results derived from this study primarily center around the prerequisites for ML applications aimed at enhancing HO processes within cellular networks. Beginning with the foundational step of defining input data for modeling, the focus extends to encompass data acquisition, delineating the sources from which data are procured, and dataset construction, which governs data suitability and novel data generation. Further exploration delves into the ML process concerning handovers, encompassing algorithms employed in modeling and the task specification directed at handover improvement.

3.1. Input Data

3.1.1. Data Acquisition

In some instances, researchers turn to simulators like Matlab to assess the effectiveness of their developed models. In [12], the mobility patterns of cellular service users were extracted to predict trajectories using statistical models and Deep Learning. The objective was to optimize HO, considering signal strength and network load balancing. To finalize the research, the model's performance was evaluated by simulating the HO using the mobility predictions, demonstrating that the prediction error was the lowest compared to existing methods [12].

Another study utilized an Automatic Tuning Optimization (ATO) algorithm with Matlab to evaluate its performance. This algorithm leverages user rate and the reference power of the received signal to adapt HCP in LTE-A and 5G Heterogeneous Networks (HetNets). The ATO algorithm was assessed with numerous macro eNBs (MeNBs) and small eNBs (SeNBs) based on 3GPP [6].

In pursuit of improving HO in 5G multi-level intraRatio Access Technology (RAT), a HO decision algorithm was developed. Matlab simulations were used, and input conditions were defined for each event. If these conditions were met for a certain Time-to-Trigger (TTT), the simulated base station (BS) would initiate the HO process. However, if the RSRP of the user equipment (UE) fell below the output condition or failed to meet the input condition after the TTT, the UE remained connected to the current BS [13].

Actual mobile network data collection has been conducted directly by cellular operators or through mobile devices using cellular applications. In [1], the serving BS collected reports containing RSRP measurements from the Mobile Stations for a centralized CMAB agent to make HO decisions. Each BS ran an algorithm to determine whether a BS needed a HO. Furthermore, a telecommunications company provided real LTE traffic data to extract network characteristics for predicting 5G traffic [14].

HO prediction is a significant area of research in 5G systems. To predict bandwidth and HO in 5G or 4G/5G networks, data captured through cellular applications, including parameters such as bandwidth, LTE neighbors, Received Signal Strength Indicator (RSSI), Reference Signal Reception Quality (RSRQ), and speed, among others, were utilized [15].

In [16], the history of radio base HOs was used to predict HO occurrences through two methods: one in vehicles with defined topologies, and another in areas with mobile people. Data were collected from the real world using Android applications installed in vehicles.

3.1.2. Dataset Construction

In ML, it is crucial that researchers have a clear understanding of the characteristics that will be fed into the model to achieve the desired results. Researchers in the field of HO and ML have identified various factors influencing HO behavior. These factors include UE measurements such as RSSI, RSRQ, user speed, distance to the BS, signal-to-noise ratio (SNR), as well as data obtained from the base radio, such as network traffic, channel capacity, error rate, and HO requests. These measurements give rise to characteristics such as HO failure (HOF), Unnecessary HO (UHO), Radio Link Failure (RLF), Too Early HO (TEHO) and Too Late HO (TLHO).

The primary decision parameters for inter-cell HO can be categorized based on speed, RSS or RSRP, cost, energy efficiency, and interference [17].

Additionally, some authors contend that reasons for HOs include a reduced signal strength insufficient to maintain the call, the user equipment moving away from the base station, the exhaustion of all the BS's capacity with pending calls, the assignment of a channel to another BS's call in the overlapping range, and interference avoidance when one user equipment item uses a channel that interferes with another user equipment item using the same channel in another cell [18].

3.2. *Machine Learning in Handover*

In the context of HO in cellular networks, the application of ML involves two essential components: "Algorithms" and "Task."

3.2.1. Algorithms

ML algorithms play a pivotal role in the modeling process. These algorithms are selected based on the learning approach and the objectives of the HO optimization. They can be categorized into three primary types.

Supervised

The choice of ML algorithms in HO research is closely tied to the available data and the desired outcomes. Supervised learning algorithms are employed when working with labeled data to predict specific behaviors.

To mitigate call losses in cellular networks stemming from suboptimal HO performance or channel allocation, researchers have applied supervised algorithms, including neural networks. In one instance, an adaptive HO threshold, based on signal-to-interference ratio and available channels, was used in a decision matrix to label data and determine HO decisions [19].

Unsupervised

The SIM-Known semantic information model aids in HO decisions by considering various cellular network contexts, such as network, application, user, device, and HO. Through semantic analysis, it categorizes HOs as network-controlled, mobile-controlled, network-assisted, or mobile-assisted [10].

Another unsupervised algorithm employed for predicting cellular network traffic, including HOs, is convolutional graphs. This approach extracts network features using real LTE traffic data, and has been instrumental in predicting traffic patterns in 5G networks, utilizing auxiliary features like time and day [7].

Reinforcement

Intelligent decision making in HO optimization leverages historical user trajectory information, often employing Double Deep Reinforcement Learning algorithms. These algorithms consider network, channel, base radio antenna beamforming, and Signal-to-Interference-plus-Noise Ratio (SINR) aspects to optimize the density of HOs and base radios [20].

In the pursuit of optimizing HO and power allocation for maximizing throughput in 5G HetNet systems, researchers have proposed models utilizing reinforcement learning techniques. Each mobile device contributes base radio selection information and power requirements to an incentive-based system, which periodically updates selection policies based on results obtained [21].

3.2.2. Task

The "Task" in ML for HO optimization comprises specific objectives or problems that ML models aim to solve.

Self-organizing

Self-organizing networks (SON) represent a key application of machine learning in HO optimization. SON aims to optimize cellular networks through three distinct phases, detection of failures, diagnosis of the causes of these failures (including issues like hardware problems and parameter misconfigurations), and recovery, to establish a fault-free network [9].

HO Control Parameters

In the realm of management HO (HM), ML models are applied to optimize HCPs. Research efforts focus on developing algorithms for the adaptive adjustment of HCPs based on various network factors, including RSRP, SINR, and UE speed. Performance indicators such as HO failure rate (HOFR) are also considered [10].

Incorrect HCP configurations can lead to an increased RLF rate. Therefore, frequent adjustments based on UE mobility help minimize UHOs. The adjustment process often involves considering TTT based on the number of HOs Performed (HOPP) within a measurement interval [6].

Self-optimizing network

Self-optimizing networks represent another significant application of machine learning in cellular networks. Self-optimizing networks processes include self-configuration, self-optimization, and self-healing within cellular networks. Self-optimizing networks initiates by collecting Key Performance Indicators (KPIs) such as retainability, HO success rate, RSRP, RSRQ, SINR, distance, and average throughput [9].

In a study focusing on 5G HetNets, an auto-tuning self-optimization approach was proposed to address mobility challenges in LTE. The auto-tuning optimization (ATO) algorithm utilizes the user's speed and the reference power of the received signal to adjust HO margin and activation time. HCPs are adapted after each measurement, reducing HOs Per Path (HOPP), HO delay, and call drop rates (CDR) [6].

4. Discussion

The proposed study outlines the prerequisites for conducting a ML investigation applied to the HO process within cellular networks. The fundamental resource for building ML models lies in the data, which enable the model to acquire knowledge and forecast future behavior for process enhancement, such as HO. Through a comprehensive review of the literature concerning the utilization of machine learning in cellular network HOs, key parameters commonly employed in this context have been identified.

To undertake a machine learning modeling for the HO process in cellular networks, we propose a structured approach, as depicted in Table 1. Firstly, it entails the collection of relevant parameters associated with HO, contingent upon the specific scenario and the accessibility of these parameters. Access channels can vary, including retrieval via the base radio for access to wireless network data or wireless network measurements. Alternatively, access may be facilitated through user devices to acquire wireless network data or user-specific information. In cases where direct access is unavailable, simulation tools such as MATLAB can be employed.

Secondly, the acquired data necessitate preprocessing, which involves data cleansing to eliminate outliers and the creation of new variables to capture data behavior during HOs, thereby enhancing their utility. Finally, this process culminates in the construction of a comprehensive dataset.

Within the machine learning process, it is imperative to precisely define the intended task for the model. In the context of this case study, the objective is to identify parameters related to HOs that can be inferred through machine learning modeling, while recognizing their reliance on input data.

Furthermore, after specifying input parameters and expected outcomes, machine learning algorithms that align with the study's requirements are selected. The choice of algorithms is intrinsically linked to the characteristics of the data, encompassing both input and output, and a range of algorithms are available to accommodate these data-specific needs.

In accordance with the steps, the application of machine learning to HO process data within cellular networks serves the purpose of discovering features that enhance the quality of service provided to users when their mobile devices necessitate a HO process. Following the development of the machine learning model, a comprehensive evaluation is imperative to gauge the effectiveness and efficiency of the proposed model.

Table 1. Requirements to apply machine learning in HO of cellular networks.

<table>
<tr><th colspan="3">Data</th></tr>
<tr><td rowspan="3">Data Acquisition</td><td>Wireless network</td><td>RSSI, RSNR, SINR, bandwidth, RSRQ, channel capacity, traffic</td></tr>
<tr><td>User equipment</td><td>Speed, mobility patterns, distance to BS</td></tr>
<tr><td>Base station</td><td>HO history, call set-up success rate, field intensity measurements, voice quality measurements</td></tr>
<tr><td rowspan="5">Dataset Construction</td><td colspan="2">HO failure (HOF)</td></tr>
<tr><td colspan="2">HO too early (TEHO)</td></tr>
<tr><td colspan="2">HO too late (TLHO)</td></tr>
<tr><td colspan="2">Unnecessary HO (UHO)</td></tr>
<tr><td colspan="2">Radio Link Failure (RLF)</td></tr>
<tr><th colspan="3">Machine Learning</th></tr>
<tr><td rowspan="3">Algorithms</td><td>Supervised</td><td>Support vector machine (SVM), artificial neural network (ANN), decision tree, random forest, Graph Convolutional</td></tr>
<tr><td>Unsupervised</td><td>k-means, Principal Component Analysis (PCA), maximization of expectations, analysis of independent components</td></tr>
<tr><td>Reinforcement</td><td>Q-learning, SARSA, Proximate Policy Optimization, Markov Chain</td></tr>
<tr><td rowspan="3">Task</td><td>Self-organization</td><td>HRQ, PPHO, type HO</td></tr>
<tr><td>HO control parameters</td><td>TTT, DCR, HM, cost HO</td></tr>
<tr><td>Network self-optimization</td><td>UHO, TEHO, TLHO, HOSR, HOSR, cost HO</td></tr>
</table>

5. Conclusions

The challenges posed by HOs in cellular networks revolve around their impact on increasing signaling load within the mobile network. This not only diminishes the quality of service delivered to users but also incurs a computational cost for the network. Detecting HO failures is both essential and intricate, given that these failures can stem from a wide array of causes, including hardware issues and suboptimal parameter configurations in cellular networks.

The application of machine learning to cellular network HOs offers the prospect of context-aware HO decisions. Such decisions are rooted in the consideration and correlation of criteria drawn from diverse data collection scenarios, encompassing perspectives from users, wireless networks, and base radio stations. Emphasis has been consistently placed on the pivotal role of high-quality data as the bedrock for training effective machine learning models. Rigorous data collection and preprocessing are critical precursors to constructing datasets that underpin specific HO-related tasks. The literature frequently underscores signal strength as a paramount parameter for enhancing the HO process, encompassing metrics such as RSSI or RSRP, alongside RSRQ, distance, speed, and network traffic. The wide spectrum of machine learning algorithms available for use is intricately intertwined with the nature of the data and the specific tasks at hand.

Furthermore, it is crucial to recognize that the successful integration of machine learning into cellular network HOs can yield substantial enhancements in the quality of service furnished to users. The predictive and optimization capabilities of machine learning can effectively reduce connection interruptions, ultimately enhancing the user experience within cellular networks. Advancements in machine learning within this domain have facilitated the conceptualization and development of solutions such as self-organizing and self-optimizing HOs. These innovations represent a forward-looking vision for self-optimizing HOs that encompass the realms of self-configuration, self-optimization, and self-healing.

In conclusion, we underscore the imperative need for comprehensive evaluations of the models devised to gauge their real-world effectiveness and efficiency. This serves as a foundational stepping stone for future research endeavors in the realm of applying self-optimizing machine learning in the context of cellular network HOs. As communication technology continues its inexorable evolution, machine learning stands as an indispensable tool for the optimization and enhancement of service quality within these pivotal networks. It is our sincere hope that this work will inspire and guide researchers and practitioners alike in their quest for innovative solutions in this ever-evolving field.

Author Contributions: Conceptualization, methodology, investigation, writing—original draft preparation, writing—review and editing, V.P.-V.; supervision, P.L.-M., F.G. and H.C. All authors have read and agreed to the published version of the manuscript.

Funding: This research received no external funding.

Institutional Review Board Statement: Not applicable.

Informed Consent Statement: Not applicable.

Data Availability Statement: Not applicable.

Acknowledgments: We are grateful for the support of the Internal Research Project without Funding PII-DETRI-2022-02 entitled "Analysis of the performance of cellular networks based on field measurements and Machine Learning techniques (Case study city of Quito)" of the Escuela Politécnica Nacional.

Conflicts of Interest: The authors declare no conflict of interest.

References

1. Yajnanarayana, V.; Rydén, H.; Hévizi, L. 5G Handover using Reinforcement Learning. In Proceedings of the 2020 IEEE 3rd 5G World Forum (5GWF), Bangalore, India, 10–12 September 2020; pp. 349–354. [CrossRef]
2. Boutaba, R.; Mohammad, A.; Salahuddin, N.; Sara, A.; Nashid, S.; Estrada-Solano, F.; Caicedo, O. A comprehensive survey on machine learning for networking: Evolution, applications and research opportunities. *J. Internet Serv. Appl.* **2018**, *9*, 16. [CrossRef]
3. Hafez, H. Mining Big Data in Telecommunications Industry: Challenges, Techniques, and Revenue Opportunity. *Int. J. Comput. Electr. Autom. Control Inf. Eng.* **2016**, *10*, 183–190.
4. Azeem, M.; Usman, M.; Fong, A. A churn prediction model for prepaid customers in telecom using fuzzy classifiers. *Telecommun. Syst.* **2017**, *66*, 603–614. [CrossRef]
5. He, Y.; Yu, F.; Zhao, N.; Yin, H.; Yao, H.; Qiu, R. Big Data Analytics in Mobile Cellular Networks. *IEEE Access* **2016**, *4*, 1985–1996. [CrossRef]
6. Alhammadi, A.; Roslee, M.; Alias, M.; Shayea, I.; Alquhali, A. Velocity-Aware Handover Self-Optimization Management for Next Generation Networks. *Appl. Sci.* **2020**, *10*, 1354. [CrossRef]
7. Sreejith, S.; Rajak, A. Study on optimization of handoff process using fuzzy logic for mobile communication. *J. Phys. Conf. Ser* **2020**, *1706*, 012161. [CrossRef]
8. Kassan, S.; Hadj–Kacem, I.; Jemaa, S.B.; Allio, S. A Hybrid machine learning based model for congestion prediction in mobile networks. In Proceedings of the 2022 IEEE 33rd Annual International Symposium on Personal, Indoor and Mobile Radio Communications (PIMRC), Kyoto, Japan, 12–15 September 2022; pp. 583–588. [CrossRef]
9. Chen, M.; Zhu, K.; Chen, B. Root Cause Analysis for Self-organizing Cellular Network: An Active Learning Approach. *Mob. Netw. Appl.* **2020**, *25*, 2506–2516. [CrossRef]
10. Vivas, F.; Caicedo, O.; Nieves, J. A Semantic and Knowledge-Based Approach for Handover Management. *Sensors* **2021**, *21*, 4234. [CrossRef] [PubMed]
11. Zaldumbide, J.; Parraga, V. Systematic Mapping Study of Literature on Educational Data Mining to Determine Factors That Affect School Performance. In Proceedings of the 2018 International Conference on Information Systems and Computer Science (INCISCOS), Quito, Ecuador, 13–15 November 2018; pp. 239–245. [CrossRef]
12. Bahra, N.; Pierre, S. A Hybrid User Mobility Prediction Approach for Handover Management in Mobile Networks. *Telecom* **2021**, *2*, 199–212. [CrossRef]

13. Kapadia, P.; Seet, B.-C. Multi-Tier Cellular Handover with Multi-Access Edge Computing and Deep Learning. *Telecom* **2021**, *2*, 446–471. [CrossRef]
14. Zhao, S.; Xiaopeng, J.; Guy, J.; Rittwik, J.; Wen-Ling, H.; Raif, R.; Manoop, T.; Syed, A.; Yi, C.; Borcea, C. Cellular Network Traffic Prediction Incorporating Handover: A Graph Convolutional Approach. In Proceedings of the 2020 17th Annual IEEE International Conference on Sensing, Communication, and Networking (SECON), Como, Italy, 22–25 June 2020; IEEE: Piscataway, NJ, USA, 2020; pp. 1–9. [CrossRef]
15. Mei, L.; Gou, J.; Cai, Y.; Cao, H.; Liu, Y. Realtime Mobile Bandwidth and Handoff Predictions in 4G/5G Networks. *arXiv* **2021**, arXiv:2104.12959. Available online: http://arxiv.org/abs/2104.12959 (accessed on 15 February 2023). [CrossRef]
16. Abdah, H.; Barraca, J.; Aguiar, R. Handover Prediction Integrated with Service Migration in 5G Systems. In Proceedings of the ICC 2020—2020 IEEE International Conference on Communications (ICC), Dublin, Ireland, 7–11 June 2020; IEEE: Piscataway, NJ, USA, 2020; pp. 1–7. [CrossRef]
17. Achhab, T.; Abboud, F.; Assalem, A. A Robust Self-Optimization Algorithm Based on Idiosyncratic Adaptation of Handover Parameters for Mobility Management in LTE-A Heterogeneous Networks. *IEEE Access* **2021**, *9*, 154237–154264. [CrossRef]
18. Dhake, T.; Jain, E.; Kabra, P.; Khan, A.; Joglekar, C. Effect of Power Adjustment on Handover in Communication Network. *Eng. Technol.* **2020**, *8*, 7.
19. Islam, M.; Hasan, M.; Begum, A. Improvement of the Handover Performance and Channel Allocation Scheme using Fuzzy Logic, Artificial Neural Network and Neuro-Fuzzy System to Reduce Call Drop in Cellular Network. *J. Eng. Adv.* **2020**, *1*, 130–138. [CrossRef]
20. Mollel, M.; Attai, I.; Metin, O.; Shubi, F.; Kisangiri, M.; Sajjad, H.; Muhammad, A.; Qammer, H. Intelligent Handover decision scheme using double deep reinforcement learning. *Phys. Commun.* **2020**, *42*, 101133. [CrossRef]
21. Guo, D.; Tang, L.; Zhang, X.; Liang, Y. Joint Optimization of Handover Control and Power Allocation Based on Multi-Agent Deep Reinforcement Learning. *IEEE Trans. Veh. Technol.* **2020**, *69*, 13124–13138. [CrossRef]

engineering proceedings

MDPI

Proceeding Paper

Optimal Analysis of Microgrid with HOMER According to the Existing Renewable Resources in the Sector of El Aromo and Villonaco, Ecuador †

Fernando Mariño [1], Víctor Tibanlombo [2,*], Jorge Medina [3] and William Chamorro [3]

1 Facultad de Ingeniería Eléctrica y Electrónica, Carrera de Electricidad, Universidad Politécnica Salesiana, Quito 170517, Ecuador; wmarinoa@ups.edu.ec

2 Facultad de Ingeniería Eléctrica y Electrónica, Departamento de Energía Eléctrica, Escuela Politécnica Nacional, Quito 170525, Ecuador

3 Facultad de Ingeniería Eléctrica y Electrónica, Departamento de Automatización y Control Industrial, Escuela Politécnica Nacional, Quito 170525, Ecuador; jorge.medinaj@epn.edu.ec (J.M.); william.chamorro@epn.edu.ec (W.C.)

* Correspondence: victor.tibanlombo@epn.edu.ec

† Presented at the XXXI Conference on Electrical and Electronic Engineering, Quito, Ecuador, 29 November–1 December 2023.

Abstract: In today's world, the integration of renewable energies is essential to meet the surging global energy demand and reduce pollution. Ecuador, with its favorable geographical location, boasts abundant renewable resources. Currently, the country has large-scale renewable energy projects such as "El Aromo" in Manabí, led by Solarpack, and Villonaco in Loja, with its wind power plant. In this study, Homer Pro software is used to simulate two microgrids with solar and wind energy in the mentioned sectors, allowing us to conduct comprehensive economic and energy analyses to determine the most viable configurations.

Keywords: distributed; Ecuador; generation; optimization; photovoltaic; renewables; wind

Citation: Mariño, F.; Tibanlombo, V.; Medina, J.; Chamorro, W. Optimal Analysis of Microgrid with HOMER According to the Existing Renewable Resources in the Sector of El Aromo and Villonaco, Ecuador. *Eng. Proc.* **2023**, *47*, 3. https://doi.org/10.3390/engproc2023047003

Academic Editor: Jaime Cepeda

Published: 6 December 2023

1. Introduction

This work presents a study of two grid-connected microgrids in different areas of Ecuador, incorporating solar and wind energy sources. The study areas are El Aromo in the province of Manabí, where the country's largest photovoltaic power station is projected, and Villonaco, located in Loja, where large-scale wind projects are being developed. As documented in [1–4], there is currently a growing interest in integrating renewable energy sources to form energy microgrids. Simulations allow us to predict behaviors and facilitate decision-making regarding microgrid characteristics.

Currently, several computer tools can be employed for the design, planning, and analysis of microgrids, including ETAP, OPAL-RT, and Sandia's Microgrid Design Toolkit (MDT). However, HOMER (hybrid optimization of multiple energy resources) Energy Plus has been the most widely used and documented tool for the design and analysis of microgrids [1–3]. It offers numerous functionalities, such as annual energy analysis, optimization, and economic analysis.

Using HOMER, we propose an economic comparison of two scenarios: one with the use of renewable energies and another without them. We conduct cost and efficiency assessments. It is important to highlight that HOMER also presents economic combinations of the simulated renewable energy systems, enabling comparison and decision-making among various configurations. This provides a more robust approach to the optimal selection of energy solutions. The following section presents details of the methodology, experiments, results, and conclusions.

2. Theorical Framework

HOMER Pro (hybrid optimization of multiple electric renewable energies) is a tool that allows us to simulate various configurations in a system through its sensitivity analysis and optimization algorithms [5].

The software incorporates two optimization algorithms: the original grid search algorithm, which simulates all possible microgrid configurations within the search space, and a derivative-free algorithm that identifies the system with the lowest cost. After these algorithms complete their processes, HOMER generates a list of all feasible configurations, ranked by net annual cost or life cycle cost. Additionally, HOMER allows for sensitivity analysis, where the user can define sensitivity variables as inputs, and the software will initiate the optimization process for each of the entered sensitivity variables [6,7].

Depending on the number of variations, HOMER systematically explores all possible combinations and optimizes them. In this specific study, sensitivity variables were not included.

2.1. Levelized Cost of Energy (LCOE)

The levelized cost of energy is a key economic performance metric, enabling us to estimate the cost of electricity production for each evaluated case [3,8–10]. It represents the average cost per kWh of useful electricity produced by the grid.

The cost of electricity production can be computed as follows:

$$LCOE = \frac{C_y - C_b H_L}{E_L} \tag{1}$$

where C_y is total annualized cost per year (USD/year), C_b is Marginal boiler cost (USD/kWh), H_L is the total thermal load served per year (kWh/year), and E_L is the total electrical load served per year (kWh/year).

H_L is the portion of the annualized cost that results from serving the thermal load. In systems such as wind or photovoltaics, which do not serve a thermal load (H_L = 0), this term is zero.

2.2. Net Present Cost (NPC)

The net present cost (or life-cycle cost) of a component is the present value of all the costs of installing and operating the component over the project's lifetime, minus the present value of all the revenues that it earns over the project's lifetime. HOMER calculates the net present cost of each component in the system and of the system as a whole [5]. It is the monetary value of the difference between income and expenditure flows, subtracting the initial investment [11,12].

2.3. Operating Cost (OC)

The operating cost is the annual value of the sum of all costs and incomes excluding the capital cost. The objective of the optimization is to seek a reduction in the operating cost of the microgrid [13,14]. The operating cost equation is as follows:

$$OC = C_y - C_c \tag{2}$$

where C_c is the total annualized capital cost per year (USD/year). The total annualized capital cost is equal to the total initial capital cost multiplied by the capital recovery factor [10].

2.4. Initial Capital Cost

Total cost of installing each element at the beginning of the project. This includes equipment installation costs, equipment costs, materials, or auxiliary systems for installation.

3. Existing Resources

Thanks to Ecuador's strategic location and the presence of the Andes Mountain range, the country boasts significant photovoltaic potential [15–17].

Irradiance and wind data, along with climatic information, are directly obtained from NASA's website and via the HOMER software once the microgrid's design location is selected [17]. To validate this data, it employs historical ground-based measurement data from weather stations located in El Aromo and Villonaco; this data is available at meteo-scinergy.epn.edu.ec (accessed on 27 June 2023).

3.1. *Solar Resource in El Aromo Microgrid*

In Figure 1, it can be observed that the months with the highest irradiance in the El Aromo sector are January, March, April, and May, reaching an approximate average of 6 kWh/m^2/day. We can observe the data of global horizontal solar irradiance in the Aromo area for each month of the year, with an annual average irradiance of 4.82 kWh/m^2/day [18].

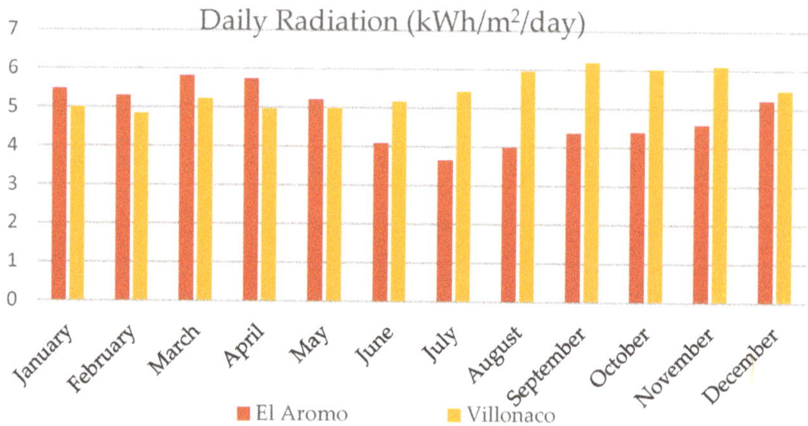

Figure 1. Daily average radiation in El Aromo and Villonaco.

In Villonaco, as shown in Figure 1, it can be observed that the months with the highest solar irradiation are August, September, October, and November, reaching average radiation levels of 6 kWh/m^2/day. As observed in the available resource data in Villonaco, there is an average annual irradiation higher than that in the Aromo sector. In Aromo, it is 4.82 kWh/m^2/day, meanwhile in Villonaco, it is 5.46 kWh/m^2/day.

3.2. *Wind Resource in El Aromo and Villonaco Microgrid*

From the NASA database, monthly average wind speed data are obtained. The graph shows the values for each month, giving an annual average wind speed of 2.78 m/s. Wind turbines start operating at 3–4 m/s and reach maximum power generation at 13–14 m/s. The integration of wind power in El Aromo may not be feasible, but results will be obtained through software optimization. In Figure 2, the months with the highest average wind speed are June to November, with speeds of 3 m/s.

The wind resource is also superior in Villonaco with an annual average wind speed of 3.02 m/s, while the average wind speed in Aromo is 2.78 m/s. In Figure 2, it can be observed that the months with the highest wind speeds in Villonaco are June, July, and August, surpassing average speeds of 3 m/s.

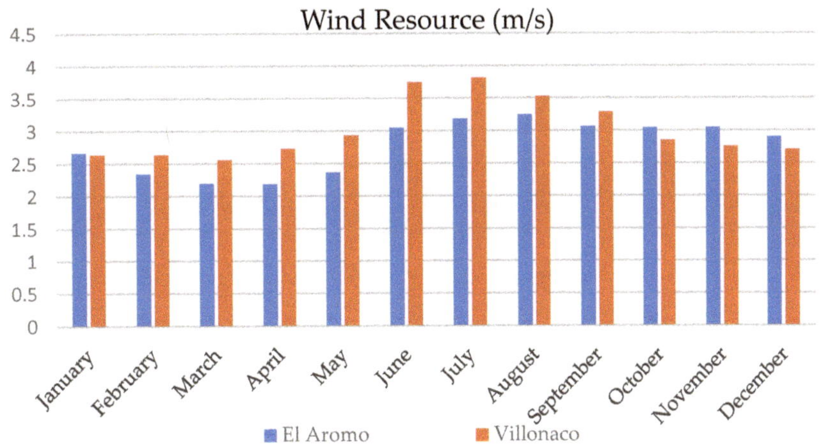

Figure 2. Wind resource in El Aromo and Villonaco.

4. Characteristics of Modules Used in the Design of Microgrids

Given that this is a comparative study between two different areas with different environmental resources, the microgrid designs will be carried out using modules with identical characteristics. This is to ensure an objective analysis focused primarily on the quantity of existing resources in each sector.

4.1. Grid

Distribution networks are a crucial component of power systems, because all the generated power needs to be distributed among users who are scattered across vast territories [16]. Residential loads are connected to the local grid through the grid module, as well as to the microgrid integrated with photovoltaic and wind generation, forming a hybrid system. The grid module allows us to incorporate real-time or scheduled energy purchases and sale rates. In this case, it has been configured using the simple tariff mode, enabling the input of constant energy purchases and sale prices, as well as the capacity to sell energy back to the distribution company or grid.

For the calculation and input of data in the simple tariff, an average value of the cost per kWh and the purchase cost based on the actual average tariff in Ecuador was estimated. The price of grid energy is USD 0.09/kWh and selling price to the grid is USD 0.07/kWh.

4.2. Photovoltaic Module (PV)

A photovoltaic module (PV) can convert solar energy into electrical energy, and the module allows us to input a matrix of photovoltaic panels and their complementary elements into the microgrid. HOMER's library provides a wide selection of panel brands and generic elements with different power generation capabilities. We use as inputs capital costs, replacement costs, and annual operation and maintenance costs for each of these elements. It is important to consider the costs associated with the photovoltaic system, which may include costs of photovoltaic panels, installation costs, wiring costs, additional accessories, and mounting costs, among others [19–22].

These additional costs may also include the power electronics parts such as, in this case, the DC-AC converter module, which has not been included in this case.

The specification of the selected solar module is shown in Table 1, and its mentioned reference costs have been tabulated in Table 2.

The nominal capacity of the photovoltaic array is 4.4 kW, which allows covering almost half of the demand.

Table 1. Properties of PV module Fromius Symo 4.5-3-S with generic PV.

Variable	Description
Type	Flat plate
Nominal capacity (kW)	4.4
Temperature coefficient	−0.41
Operation temperature (°C)	45
Efficiency (%)	17.3
Manufacturer	Fromius

Table 2. Costs of PV module Fromius Symo 4.5-3-S with generic PV.

Capacity (kW)	Capital (USD)	Replace (USD)	O&M (USD/year)
1	2200	2200	10

4.3. Storage Module

In photovoltaic systems, batteries are mainly used as energy storage systems due to the temporal displacement that may exist between generation periods (during the day) and consumption periods (during the night), allowing the operation of loads when the photovoltaic generator alone cannot generate sufficient power to supply consumption [23]. An idealized battery has been integrated into the photovoltaic system through the storage module, which allows for the independent configuration and sizing of energy and power. This type of battery operates based on a flat capacity curve.

The specification of the selected storage module is shown in Table 3, and its mentioned reference costs have been tabulated in Table 4.

Table 3. Properties of storage module RedT5kW-20kWh energy storage.

Variable	Description
Nominal voltage (V)	48
Nominal capacity (kWh)	20
Nominal capacity (Ah)	417
Round-trip efficiency (%)	75
Maximum charge current (A)	105
Maximum discharge current (A)	105
Service life (years)	25
Efficiency (kWh)	876,000

Table 4. Costs of storage module RedT5kW-20kWh energy storage.

Capacity (N°)	Capital (USD)	Replace (USD)	O&M (USD/year)
1	750	750	20

4.4. Converter Module

An important decision must be made regarding whether to establish power consumption in alternating current (AC) or direct current (DC), considering that the standard electrical grid operates on AC, sinusoidal at 60Hz in Ecuador, and the majority of consumer appliances and industrial equipment are designed to function with AC [24]. The converter module becomes imperative for systems necessitating the seamless integration of both AC and DC components, as is the case with integrating the photovoltaic array into the broader microgrid. This module empowers us to precisely define costs and fine-tune the parameters for both the inverter and rectifier, ensuring optimal performance and efficiency. The specifications of the selected converter module are shown in Table 5, and its reference costs have been tabulated in Table 6.

Table 5. Properties of converter module SolarX X3-hybrid5.

Variable	Description
Capacity (kW)	5
Efficiency (%)	97.7
MPPT Efficiency (%)	99
Minimum voltage of MPPT range (Vdc):	180
Maximum voltage of MPPT range (Vdc):	950

Table 6. Costs of converter module SolarX X3-hybrid5.

Capacity (kW)	Capital (USD)	Replace (USD)	O&M (USD/year)
5	400	400	10

4.5. Wind Turbine Module

In wind installations, particularly in smaller and medium-sized setups, achieving extremely high efficiency might not hold the same level of importance, given the freely available nature of wind energy (at least for now). However, optimal performance still holds significance as it facilitates the reduction of rotor diameter, leading to cost savings and addressing construction and mechanical complexities [23–25].

Within the microgrid context, a generic 3 kW wind turbine has been seamlessly integrated to cater to an additional segment of the energy demand. The specifications of the selected solar module are shown in Table 7, and its reference costs have been tabulated in Table 8.

Table 7. Properties of wind turbine module G3 3 kW.

Variable	Description
Capacity (kW)	3
Manufacturer	Generic

Table 8. Costs of wind turbine module G3 3 kW.

Capacity (kW)	Capital (USD)	Replace (USD)	O&M (USD/year)
3	18,000	18,000	180

Figure 3 illustrates the power characteristic curve generated in relation to wind speed for the chosen generator in the simulation.

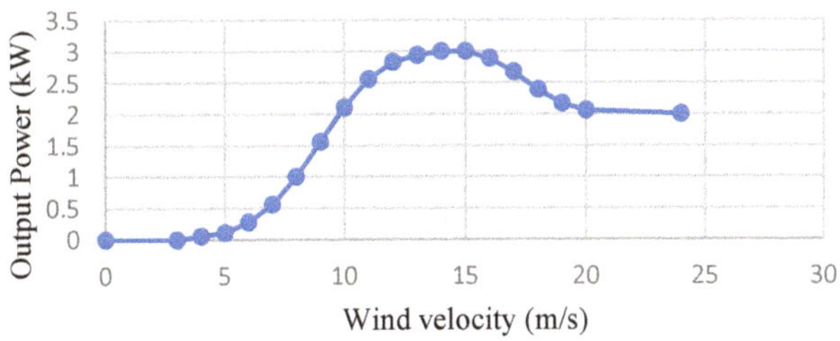

Figure 3. Power characteristic curve of wind turbine module G3.

5. Demand Characteristics

An important strength of this research on electrical consumption projections lies in its distinct seasonality. For electricity producers, understanding which months of the year experience higher or lower electricity consumption is crucial for scheduling maintenance activities on machinery [26].

There are three types of users, each with their own distinct characteristics: residential, commercial, and industrial [27].

The simulated demands will focus on the residential type, utilizing their characteristic load curve but based on the average electrical consumption data for each sector.

5.1. El Aromo's Demand

The grid-connected microgrid simulated in the El Aromo sector features a distinct primary residential load with a demand of 9.97 kWh/day. This demand value corresponds to the average residential demand on the Ecuadorian coast, including the approximate monthly consumption of 100 kWh for an induction cooker [27].

As illustrated in Figure 4, the daily residential load profile in El Aromo displays a distinct demand pattern, with pronounced peaks during the late afternoon and evening hours, coinciding with the times when most consumers are present in their households. This characteristic demand shape accurately mirrors the consumption behavior prevalent in this region.

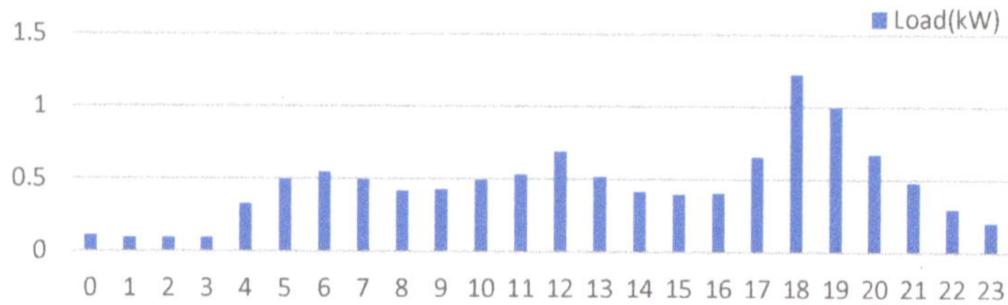

Figure 4. Daily load profile of El Aromo.

5.2. Villonacos's Demand

The demand profile in Villonaco showcases a distinct residential primary load with a daily demand of 8.40 kWh. This demand value aligns with the average residential consumption in the Ecuadorian highlands, encompassing an estimated 100 kWh/month usage from an induction cooker [27].

6. Results

Displayed in Figures 5 and 6 are schematic diagrams showcasing the simulated hybrid microgrid located within Ecuador's Manabí province, specifically in the El Aromo sector. This complex setup seamlessly integrates all the components. Furthermore, an analogous schematic diagram for Villonaco is provided, reflecting a similar configuration of elements. It's worth noting that the disparities primarily pertain to load characteristics and the distinct energy resources accessible in each respective area.

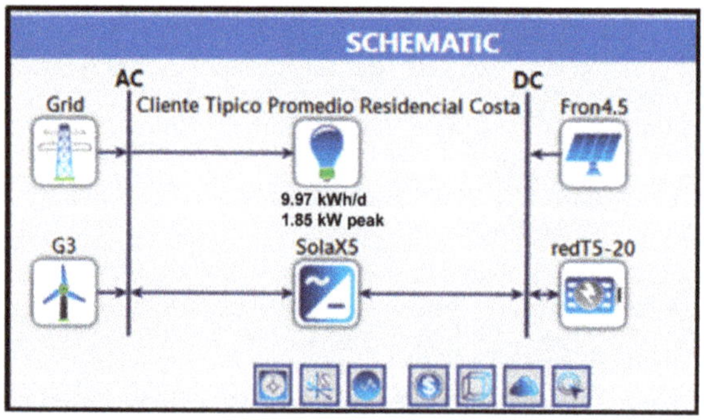

Figure 5. El Aromo microgrid configuration.

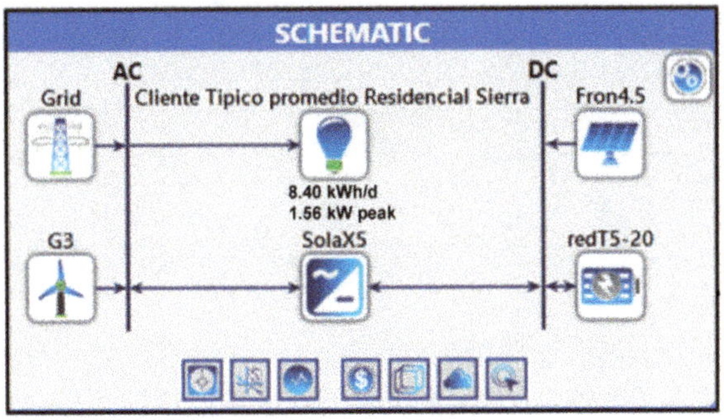

Figure 6. Villonaco microgrid configuration.

6.1. El Aromo Microgrid

The presented case involves a user solely connected to the electrical grid, with an initial capital requirement of USD 0. The annual cost, or the yearly expense that the user would pay to the distribution company for consumption amounts to USD 327.40, equivalent to a monthly fee of USD 27.28. This scenario does not consider the integration of any other type of power generation into the system. Below, in Table 9 is shown the data of sensitivity analysis of this case.

Table 9. Table of sensitivity analysis for El Aromo.

Grid (kW)	LCOE (USD)	NPC (USD)	Operating Cost (USD/year)	Initial Capital Cost (USD)
1	0.09	4,232,505	180	0

Optimization Results

As depicted in the optimization table shown in Table 10, the annual cost for energy consumption solely sourced from the grid amounts to USD 327.40, translating to a monthly payment of USD 27.28. For a residential load characterized as proposed, the most viable choice emerges in the third row. This selection involves integrating the grid with a 4.5 kW photovoltaic panel array (from 4.5 kW) and a 5 kW DC-AC SolarX5 converter. This integration necessitates an initial capital investment of USD 10,080 and yields a negative operational cost of USD −158.62, indicating surplus energy being sold back to the distribution entity. This configuration not only facilitates the long-term recovery of the initial investment but also encompasses the compelling environmental advantages associated with renewable energy utilization.

Table 10. Optimization analysis table displaying possible microgrid combinations for El Aromo sector.

From 4.5 kW	G3	redT5-20	Grid (kW)	SolarX5 (kW)	LCOE (USD)	NPC (USD)	Operating Cost (USD/yr)	Initial Capital (USD)
			1		0.09	4232.505	327.4028	0
		1	1	1	0.1217797	5752.891	356.0538	1150
1			1	1	0.0733681	8029.5	−158.6151	10,080
1		1	1	1	0.08226459	9020.611	−139.9641	10,830
	1		1		0.5696319	26,854.04	684.8985	18,000
	1	1	1	1	0.599184	28,374.42	713.5494	19,150
1	1		1	1	0.2764108	30,664.86	199.9506	28,080
1	1	1	1	1	0.2847995	31,655.97	218.6015	28,830

6.2. Villonaco Microgrid

The annual cost of energy consumption obtained solely from the electrical grid with a fixed rate of USD 0.09 per kWh in Villonaco amounts to USD 275.94, equivalent to a monthly cost of USD 22.99. Below, in Table 11 is shown the data of sensitivity analysis of this case.

Table 11. Table of sensitivity analysis for Villonaco.

Grid (kW)	LCOE (USD)	NPC (USD)	Operating Cost (USD/year)	Initial Capital Cost (USD)
1	0.09	3567	275.94	0

Optimization Results

Analyzing Table 12, it can be concluded that the most viable configuration option integrates the electrical grid with the photovoltaic panel array from 4.5 kW and the DC-AC converter SolarX5 5kW, requiring an initial capital investment of USD 13,600 and resulting in a negative operating cost of USD -283.99.

Table 12. Optimization analysis table displaying possible microgrid combinations for Villonaco sector.

From 4.5 kW	G3	redT5-20	Grid (kW)	SolarX5 (kW)	LCOE (USD)	NPC (USD)	Operating Cost (USD/yr)	Initial Capital (USD)
			1		0.09	3567.219	275.94	0
		1	1	1	0.1276727	5087.612	304.5916	1150
1			1	1	0.0851245	9928.711	−283.9902	13,600
1		1	1	1	0.09345291	10,919.85	−265.3368	14,350
	1		1		0.6498159	26,081.54	625.1429	18,000
	1	1	1	1	0.6840639	27,601.94	653.7945	19,150
1	1		1	1	0.2724693	32,463.09	66.76398	31,600
1	1	1	1	1	0.2804584	33,458.66	85.75963	32,350

This implies that energy is being sold back to the distribution company, leading to a faster recovery of the initial investment cost in the long term compared to the El Aromo microgrid.

7. Conclusions

The optimal design for the proposed microgrids has been successfully accomplished, providing insightful outcomes that pinpoint the most cost-effective and energy-efficient configurations when integrated with the local distribution utility in each specific case.

The optimization results from HOMER lead us to a firm conclusion: the most optimal setup involves a grid integration strategy incorporating the from 4.5 kW photovoltaic panel array along with the SolarX5 (5kW) DC-AC converter. This optimal arrangement is consistent for both the El Aromo and Villonaco microgrid scenarios.

It can be deduced that the payback period for the initial investment cost, in the context of the proposed average residential consumption within the coastal and mountainous regions of Ecuador, spans around 20 years. However, it's crucial to note that this initial investment cost may pose a considerable hurdle for an average household. To reduce this payback period and promote the adoption of renewable energy, strategies such as government subsidies, accessible financing, public education, and large-scale projects can be implemented, along with improving energy efficiency to make the investment more attractive and profitable in both the short and long term.

Author Contributions: Conceptualization, F.M. and V.T.; methodology, J.M.; software, V.T.; validation, F.M., V.T., J.M. and W.C.; formal analysis, V.T.; investigation, F.M.; data curation, J.M. and W.C.; writing—original draft preparation, V.T.; writing—review and editing, V.T., J.M. and W.C. All authors have read and agreed to the published version of the manuscript.

Funding: This research received no external funding.

Institutional Review Board Statement: Not applicable.

Informed Consent Statement: Not applicable.

Data Availability Statement: Not applicable.

Conflicts of Interest: The authors declare no conflict of interest.

References

1. Restrepo, D.; Restrepo-Cuestas, B.; Trejos, A. Microgrid Análisis Using HOMER: A Case Study. DYNA. Available online: http://www.scielo.org.co/pdf/dyna/v85n207/0012-7353-dyna-85-207-129.pdf (accessed on 12 June 2023).
2. Swarnkar, N.; Gidwani, L. Analysis of hybrid energy system for supply residential electrical load by HOMER and RETScreen: A case in Rajasthan, India. In Proceedings of the IEEE International Conference on Recent Advances and Innovations in Engineering (ICRAIE), Jaipur, India, 23–25 December 2016.
3. Garni, H.; Awashti, A. Techno-economic feasibility analysis of a solar PV grid-connected system with different tracking using HOMER software. In Proceedings of the 5th IEEE International Conference on Smart Energy Grid Engineering, Oshawa, ON, Canada, 14–17 August 2017; pp. 217–223.
4. Zahboune, H.; Zouggar, S.; Krajacic, G.; Varbanov, P.S.; Elhafyani, M.; Ziani, E. Optimal hybrid renewable energy design in autonomous system using Modified Electric System Cascade Analysis and Homer software. *Energy Convers. Manag.* **2016**, *126*, 909–922. [CrossRef]
5. Bienvenido A HOMER. Available online: https://www.homerenergy.com/products/pro/docs/latest/index.html (accessed on 12 July 2023).
6. Jahangir, M.H.; Cheraghi, R. Economic and environmental assessment of solar-wind-biomass hybrid renewable energy system supplying rural settlement load. *Sustain. Energy Technol. Assess.* **2020**, *42*, 100895. [CrossRef]
7. Santiago, C.N.B.; Sautto, A.M.A. Generación De Energía Eléctrica En Aerogeneradores: Evidencia Empírica. In *Experiencias en la Modelación de la Toma de Decisiones en la Salud Humana, Medio Ambiente Y Desarrollo Humano Tomo 1*; Universidad de Granada: Granada, Spain, 2015; Volume 7, pp. 147–160. [CrossRef]
8. Paredes, A. Serie Ponencias: Perspectivas de Sustentabilidad México [Conference Series: Sustainability Perspectives Mexico]. pp. 119–134. Available online: https://www.researchgate.net/profile/Luis_Cervera-Gomez2/publication/322701637_Diagnostico_del_recurso_solar_y_el_desarrollo_economico_y_social_del_Valle_de_Juarez_Chihuahua_Pp_59/links/5a6a2c18a6fdccf88497c8ec/Diagnostico-del-recurso-solar-y-el-desarroll (accessed on 12 June 2023).
9. La, P.D.E.; Residual, B.; Proceso, D.E.L.; Palma, D.E. Potencial de La Biomasa Residual del Proceso de Extracción de Palma, 2021; pp. 1–101. Available online: https://repositorio.cuc.edu.co/bitstream/handle/11323/8286/Potencial%20de%20generaci%C3%B3n%20de%20energ%C3%ADa%20el%C3%A9ctrica%20a%20partir%20de%20la%20biomasa%20residual%20del%20proceso%20de%20extracci%C3%B3n%20de%20palma%20de%20aceite%20en%20la%20zona%20norte%20de%20Colombia.pdf?sequence=1&isAllowed=yhtml (accessed on 12 July 2023).
10. Costo Nivelado de Energía. Available online: https://www.homerenergy.com/products/pro/docs/latest/levelized_cost_of_energy.html (accessed on 12 July 2023).
11. Córdoba Padilla, M. Formulación y Evaluación de Proyectos. Colombia: Ecoe Ediciones, 2011. ProQuest Ebrary. Available online: http://repositorio.uasb.edu.bo:8080/bitstream/54000/1206/1/C%C3%B3rdoba-evaluaci%C3%B3n%20d%20eproyectos%202da%20ed.pdf (accessed on 12 July 2023).
12. Mosquera, F. Localización óptima de plantas virtuales de generación en sistemas eléctricos de potencia basados en flujos óptimos de potencia. *I+D Tecnol.* **2020**, *16*, 5–16. [CrossRef]
13. Costo Operacional. Available online: https://www.homerenergy.com/products/pro/docs/latest/operating_cost.html (accessed on 12 July 2021).
14. Olarte, D.M.; Vásquez, M.M. Plan de Negocio para la Puesta en Marcha de una Empresa de Financiamiento de Proyectos de Energía Renovable Mediante Contratos de Suministro y venta de Electricidad a Largo Plazo (PPA)—DMM Soluciones Financieras Integrales Manuel Mejía Vásquez. Trabajo de Grado. Bachelor's Thesis, Universidad EIA, Las Palmas, Colombia, 2016.
15. Moragues, J.; Rapallini, A. Instituto Argentino de La Energía. Energía Eólica. 2003, p. 22. Available online: http:///docplayer.es/10404081-Energia-eolica-jaime-moragues-y-alfredo-rapallini.html (accessed on 12 June 2023).
16. Burneo Nieto, J.P. Análisis de Sensibilidad para la Integración de Nuevos Sistemas de Generación Eléctrica en el Ecuador. Basados en Energía Eólica. Master's Thesis, Universidad Espíritu Santo, Guayaquil, Ecuador, 2016.
17. POWER Data Access Viewer. Available online: https://power.larc.nasa.gov/data-access-viewer/ (accessed on 12 July 2023).
18. Holguín, W.D.J.; Henao, E. Desarrollo De Proyecto Pv Usando El Software Homer Pro. 2019. Available online: https://repositorio.itm.edu.co/bitstream/handle/20.500.12622/419/HolguinBoteroWilliamdeJesus2018.pdf?sequence=1&isAllowed=y (accessed on 12 July 2023).
19. Juárez Cervantes, J.D. Sistemas de Distribución de Energía Eléctrica. 2002, Volume 52. Available online: https://core.ac.uk/download/pdf/48392416.pdf (accessed on 13 June 2023).
20. Roé, J.C. Diseño e Implementación de un Inversor Multinivel para Sistemas Fotovoltaicos Conectados a Red. Master's Thesis, Universitat Politècnica de Catalunya, Barcelona, Spain, 2010.

21. Krenzinger, A.; Prieb, C.W.M. Clasificación y Selección de Módulos Fotovoltaicos para una Central Conectada a la Red. *Av. En Energ. Renov. Y Medio Ambiente* **2005**, *9*, 4–19. Available online: http://sedici.unlp.edu.ar/bitstream/handle/10915/82225/Documento_completo.pdf-PDFA.pdf?sequence=1&isAllowed=y%0Ahttp://asades.org.ar/modulos/averma/trabajos/2005/2005-t004-a004.pdf (accessed on 12 June 2023).
22. Alonso, M. Master en Energias Renovables y Mercado Energetico. Ciemat. Available online: http://api.eoi.es/api_v1_dev.php/fedora/asset/eoi:45340/componente45338.pdf (accessed on 12 June 2023).
23. Abella, M.A.; Romero, F.C.; Fotovoltaico, S.D.B. Máster en Energías Renovables y Mercado Energético: Energía Solar Fotovoltaica. Escuela de Organización Industrial. *Dimens. Sist. Fotovoltaicos* **2017**, *1*, 73.
24. Windräder, K. Introducción a La Teoría de Las Turbinas Eólicas. 2009. Available online: https://www.studocu.com/es-ar/document/universidad-nacional-de-mar-del-plata/calculo-de-elementos-de-maquinas/introduccion-a-la-teoria-de-las-turbinas-eolicas-amics-21/4864468 (accessed on 13 July 2023).
25. Fuentes, C.A.; Aguilar, R. Proyección de Series de Tiempo para el Consumo de la Energía Eléctrica a Clientes Residenciales en Ecuador. *Rev. Tecnol. ESPOL-RTE* **2016**, *29*, 1–21.
26. Bayas, L.; Jaramillo, M.; Betancourt, E.; Reinoso, J.P. La tarifa horaria en el Ecuador como incentivo de eficiencia energética. *Rev. Politec.* **2009**, *25*, 50–68.
27. MERNNR. Plan Maestro de Electricidad. Minist. Energía y Recur. No Renov. Energía y Recur. No Renov. Available online: https://www.recursosyenergia.gob.ec/plan-maestro-de-electricidad/ (accessed on 12 July 2023).

engineering proceedings

MDPI

Proceeding Paper

Private LoRaWAN Network Gateways: Assessment and Monitoring in the Context of IIoT-Based Management †

Oscar Torres Sanchez [1,2,*], Duarte Raposo [3], André Rodrigues [4], Fernando Boavida [1,2] and Jorge Sá Silva [5]

1 Centre of Informatics and Systems, University of Coimbra (CISUC), 3030-290 Coimbra, Portugal; boavida@dei.uc.pt
2 Department of Informatics Engineering, University of Coimbra, 3030-290 Coimbra, Portugal
3 Instituto de Telecomunicações, 3810-193 Aveiro, Portugal; dmgraposo@av.it.pt
4 Polytechnic Institute of Coimbra, Coimbra Business School, Quinta Agrícola-Bencanta, 3045-231 Coimbra, Portugal; andre@iscac.pt
5 Institute for Systems Engineering and Computers (INESC), University of Coimbra, 3030-790 Coimbra, Portugal; sasilva@deec.uc.pt
* Correspondence: otorres@dei.uc.pt
† Presented at the XXXI Conference on Electrical and Electronic Engineering, Quito, Ecuador, 29 November–1 December 2023.

Abstract: In the ever-evolving construction industry, the incorporation of the Industrial Internet of Things (IIoT) through Low-power Wireless Area Networks (LPWAN), such as LoRaWAN, has emerged as a practical solution for addressing the challenges posed by the limited 5G cellular coverage found in solutions like NB-IoT and LTE-M, especially when deployed in remote locations. Open-source LPWAN platforms like The Things Network (TTN) and ChirpStack have played a pivotal role in fostering the adoption of LoRa technology by providing a mature and cost-effective ecosystem that facilitates efficient device resource management. Within this context, maintaining continuous surveillance of LPWAN network gateways becomes critically important, requiring a meticulous examination of status indicators and an evaluation of the communication quality. This paper introduces a structured approach for extracting data from TTN to create a comprehensive gateway monitoring system. The methodology encompasses various aspects, including ensuring seamless server connectivity, specifically focusing on efficient information management and integration of real-world construction data. This foundational work sets the stage for a more in-depth exploration of the diverse management components within the network ecosystem.

Keywords: private TTN networks; gateway monitoring; IIoT; LoRaWAN

Citation: Sanchez, O.T.; Raposo, D.; Rodrigues, A.; Boavida, F.; Sá Silva, J. Private LoRaWAN Network Gateways: Assessment and Monitoring in the Context of IIoT-Based Management. *Eng. Proc.* **2023**, *47*, 4. https://doi.org/10.3390/engproc2023047004

Academic Editor: Soraya Lucia Sinche Maita

Published: 4 December 2023

1. Introduction

In the context of Industry 4.0, the construction industry is actively seeking innovative solutions to overcome operational challenges. Adopting the IIoT, particularly through private LoRaWANs, can revolutionize the industry by addressing communication, monitoring, and management issues [1]. However, effective gateway monitoring and management are essential to leverage private LoRaWANs' benefits fully [2,3]. Monitoring LoRaWAN private gateways can bring advantages such as increased network reliability, proactive maintenance, data integrity, and optimal LoRaWAN network performance [4,5]. Furthermore, these methods can detect connectivity problems and enable corrective actions for seamless data flow, enhancing the infrastructure's reliability, which is vital for IIoT devices. This work aims to provide a method for assessing and effectively monitoring gateways in private networks, specifically designed for an IIoT framework centered on industrial machinery management [6,7]. This endeavor represents a significant step toward addressing the pressing operational challenges faced by the construction industry in its quest for innovation and enhanced efficiency in the digital era of Industry 4.0.

Therefore, the contributions of this work are:

- The implementation of a method for active monitoring of gateways in private LoRaWAN networks based on the TTN platform.
- A data management agent for collecting, structuring, and sending data from the TTN server to third-party applications.
- Analysis of data extracted from monitoring the communication between private gateways and connected devices in a real industrial environment.

This manuscript is structured as follows: Section 2 reviews relevant literature; Section 3 discusses the proposed network monitoring architecture; Section 4 presents some evaluation results within the gateway scope; and Section 5 concludes and paves the way for future work.

2. Related Works

Table 1 depicts an analysis of a few recent related works, and, despite advances in the area, these studies have not implemented private gateway monitoring in the way this article addresses.

Table 1. Comparison of related works and approaches for monitoring and management of Private LoRaWAN Gateways.

Reference	Year	Contribution	Limitation and Challenges
[1]	2023	A new solution for real-time integrated monitoring of construction machinery in the construction industry.	The paper does not provide a detailed analysis of the solution's performance in terms of operational efficiency, including, e.g., gateways.
[8]	2022	IIoT solution for a factory in the Manaus Industrial Complex that meets specific requirements such as MQTT message publishing gateway in a private broker, end-devices with adequate processing/storage capacity.	The system's performance is evaluated based on a limited number of transmitted packets. Furthermore, the work needs to discuss the scalability of the system and its ability to handle many end devices.
[3]	2022	It provides original results comparing the performance of private and public LoRaWAN deployment options.	The paper does not provide an in-depth analysis of the deployment process for each approach.

3. IIoT-Based Management Architecture

As depicted in Figure 1, the architecture establishes a construction machinery monitoring architecture rooted in the paradigm of IIoT. This structure revolves around end devices interfacing with machinery through industrial communication protocols. These devices employ LoRaWAN technology to transmit data extracted from machinery to dedicated management sites. Notably, the architecture relies on a private LoRaWAN network, emphasizing that gateways' creation, configuration, customization, and administration require oversight from network administrators [8,9].

The Things Network, in its community version, provides a long-range, low-power communication framework finely tuned to the needs of Internet of Things (IoT) devices and applications that utilize LoRaWAN. This platform embodies an open and collaborative infrastructure, empowering users to establish personalized IoT networks in the various locations that the construction industry demands. Using LoRaWAN private networks in TTN allows the management of their gateways through procedures for the creation, configuration, deletion, monitoring, validation, and user authentication, among others.

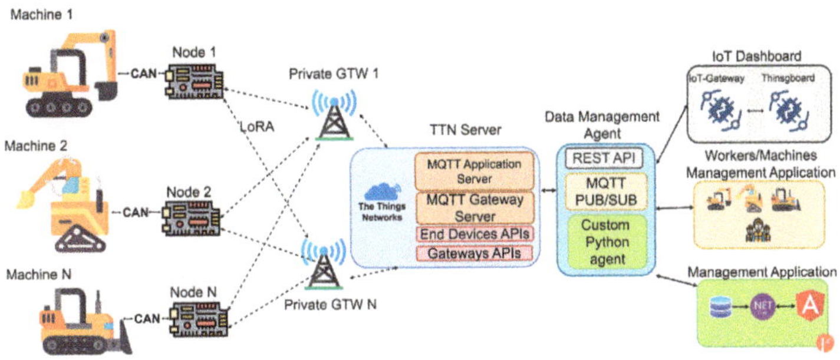

Figure 1. IIoT-based monitoring architecture established over a LoRaWAN private network.

3.1. TTN Server and Integrators

The function of integrating devices with the consumer's application is carried out by TTN integrators, serving as intermediaries, generating connections between devices' events/messages and third-party applications. This role assumed by integrators facilitates the seamless transmission of data acquired by IoT devices to designated management or visualization platforms. The integration can be performed by utilizing Application Programming Interfaces (APIs) and MQTT connectors provided by TTN as part of their services. The choice of the integrator towards TTN can influence the latency, reliability, adaptability, transformation, communication protocol, and connection of data to the platforms; therefore, the type of integrator must be related to the monitoring needs of the different events and messages available for each device or gateway.

3.1.1. MQTT Application and Gateway Server

In the TTN architecture, the MQTT Application Server and MQTT Gateway Server play distinct yet crucial roles. The MQTT Application Server connects IoT devices and applications, routing converted device data to specific applications and managing data subscriptions and publications from external applications. On the other hand, the MQTT Gateway Server interfaces with network-connected gateways, receiving, processing, and distributing gateway-transmitted data to TTN, ensuring smooth data flow in TTN's IoT ecosystem. Furthermore, gateways occasionally send status updates to the MQTT Gateway Server after long periods; in the context of monitoring, it may be less effective to just rely on this source of status information.

Therefore, to monitor the private gateways, TTN's MQTT Application Server grants access to crucial IoT device uplink data, like RSSI and SNR, reflecting connection quality. Meanwhile, the MQTT Gateway Server provides gateway performance data, availability, and connectivity status. It also enables gateway management and coordination for infrastructure control [10,11].

3.1.2. TTN Application and Gateway APIs

TTN's application and gateway APIs are available via HTTP GET and POST requests. GET requests to specific URLs offer insights on device events and allow tailored device commands. Similarly, Gateway API GET requests provide gateway details, including status and performance metrics. Conversely, POST requests enable command distribution and gateway data retrieval. For gateway health monitoring, GET requests to relevant URLs yield real-time data on availability, including packet timestamps, connection status, version, model, and more. These data support alert systems and visualizations, ensuring prompt issue detection and optimal gateway operation in TTN [10].

3.2. Data Management Agent

The data management agent is pivotal in facilitating structured connectivity and oversight of acquired parameters via MQTT or API queries. This agent encompasses a publication and subscription service for gathering data from end devices. For gateway monitoring, it archives device-specific information accompanying uplink messages, including sender identity, receiving gateway identifier, and metrics such as RSSI, SNR, bandwidth, spreading factor, data rate index, and frequency. Moreover, the agent encompasses a service to query TTN APIs, offering insights into status, protocol, connection history, update frequency, message count since the last update, device model, and utilized frequencies. These data agent programmed with Python allows the management of TTN in a customized and non-standardized way. Therefore, it allows us to adapt to different sources and information from TTN, perform more complex processes, and manage the multiple destinations requiring the TTN's components data. This agent's role encompasses rectifying erroneous, atypical, or missing data, precluding disruptive application queries, and ensuring gateway monitoring stability. It optimizes database storage by retaining pertinent gateway monitoring data, excluding redundant information in the MQTT server and API queries. Crucially, this agent safeguards against the gateway status information gap by autonomously soliciting the current gateway status during device communication without relying solely on incoming messages.

4. Gateways Monitoring Assessment

This section shows the results obtained through the monitoring of private gateways. These gateways have been installed to collect data from a construction scenario in Vilar Formoso, Portugal, so the data represents the operation in a real IIoT scenario. Two private gateways were installed, a LORIX One LoRaWAN base station at the Vilar Formoso construction site and a LtAP LR8 LTE Mikrotik kit (MikroTik, Riga, Latvia) in a semi-fixed station in a crusher 12 km from the initial station at Porto de Ovelha, Portugal. These were intended to have the best coverage in a more than 30 km construction line. Therefore, due to the locations of the private gateways and their difficult access, creating an effective monitoring system is mandatory. First, an alert system was created to receive an email notification when a gateway was disconnected. In addition, this system displayed an updated status of the gateways to the construction operators and network management, allowing them to verify whether the gateways were operational or not.

In an actual industrial setting, the dataset includes transmission data from machines like a Multifunction, a Jaw Crusher, a bulldozer Dossan DL200, and a Drill Roc D7. The system uses the data management agent described in previous sections, which systematically collects and organizes device-specific data and gateway information, and the datasets comprise around two months of tests. As part of the monitoring tasks, the distribution of messages at each gateway is required; it is important to evaluate if the allotment and location of the gateways around the construction site are correct. Figure 2 shows message distribution among gateways and each machine, enabling analysis of gateway usage for each machine in our management platforms. In general, the use of the gateway far from the main site (Crusher Station) is greater for most machines, which may be due to its location and where the machines work. Another vital aspect of gateway monitoring is assessing communication quality between the machines and gateways. This dataset allows in-depth analysis for identifying issues related to LoRaWAN network coverage that affect the construction zones communication and eventually the loss of system reliability. Figure 3 demonstrates how gateway monitoring, combined with device in-formation, enables the evaluation of the average RSSI for each gateway–machine pair, aiding in the detection of abnormal data points. This insight helps in defining strategies to prevent future device-to-gateway communication issues. Finally, Figure 4 shows the variation in the communication quality between the gateways and machines over the distance from the antenna. It can be observed that the central station tolerates longer-distance communications, with a noticeable degradation when communication reaches 15 km. In contrast, the Crusher station does not tolerate longer communication distances, presenting a degradation and a maximum tolerance below 9 km of distance. Despite receiving a higher quantity of messages, its tolerance for longer distances can be diminished by its location in a working area, which typically has a less accessible topography.

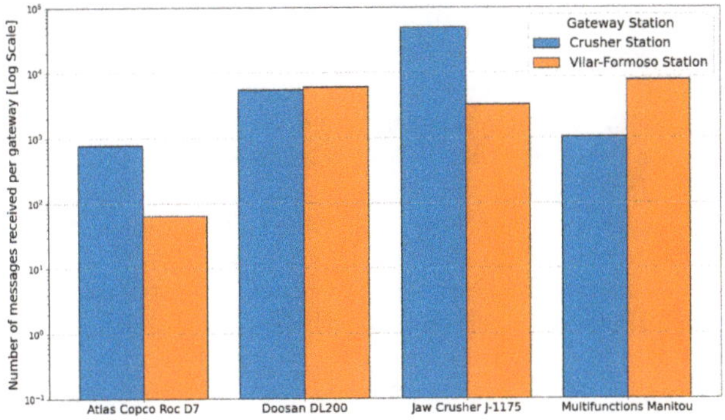

Figure 2. Messages per Gateway and Machine.

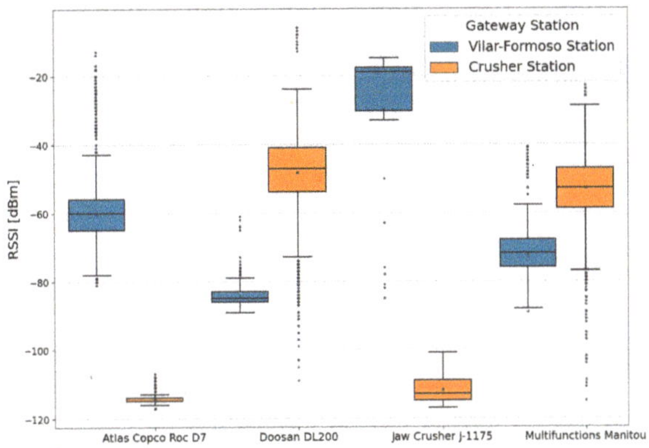

Figure 3. RSSI values by Gateway and Machine.

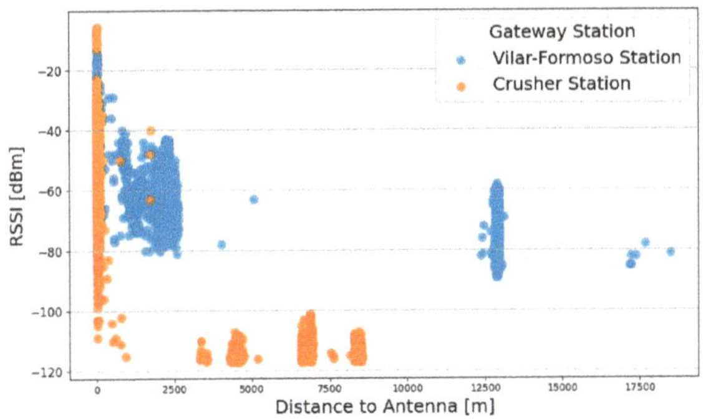

Figure 4. RSSI values by Gateway station considering distance to antenna.

5. Conclusions and Future Works

Integrating IIoT technology, such as implementing networks like TTN, holds transformative potential for revolutionizing equipment management and utilization strategies within the construction domain. Providing real-time connectivity, monitoring capabilities, and comprehensive data collection engenders more informed decision-making processes, optimization of resources, and operational workflows. The continuous monitoring of gateways emerges as a pivotal imperative to ensure connectivity and check performance standards within the TTN network. Effective management of gateway status and communication quality remains essential to uphold network stability and reliability. This study demonstrates an efficacious approach to resource monitoring, particularly on gateways, by harnessing data acquisition via MQTT or relevant requests directed at the TTN server. Moreover, the work establishes prerequisites for data cleaning, organization, and compaction, optimizing data transmission to visualization platforms and third-party entities. This system belongs to a complete scheme for IIoT-integrated management for the construction industry; hence, this work establishes a foundation for systematic and orderly data collection, laying the groundwork for future in-depth analyses. This study mainly shows the data obtained from monitoring the communication performance, mainly from uplink messages, as well as the status of the gateways. However, it does not consider monitoring the active management of resources and gateways, such as creation, update, deletion, purge, and authentication. Future work will consider the implementation of some of these features. Also, future implications can encompass a comprehensive assessment of the integrated device, machinery, operator, and network management; it provides a structured approach for systematically acquiring data to facilitate more profound investigations in subsequent research and works.

Author Contributions: Conceptualization, O.T.S., D.R., A.R. and J.S.S.; methodology, O.T.S., D.R., A.R. and J.S.S.; software, O.T.S.; data curation, O.T.S.; writing—original draft preparation, O.T.S.; writing—review and editing, D.R., A.R., F.B. and J.S.S.; supervision, A.R., F.B. and J.S.S. All authors have read and agreed to the published version of the manuscript.

Funding: This research was funded by the FCT—Foundation for Science and Technology, I.P./MCTES through national funds (PIDDAC), within the scope of CISUC R&D Unit—UIDB/00326/2020 or project code UIDP/00326/2020. Óscar Torres wishes to acknowledge the Portuguese funding institution FCT—Foundation for Science and Technology—for supporting his research under the Ph.D. grants. The authors also acknowledge INESC Coimbra under the project 2023—PDS and VDS.

Institutional Review Board Statement: Not applicable.

Data Availability Statement: Not applicable.Data Availability StatementThe data presented in this study are available on request from the corresponding author. The data are not publicly available due to it is obtained from CONDURIL-ENGENHARIA, S.A. machine monitoring and an access request is required.

Conflicts of Interest: The authors declare no conflict of interest.

References

1. Sanchez, O.T.; Raposo, D.; Rodrigues, A.; Boavida, F.; Marculescu, R.; Chen, K.; Silva, J.S. An IIoT-Based Approach to the Integrated Management of Machinery in the Construction Industry. *IEEE Access* **2023**, *11*, 6331–6350. [CrossRef]
2. Sisinni, E.; Mahmood, A. *Wireless Communications for Industrial Internet of Things: The LPWAN Solutions*; Wireless Networks and Industrial IoT; Springer: Cham, Switzerland, 2021; pp. 79–103. [CrossRef]
3. Fujdiak, R.; Mikhaylov, K.; Pospisil, J.; Povalac, A.; Misurec, J. Insights into the Issue of Deploying a Private LoRaWAN. *Sensors* **2022**, *22*, 2042. [CrossRef]
4. Almuhaya, M.A.M.; Jabbar, W.A.; Sulaiman, N.; Abdulmalek, S. A Survey on LoRaWAN Technology: Recent Trends, Opportunities, Simulation Tools and Future Directions. *Electronics* **2022**, *11*, 164. [CrossRef]
5. Mendes, B.; Plessis, S.D.; Passos, D.; Correia, N. Framework for the Integration of Transmission Optimization Components into LoRaWAN Stack. *Lect. Notes Netw. Syst.* **2022**, *461*, 421–431. [CrossRef]
6. Sherazi, H.H.R.; Grieco, L.A.; Imran, M.A.; Boggia, G. Energy-Efficient LoRaWAN for Industry 4.0 Applications. *IEEE Trans. Ind.* **2021**, *17*, 891–902. [CrossRef]
7. Bahri, N.; Saadaoui, S.; Tabaa, M.; Sadik, M.; Medromi, H. Wireless Technologies and Applications for Industrial Internet of Things: A Review. *Adv. Intell. Syst. Comput.* **2021**, *1188*, 505–516. [CrossRef]
8. Bessa, A.R.R.; Santos, M.S.; Linhares, J.E.B.S.; Valadão, M.D.M.; Amoedo, D.A.; Mattos, E.V.C.U.; Pereira, A.M.C.; Torres, L.M.; Valney, M.N.; Santos, A.M.; et al. Design. Implementation and Evaluation of a Private LoRaWan Network for Industrial Internet of Things (IIoT) Applications. In Proceedings of the 2022 IEEE International Conference on Consumer Electronics—Taiwan, ICCE, Taipei, Taiwan, 6–8 July 2022; pp. 513–514. [CrossRef]

9. Ahmed, S.T.; Annamalai, A. On Private Server Implementations and Data Visualization for LoRaWAN. In Proceedings of the 13th IEEE Symposium on Computer Applications and Industrial Electronics, ISCAIE, Penang, Malaysia, 20–21 May 2023; pp. 342–347. [CrossRef]
10. Barros, T.G.F.; Da Silva Neto, E.F.; Neto, J.A.D.S.; De Souza, A.G.M.; Aquino, V.B.; Teixeira, E.S. The Anatomy of IoT Platforms—A Systematic Multivocal Mapping Study. *IEEE Access* **2022**, *10*, 72758–72772. [CrossRef]
11. Akshatha, P.S.; Kumar, S.M.D.; Venugopal, K.R. MQTT Implementations, Open Issues, and Challenges: A Detailed Comparison and Survey. *Int. J. Sens. Wirel. Commun. Control* **2022**, *12*, 553–576. [CrossRef]

engineering proceedings

MDPI

Proceeding Paper

Visual State Estimation for False Data Injection Detection of Solar Power Generation †

Byron Alejandro Acuña Acurio [1,*,‡], Diana Estefanía Chérrez Barragán [1,‡], Juan Camilo López [2,‡], Felipe Grijalva [3,‡], Juan Carlos Rodríguez [4,‡,§] and Luiz Carlos Pereira da Silva [1,‡]

1 Faculdade de Engenharia Elétrica e de Computação (FEEC), Universidade Estadual de Campinas (UNICAMP), Campinas 13083-852, SP, Brazil; diana.cherrez@gmail.com (D.E.C.B.); lui@unicamp.br (L.C.P.d.S.)
2 Department of Electrical Engineering, Mathematics and Computer Science (EEMCS), University of Twente, 7522 Enschede, Overijssel, The Netherlands; j.c.lopezamezquita@utwente.nl
3 Colegio de Ciencias e Ingenierías "El Politécnico", Universidad San Francisco de Quito USFQ, Quito 170157, Ecuador; fgrijalva@usfq.edu.ec
4 Analog Devices Inc., Wilmington, MA 01887, USA; juancarlos.rodriguez@analog.com
* Correspondence: byronacunia@gmail.com
† Presented at the XXXI Conference on Electrical and Electronic Engineering, Quito, Ecuador, 29 November–1 December 2023.
‡ These authors contributed equally to this work.
§ The opinions expressed in this publication are those of the authors. They do not purport to reflect the opinions or views of the ADI, its subsidiaries or employees.

Abstract: As the penetration level of solar power generation increases in smart cities and microgrids, an automatic energy management system (EMS) without human supervision is most communly deployed. Therefore, assuring safe and reliable data against cyber attacks such as false data injection attacks (FDIAs) has become of utmost importance. To address the aforementioned problem, this paper proposes detecting FDIAs considering visual data. The aim of visual state estimation is to enhance the resilience and security of renewable energy systems. This approach provides an additional layer of defense against cyber attacks, ensuring the integrity and reliability of solar power generation data and facilitating the efficient and secure operation of EMS. The proposed approach uses a modified VGG-16 neural network model to obtain an intermediate representation that provides textual and numerical explanations about the visual weather conditions from sky images. Numerical results and simulations corroborate the validity of our proposed approach. The performance of the modified VGG-16 neural network model is also compared with previous state-of-the-art machine learning models in terms of accuracy.

Keywords: solar power generation; false data injection attacks; computer vision; statistical approach

Citation: Acuña Acurio, B.A.; Chérrez Barragán, D.E.; López, J.C.; Grijalva, F.; Rodríguez, J.C.; da Silva, L.C.P. Visual State Estimation for False Data Injection Detection of Solar Power Generation. *Eng. Proc.* **2023**, *47*, 5. https://doi.org/10.3390/engproc2023047005

Academic Editor: Jaime Cepeda

Published: 4 December 2023

1. Introduction

False data injection attacks (FDIAs), such as data poisoning or noise injection, can significantly affect the decision-making process of energy management system (EMS) applications, for instance, voltage regulation [1]. FDIAs are destructive to EMS [2], because the attacker can manipulate the meter readings by injecting additional false data, causing system instability [3] and even cascading failures leading to massive blackouts [4]. Therefore, various methodologies have been developed over the past decade to defend against such attacks. The existing methodologies can be divided into (i) protection-based and (ii) detection-based approaches [5]. The protection-based approaches [6,7] are based on protecting specific sensors, but these methods have two drawbacks: (i) protecting data will reduce the amount of measured data, and (ii) the protection mechanism can not ensure that the data are always safe. On the other hand, FDIA detection methods traditionally were model-driven approaches. In [8], the authors proposed a method based on Kalman filters to detect FDIA in power grids. A multiobjective optimal detection scheme based on the parity method was proposed in [9], which only applies to DC microgrids but has poor adaptability to the current mainstream AC power grid. Although high detection accuracy is shown when tested on traditional power grid scenarios (without renewable energy resources), the literature does not answer whether the existing detection methods can be applied to microgrids with a high share of solar power generation plants. To the best of the authors' knowledge, the present work is the first attempt to apply computer vision techniques to the FDIA detection problem. However, for a detailed survey of FDIA detection strategies, the reader is referred to [10].

The fundamental problem in FDIA detection methods is to identify tampered measurement z_a reported from a smart metering system, which can be expressed as:

$$z_a = z + a. \tag{1}$$

In this case, a is a nonzero attack value added to the true measurement z of the solar power generation system. Facing this problem, this paper proposes an FDIA detection approach for solar power generation based on image processing of visual weather conditions using convolutional neural networks and transfer learning. To do this, a VGG-16 architecture pretrained on the ImageNet dataset was used. This work expands the theoretical explanation of transfer learning techniques to facilitate reproducibility by newcomers to this field. The main contributions of this work are:

- A novel deep learning architecture that can detect FDIA in solar power generation measurements based on sky images,
- A detailed step-by-step process to perform transfer learning from an object classification domain to the FDIA detection domain in solar power generation.

The remainder of this paper is organized as follows. After this introduction, Section 2 explains each stage of the proposed methodology, step by step. In Section 3, the experimental setup is presented. Section 4 shows the numerical results. Finally, Section 5 presents the conclusions of this work.

2. Proposed Approach

This section describes the main components of our proposed method: (i) a transfer learning procedure using a modified VGG-16 convolutional neural network, (ii) an intermediate representation with a support vector regressor, and (iii) a binary hypothesis test.

2.1. Transfer Learning Procedure

The idea of transfer learning is to use a network previously trained with a large amount of data from a specific task and reuse it in a new task. In this case, a pretrained architecture named visual geometry group 16 (VGG-16) [11] was used as a starting point for our proposed FDAI approach. VGG-16 is a deep convolutional neural network (CNN) consisting of 16 layers with 1.2 million parameters. VGG-16 was initially pretrained using the ImageNet dataset that contains around 16 million images [11] to perform the classification task for 1000 different categories. In this work, the aforementioned VGG-16 architecture of the network was modified, discarding the fully connected and softmax blocks that are highlighted in red in Figure 1.

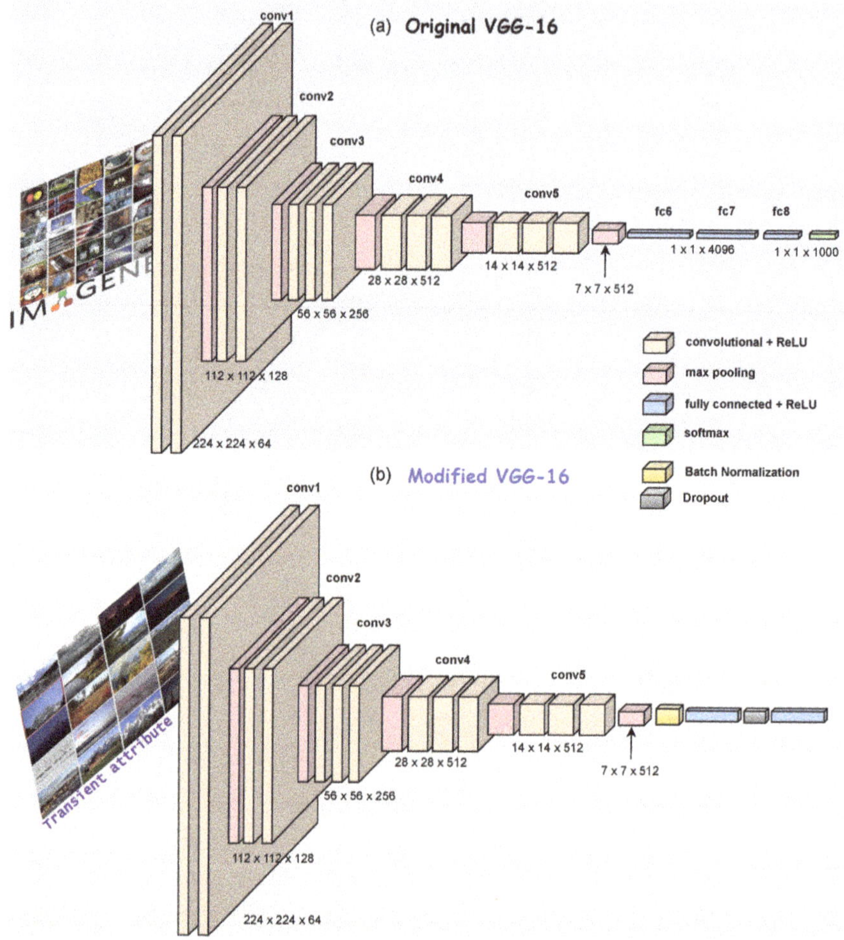

Figure 1. (**a**) Original VGG-16 [11] . (**b**) Modified VGG-16 employed to perform a transfer learning task.

The convolutional and pooling layers shown in Figure 1 work as feature extractor layers; consequently, these layers have been frozen. The removed fully connected layers were replaced with the following layers in an ordered fashion: (i) a new batch normalization layer, (ii) a fully connected layer of 1024 neurons with a rectified linear unit (ReLU) as an activation function, (iii) a dropout layer with a rate set to 10% to avoid overfitting, and (iv) finally, a fully connected layer of 40 neurons with a sigmoid activation function. The original VGG-16 was trained to perform image classification (source domain $\mathcal{D}_s$). A domain $\mathcal{D}$ is the subject that performs learning. It consists of two parts: data $(\mathcal{X}, \mathcal{Y})$ and the distribution $P(x, y)$ that generates such data for any sample (x_i, y_i) in the available data $x_i \in \mathcal{X}, y_i \in \mathcal{Y}$, as follows $\mathcal{D} = \{\mathcal{X}, \mathcal{Y}, P(x, y)\}$.

In this case, $\mathcal{D}_s$ was obtained using the ImageNet dataset that contains around 16 million images $\mathcal{X}_s$ and 1000 different categories $\mathcal{Y}_s$. With transfer learning, it is possible to use the aforementioned modified VGG-16 for a new target domain $\mathcal{D}_t = \{x_j, y_j\}_{j=1}^{N_t}$. Hence, in order to retrain the pretrained VGG-16 from the source domain to the target domain, the "*transient attribute* dataset" $(\mathcal{X}_t, \mathcal{Y}_t)$ publicly available in [12] was used.

The *transient attribute* dataset $(\mathcal{X}_t, \mathcal{Y}_t)$, has 8571 images $\mathcal{X}_t$ from 101 webcams, all annotated with 40 attribute labels $\mathcal{Y}_t$ of different types that contain values from 0 to 1.

The 40 attributes are, for instance, lighting, the season of the year (winter, summer, etc.), weather (sunny, warm, cloudy, etc.), subjective impressions (beautiful, gloomy, soft, etc.), and some additional attributes such as dirty/polluted, busy, lush vegetation, etc. (see Figure 2 for some examples).

Figure 2. Example of an image from "transient attribute database $\mathcal{X}_t, \mathcal{Y}_t$", which is publicly available in [12] with their attribute label.

To perform the aforementioned transfer learning task, a finetuning of network weights was performed over all the layers of the modified VGG-16, since the images of the *transient attribute* dataset used for transfer learning are quite different from the ImageNet dataset used in the original VGG-16. In this work, the Adam optimization algorithm was used. Unlike the stochastic gradient descent (SGD), Adam can vary the learning rate throughout the training process to obtain a better performance model. The learning rate controls the variation of the network weights for each training epoch [13]. In this work, the initial learning rate at was set at 0.001. If there was no network performance improvement during the training epochs, the learning rate was modified to 0.0005. This adjustment aimed to facilitate improvements until the final stages of training when there was no progress concerning its validation through the mean absolute error (MAE) over 15 consecutive epochs. Finally, the training process was limited to a maximum of 250 epochs. Given that the target domain $\mathcal{D}_t$ is a 40 multi-output regressor, where each output ranges from 0 to 1, in this work, it was found that the most suitable loss function for performing the transfer learning task was the mean absolute error (MAE).

2.2. Intermediate Representation

Depending on the season of year and weather conditions, the modified VGG-16 neural network that was fitted using the aforementioned transfer learning approach looks for very different points in the image, as shown in the heat maps of Figure 3.

Figure 3. A heat map shows that the modified VGG-16 neural network looks for very different points in the image, depending on the weather conditions.

Therefore, the convolutional layers of the modified VGG-16 work as a feature extractor, as shown in Figure 4.

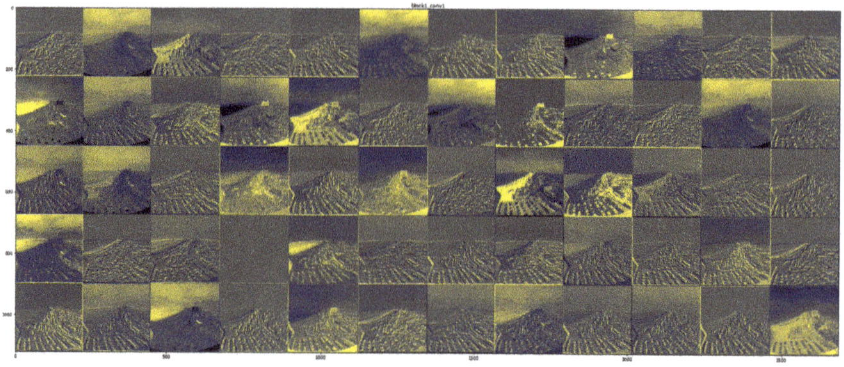

Figure 4. Example of features extracted by convolutional layers.

The output of the modified VGG-16 is 40 attributes, as shown in the correlation matrix in Figure 5.

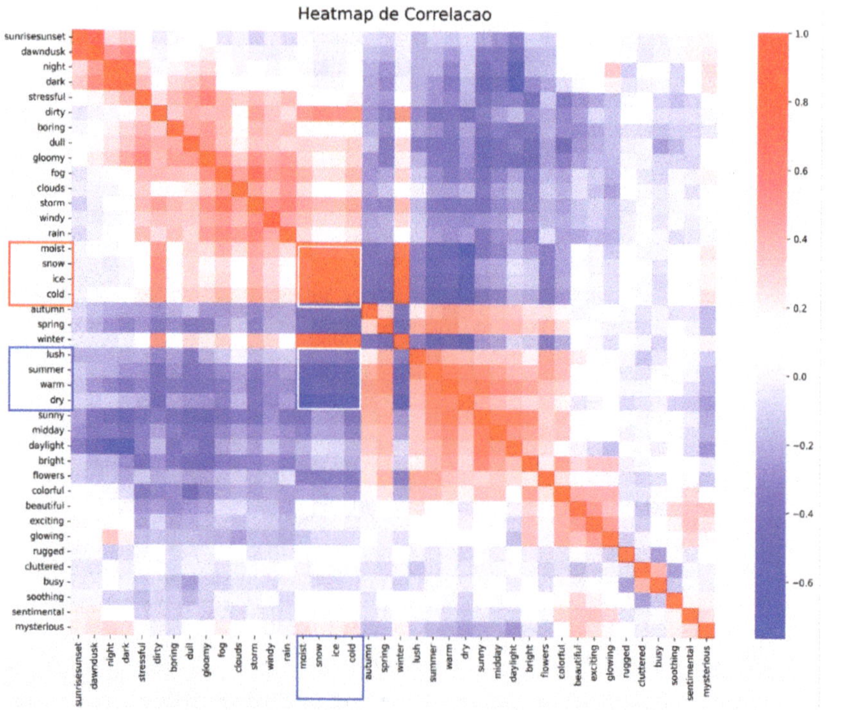

Figure 5. Correlation Matrix of the 40 attributes obtained from the modified VGG-16 from the images.

This work used these 40 attributes as an intermediate representation to perform the state estimation of solar power generation. The intermediate representation captures a numerical description of the variability in weather conditions that are very distinctive over time. To do this, historical sky images collected from the same location as the solar power generation plant under different conditions over one month were used to create a new dataset $\mathcal{D}_{sky}$ of the historical intermediate representations $\mathcal{X}_{sky}$ and the historical solar power generation $\mathcal{Y}_{sky}$. Then, a support vector regressor (SVR) with Radial Basis Function (RBF) kernel was trained using $\mathcal{D}_{sky}$ to receive the intermediate representation and return the most likely state of solar power generation. It was empirically observed that 500 sky images were enough to fit the SVR.

2.3. Binary Hypothesis Test

To detect a false data injection attack on solar power generation measurements, an inequality chi-square test χ^2 was employed. For this, an estimated chi-square value $\hat{\chi}^2$ was computed using the observed values and estimates, as follows:

$$\hat{\chi}^2 = \sum_{i=1}^{m} \frac{(O_i - E_i)^2}{E_i}, \tag{2}$$

where O represents the solar power generation measurements, E represents the estimated values of the solar power generation obtained from the SVR, and m is the number of measures over one day. To compute the chi-square probability distribution value χ^2, the degrees of freedom $k = m - n$ were defined, where n is the state variable, in this case $n = 1$. The estimated chi-square value $\hat{\chi}^2$ was compared with the value of χ^2 for a given degree of freedom k and significance level α. The significance level α is the upper bound on the probability that a Type I error will occur after performing a hypothesis test. A Type I error occurs when the null hypothesis is correct but is rejected. In this work, a significance level of α =0.03 was used, which indicates that there is a 3% chance that there are erroneous data or a confidence level of 97%. Consequently, the following chi-square test can be performed for the false data injection detection of solar power generation using visual state estimation:

- If $\hat{\chi}^2 >= \chi^2_{k,\alpha}$, false data injection attacks are suspected;
- If $\hat{\chi}^2 < \chi^2_{k,\alpha}$, false data injection attacks are not suspected.

3. Experimental Setup

Images captured by the webcams have different sizes, requiring a preprocessing step before using the aforementioned approach. Initially, all images were resized to dimensions of 200 × 200 with 3 color channels (RGB). Subsequently, for compatibility with the modified VGG16, the images were transformed into tensors of dimensions 200 × 200 × 3.

The *transient attribute* dataset [12] with 8571 images was used to perform the transfer learning stage, to finetune the modified VGG-16. The *transient attribute* dataset was divided between testing and training data according to the original paper [12], but a double-stratified k-fold cross-validation approach was used to validate the modified VGG-16 during the training stage. Under this approach, the entire training dataset was divided into $k = 10$ fold, one fold for validation and the remaining folds for training. This procedure allows us to obtain a more realistic idea of the performance of the model [13].

The modified VGG-16 neural network was trained with 250 epochs. However, empirically, it was observed that a good fit was obtained with only 25 epochs, as shown in the results section on the biases–variance curve, MAE metric, and R2 curve (see: Figure 6). Some data augmentation techniques were used in the finetuning stage of the modified VGG-16. The data augmentation stage aims to generate new images from the original ones. To achieve this goal, the original images were modified to generate new instances using the following set of transformations, which are:

- *Rotation range*: This is the degree range for random rotations. In this study, a range between −5 and +5 degrees was used;
- *Width shift range*: The original image is randomly shifted by a proportional percentage of the original image width. In this study, this parameter was set at 0.2, i.e., 20%;
- *Height shift range*: This is similar to the previous transformation, but it uses the height of the image to perform the shift. In all the experiments, this parameter was set to 0.2, i.e., 20%;

- *Zoom range*: This transformation generates a random zoom. In this work, a range between 90% and 110% of the original image was used;
- *Horizontal flip*: This randomly flips the image horizontally;
- *Vertical flip*: This randomly flips the image vertically;
- *Brightness range*: This increases or decreases the brightness of the image. Hence, this parameter was set in a range between 0.9 and 1.1;
- *Fill mode*: In all the experiments, the nearest approach fill mode was used, which fills points outside the boundaries of the image with similar information to that of the boundaries,

These transformations are applied to generate "on-demand" new images in the fine-tuning stage of the VGG-16.

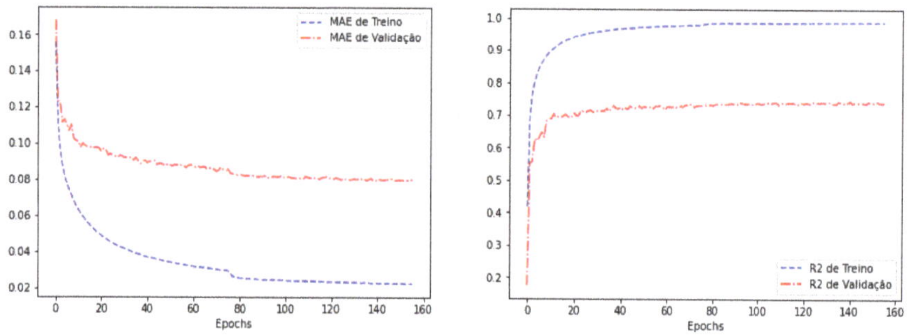

Figure 6. Numerical results of the training modified VGG-16 neural network.

Svr Training Details

For reproducibility purposes, the publicly available dataset of sky images in [14] was used to create the new dataset $\mathcal{D}_{sky}$ of historical intermediate representations $\mathcal{X}_{sky}$ and historical solar power generation $\mathcal{Y}_{sky}$. This new dataset $\mathcal{D}_{sky}$ was employed to train the support vector regression model.

4. Results and Discussion

Our proposed modified VGG-16 was compared to similar previous work [12]. To ensure a fair comparison the proposed model was trained using the same dataset that was reported in [12]. The results of our model are presented in Table 1. Figure 6 shows the training and validation learning curves of our proposed modified VGG-16 neural network model. Both the training and validation loss curves decreased after the modified VGG-16 model training began. This can be attributed to the transfer learning, indicating that the convolutional layer of the modified VGG-16 already had a high level of feature extraction, because these layers were pretrained. Although occasional fluctuations can be observed in the training loss, it is clear that the overall trend is a continuous decrease in the mean absolute error (MAE) during training.

Table 1. Comparison of the proposed modified VGG-16 neural network with previous works [12].

Modelo	MSE
SVM [12]	0.070
log reg [12]	0.093
SVR [12]	0.043
Proposed modified VGG-16	0.0319

The results of the intermediate representation show that the best-performing approach was accomplished using the proposed modified VGG-16 neural network as shown in Table 1.

The results indicate that, inside the critical region, the estimated values and the groundtruth values using $\alpha = 0.03$, as suggested in Section 2.3 for false data injection detection, can be considered the same, as shown in Figure 7.

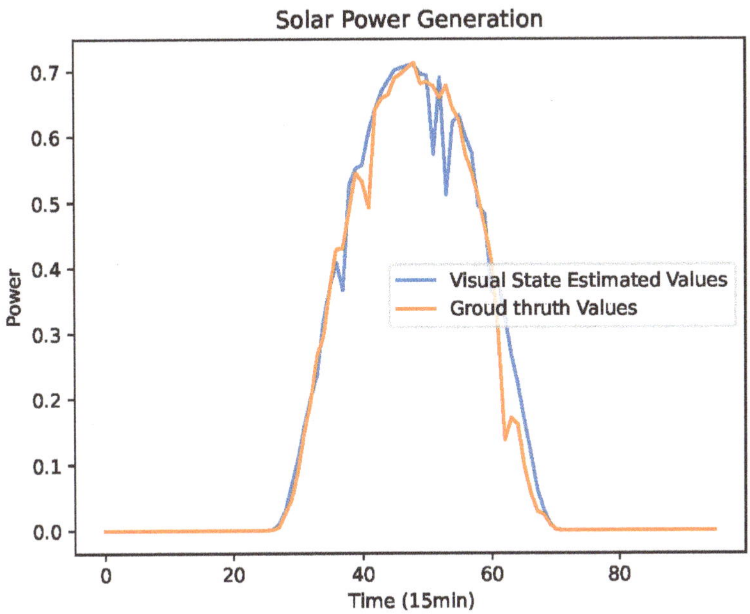

Figure 7. Comparison of the visual state estimated values and ground truth.

5. Conclusions

This work explores the false data injection detection in solar power generation from sky images, using a modified VGG-16 neural network to obtain an intermediate representation that can be used to estimate power generation with a support vector regressor. Results comparing the estimated values and the ground truth did not reveal a significant difference without a false data injection attack. A measurement discrepancy was detected when a data injection attack was performed in a random measurement. Our proposed approach overcomes the previous work [12] in terms of performance. The proposed approach is flexible and can be easily adapted to different solar power generation systems.

6. Future Works

As a future work, it could be interesting to study actions once a false data injection attack is detected; for instance, an appropriate mitigation strategy can be implemented. This might involve isolating the affected components, recalibrating sensors, restoring valid data from backups, or even triggering an automated response to neutralize the attack.

Author Contributions: Conceptualization, B.A.A.A. and D.E.C.B.; methodology, B.A.A.A.; software, D.E.C.B.; validation, L.C.P.d.S., J.C.L., F.G. and J.C.R.; formal analysis, D.E.C.B.; investigation, B.A.A.A.; resources, L.C.P.d.S.a; data curation, F.G. and J.C.R.; writing—original draft preparation, B.A.A.A.; writing—review and editing, D.E.C.B.; visualization, D.E.C.B.; supervision, J.C.L.; project administration, L.C.P.d.S.; funding acquisition, L.C.P.d.S. All authors have read and agreed to the published version of the manuscript.

Funding: This work has been supported by the following Brazilian Research Agencies: FAPESP, CAPES, CNPq, INEP, FINEP, etc. The authors are funded by the grant #2022/16881-5, #2020/03069-5, #2021/11380-5, and #2016/08645-9, Centro Paulista de Estudos da Transição Energética (CPTEn), São Paulo Research Foundation (FAPESP). This work was also developed under the Electricity

Sector Research and Development Program PD-00063-3058/2019 - PA3058: "MERGE - Microgrids for Efficient, Reliable and Greener Energy", regulated by the National Electricity Agency (ANEEL in Portuguese), in partnership with CPFL Energia (Local Electricity Distributor). This work was also supported by the Universidad San Francisco de Quito through the Poli-Grants Program under Grant 17993.

Institutional Review Board Statement: Not applicable.

Informed Consent Statement: Not applicable.

Data Availability Statement: This work used sky images dataset publicly available in [14] and transient attribute database, which is publicly available in [12].

Conflicts of Interest: The authors declare no conflict of interest.

References

1. Musleh, A.S.; Chen, G.; Dong, Z.Y. A survey on the detection algorithms for false data injection attacks in smart grids. *IEEE Trans. Smart Grid* **2019**, *11*, 2218–2234. [CrossRef]
2. Tan, R.; Nguyen, H.H.; Foo, E.Y.; Yau, D.K.; Kalbarczyk, Z.; Iyer, R.K.; Gooi, H.B. Modeling and mitigating impact of false data injection attacks on automatic generation control. *IEEE Trans. Inf. Forensics Secur.* **2017**, *12*, 1609–1624. [CrossRef]
3. Acurio, B.A.A.; Barragán, D.E.C.; Amezquita, J.C.L.; Rider, M.J.; Da Silva, L.C.P. Design and Implementation of a Machine Learning State Estimation Model for Unobservable Microgrids. *IEEE Access* **2022**, *10*, 123387–123398. [CrossRef]
4. Ameli, A.; Hooshyar, A.; El-Saadany, E.F.; Youssef, A.M. Attack detection and identification for automatic generation control systems. *IEEE Trans. Power Syst.* **2018**, *33*, 4760–4774. [CrossRef]
5. Chaojun, G.; Jirutitijaroen, P.; Motani, M. Detecting false data injection attacks in AC state estimation. *IEEE Trans. Smart Grid* **2015**, *6*, 2476–2483. [CrossRef]
6. Yang, Q.; Yang, J.; Yu, W.; An, D.; Zhang, N.; Zhao, W. On false data-injection attacks against power system state estimation: Modeling and countermeasures. *IEEE Trans. Parallel Distrib. Syst.* **2013**, *25*, 717–729. [CrossRef]
7. Wang, Z.; He, H.; Wan, Z.; Sun, Y. Detection of false data injection attacks in ac state estimation using phasor measurements. *IEEE Trans. Smart Grid* **2020**. [CrossRef]
8. Manandhar, K.; Cao, X.; Hu, F.; Liu, Y. Detection of faults and attacks including false data injection attack in smart grid using Kalman filter. *IEEE Trans. Control Netw. Syst.* **2014**, *1*, 370–379. [CrossRef]
9. Tan, S.; Xie, P.; Guerrero, J.M.; Vasquez, J.C. False data injection cyber-attacks detection for multiple dc microgrid clusters. *Appl. Energy* **2022**, *310*, 118425. [CrossRef]
10. Husnoo, M.A.; Anwar, A.; Hosseinzadeh, N.; Islam, S.N.; Mahmood, A.N.; Doss, R. False data injection threats in active distribution systems: A comprehensive survey. *arXiv* **2022**, arXiv:2111.14251v2.
11. Simonyan, K.; Zisserman, A. Very deep convolutional networks for large-scale image recognition. *arXiv* **2014**, arXiv:1409.1556.
12. Laffont, P.Y.; Ren, Z.; Tao, X.; Qian, C.; Hays, J. Transient Attributes for High-Level Understanding and Editing of Outdoor Scenes. *ACM Trans. Graph.* **2014**, *33*, 4. [CrossRef]
13. Goodfellow, I.; Bengio, Y.; Courville, A. *Deep Learning*; MIT Press: Cambridge, MA, USA, 2016.
14. Nie, Y.; Li, X.; Scott, A.; Sun, Y.; Venugopal, V.; Brandt, A. SKIPP'D: A SKy Images and Photovoltaic Power Generation Dataset for short-term solar forecasting. *Sol. Energy* **2023**, *255*, 171–179. [CrossRef]

engineering proceedings

MDPI

Proceeding Paper

Power System Dynamic Data Generation Based on Monte Carlo Simulations for Machine Learning Applications †

Jaime Cepeda

Departamento de Energia Electrica, Facultad de Ingenieria Electrica y Electronica, Escuela Politecnica Nacional, Quito 170525, Ecuador; cepedajaime@ieee.org

† Presented at XXXI Conference on Electrical and Electronic Engineering, Quito, Ecuador, 29 November–1 December 2023.

Abstract: A problem with applying machine learning for analyzing power system dynamics is the lack of specific datasets. In this realm, defining a strong methodology to obtain power system dynamic data is an important task prior to the application of any machine learning tool. This is particularly important considering the current growing research in the field of self-healing grids. Thus, this paper presents a well-defined stochastic methodology that can be used to generate dynamic data that can afterwards be analyzed using machine learning tools. The proposed method is based on a Monte Carlo simulation and this paper presents the procedure to perform it.

Keywords: dynamic data; Monte Carlo simulation; probability distribution functions

1. Introduction

Traditionally, power system security assessments have been studied using deterministic methodologies, which consider some extreme operating conditions (different load levels) and selected critical contingencies (associated with some fault types and fault locations). This type of study ignores the stochastic or probabilistic nature of real power systems, and therefore certain severe events that could lead a system to potential insecure conditions may be ignored [1]. Since the huge volume of uncertainties greatly influences the power system dynamic response, it is necessary to apply mathematical tools which allow for the consideration of the most probable scenarios. One of the main classes of probabilistic techniques is the Monte Carlo (MC)-based simulation, which provides the possibility of obtaining more realistic results, mainly for complex system analysis [1,2], since it avoids using surrogate models. The MC method allows for simulating phenomena with significant involved levels of uncertainty, which correspond to electric power systems' operational conditions. In addition to the previously stated fact, the recent spike of interest in machine learning and artificial intelligence (AI) methods that can be applied to power system dynamics has given rise to a major issue for scholars and scientists: how to objectively compare the machine learning and data mining schemes and algorithms for stability prediction (i.e., classification or regression)? [3–6]. This problem has generated even more interest with the recent design of smart grid applications to perform timely self-healing and adaptive reconfiguration actions based on system-wide analysis, with the objective of reducing the risk of power system blackouts. A real-time dynamic vulnerability assessment (DVA) must be performed to decide and coordinate the appropriate corrective control actions, depending on an event's evolution [3,4]. For this aim, the application of machine learning and AI tools constitutes a very attractive alternative. Thus, emerging technologies such as phasor measurement units (PMUs) and wide area monitoring systems (WAMSs) offer a new benchmark for performing post-contingency DVAs that could be used to trigger system integrity protection schemes (SIPSs) or activate wide area controls to implement enhanced corrective control actions [3–6]. In connection to this, the fast detection of critical conditions (from the shortest possible data record prior to, during and after a

Citation: Cepeda, J. Power System Dynamic Data Generation Based on Monte Carlo Simulations for Machine Learning Applications. *Eng. Proc.* **2023**, 47, 6. https://doi.org/10.3390/engproc2023047006

Academic Editor: Jackeline Abad

Published: 4 December 2023

fault) is instrumental in response-based special stability controls. Such a scheme should be as agnostic as possible to the type of instability that may occur within, say, 5 to 10 s after the fault clearing or the triggering event. Therefore, accurate and fair tests need to be conducted to determine the specific tool behavior. Nevertheless, the problem found in the recent literature is that most authors that are performing research on this important topic do not have specific datasets to test their models in such a way to obtain fair comparisons. In this vein, especially since the emergence of deep learning, they claim that they have "the best algorithm" ever, in terms of accuracy and pre-emption time. However, while authors have used well-known test systems, the dataset is never the same, and one of the reasons for this is the lack of a well-defined methodology for data generation. Thus, it is becoming more and more difficult to separate well-founded claims from fake claims and erroneous claims that have been made in good faith by authors. Based on the mentioned problem, this paper presents a well-defined stochastic methodology that can be used to generate sufficiently confident dynamic data that can be used for DVA applications based on machine learning. The method is based on Monte Carlo simulations, which are stochastic processes that are commonly used to perform probabilistic assessments. For instance, a Monte Carlo simulation is used in [7] to emulate the effect of high-speed wind on a test system and determine the failure state of a transmission using a reduced fragility function that was proposed to evaluate the resilience of the system. Also, a Monte Carlo simulation was combined with k-means clustering and an elbow plot to determine feasible scenarios of photovoltaic and wind generation integration into the power system in [8] for solving the probabilistic optimal power flow problem. A Monte Carlo simulation is also used in [9] to simulate the vehicle to grid (V2G) systems, including the process of power flow and the battery state, for charging and discharging (SoC) electric vehicles (EVs). However, the use of Monte Carlo simulations for power system dynamic data generation oriented to DVAs has not been explored in these works.

Therefore, the main objective of applying a Monte Carlo simulation in this paper is to generate post-contingency system static or dynamic data considering a great number of the probable scenarios and contingencies, even those that could push the system to potential cascading events and subsequent blackouts. Then, the number of MC repetitions cannot be directly determined by the convergence relations (such as those presented in [2]). Thus, the number of iterations will depend on the complexity of the system and the attainment of a dataset that is sufficiently representative to reveal post-contingency system vulnerability patterns and statuses. In connection to this, it is suggested that the number of generated vulnerable cases represents at least 20% of the total number of cases. This value has been empirically determined through several experiments that were carried out during the development of the present research.

The rest of the paper is organized as follows: Section 2 describes the concept of Monte Carlo simulations. In Section 3, the general procedure to perform a Monte Carlo simulation for power system dynamic data generation oriented to DVAs is presented. Section 4 provides numerical results to demonstrate the feasibility and performance of the proposal applied to different machine-learning-based DVA methods. Finally, a discussion of the proposed approach as well as the conclusions and outlook for future research are summarized in Section 5.

2. Monte Carlo Method as a Data Generation Technique

The Monte Carlo method is a repetitive procedure that consists of evaluating, at each repetition, the system response through the uncertainty function h, using a set of input variables (**z**) which are generated randomly from their probability distribution functions (PDFs), in order to obtain numerical random output values (**v**) [2].

$$\mathbf{v} = \mathrm{h}(\mathbf{z}) \quad (1)$$

Thus, the several MC outputs (i.e., one output per iteration) are usually aggregated in order to obtain statistical results (typically represented by the expected value and standard deviation) [1,2]. However, this paper does not specifically apply MC-based simulations with the aim of obtaining statistical information. The main objective of using this probabilistic technique is to generate post-contingency system static or dynamic data considering a great number of the probable scenarios and contingencies, even those that could push the system to potential cascading events and subsequent blackouts.

3. Materials and Methods

To perform MC-based simulations, several input data are required. These data depend on the objective of the simulation, and they are usually represented by their corresponding probability distribution functions (PDFs). For the proposed objective, the probabilistic models of power system random variables, as well as the grid topology, are structured for a short-term planning horizon to reflect the behavior of the system as realistically as possible [2]. In connection to this, the basic data to be considered in the MC-based simulation, which must be previously prepared by the system operator, are the short-term forecasting of nodal loads, the short-term unit commitment, the short-term system topology, and the random generation of N-1 contingencies.

3.1. Short-Term Forecasting of Nodal Loads

Electric power demand and energy consumption correspond to non-linear time functions that present different values in each grid load node. This peculiarity is because of the nature and diversity of customers (i.e., residential, commercial or industrial). In connection to this, the hourly behavior of each load (load chronology) is characterized by different daily load curves which resemble customers' own usage of electricity [10]. Therefore, the analysis of load forecasting must reflect this behavior in each of the nodal loads. Another aspect to be considered is the horizon of operational planning. There are three types of operational planning horizons: long-term, mid-term, and short-term [2,10]. Short-term load forecasting is one of the operator's common tasks due to it being employed as an input of the short-term economic dispatch. Since the horizon of interest for dynamic analysis is limited to the short term, which embraces a forecasting timeframe of one day to several weeks [10], three different daily load curves (i.e., typical residential, typical industrial and typical commercial) can be used. These typical load curves represent the results of the corresponding short-term load forecasting performed by the ISO. Figure 1a illustrates the load curves for various end-user classes, which can be used as nodal loads in test power systems for generating dynamic data. It is worth mentioning that whilst the active power demand is determined based on the daily load curves depicted in Figure 1a, the reactive power demand will be estimated based on the assumption that the power factor keeps a quasi-invariant behavior. Considering that, regardless of the load forecasting method, the actual system load behavior will always differ from that obtained from forecasting; the resulting uncertain behavior of the load must be considered. Thus, the forecasting uncertainty has been included in the formation of the hourly nodal loads via normal (or Gaussian) distribution functions (2).

$$\mathrm{f}(x) = \frac{1}{\sigma\sqrt{2\pi}} e^{-\frac{(x-\mu)^2}{2\sigma^2}} \tag{2}$$

where μ is the mean or expectation of the distribution, σ is its standard deviation, and its variance is therefore σ^2, whereas μ is defined depending on the base case load data and the corresponding load curve of Figure 1a and σ is introduced as a pre-defined parameter (which will depend on a historically based statistical analysis in a real system). Figure 1b presents a family of the described normal PDFs with a mean of 120 MW and different values for the parameter σ.

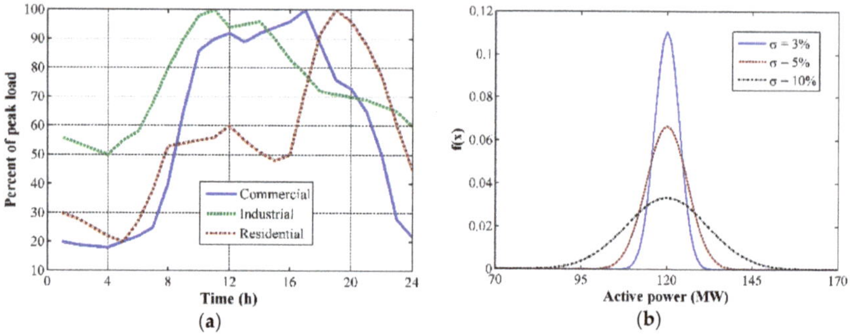

Figure 1. Stochastic load model: (**a**) class-specific daily load curves and (**b**) the family of normal distribution functions with parameter σ.

3.2. Short-Term Unit Commitment and Short-Term System Topology

The determination of the optimal scheduling of generation units to meet the demand that is forecasted over a future short-term period is a complex operational planning problem. This problem is the so-called "unit commitment", and it consists of economically optimizing the schedule of available generating units, subject to specific constraints [11].

There are two types of typical constraints: the "coupling" constraints and the "local" constraints [11], and both are associated with uncertainties [2]. The coupling constraints refer to the amount of required power, and their associated uncertainties are due to possible changes in the market dispatching rules and primary energy sources' availability. Likewise, the local constraints deal with technical limitations which might influence the unit operation status (due to maintenance plans or the failure of components). Since the solution of unit commitment comprises a complex optimization problem (solved by different approaches [11]), it is assumed that this task has been previously carried out by the operator. So, there already exists a tentative short-term schedule of dispatching units, which also includes the available spinning reserve. This tentative schedule will be used as the input for the methodology proposed in this paper.

Short-term operational planning also concerns the determination of grid topology, considering all the available transmission elements and the execution of a procedure called "network topology optimization" [12]. The objective is to use the available assets to increase grid flexibility and efficiency.

3.3. Random Generation of N-1 Contingencies

The N-1 contingencies of system elements are assumed to be independent events and are generated randomly. The set of possible independent contingencies includes generator outages, load events and three-phase short circuits in transmission lines, followed by the outage of the faulted line. The short circuits are randomly applied at different locations on the transmission lines. It is worth mentioning that the possible selected contingencies might be adapted to the system under study to consider the most probable and stressful perturbations. The probability of the fault type obeys a discrete probability distribution, which depends on a system's historical data, as shown in Equation (3) [13].

$$P_i = \frac{f_i}{\sum_{j=1}^{M} f_j} \tag{3}$$

where P_i is the probability of the i-th fault type, f_i is the number of the i-th fault type and M is the number of considered fault types. The probability of the fault's location and the amount of load to be modified in each load event is determined by adequate PDFs that are also defined by historical statistical data. In test power systems, the random generation of contingencies can be based on two types of distribution functions: uniform (random faulted element selection and random short circuit location) and Weibull (random load step size). Based on the consideration that the occurrence of any contingency is equally probable (this aspect does not necessarily need to be satisfied in a real system, where the selection of the appropriate PDF will depend on statistics), the random faulted element selection is based on a uniform PDF (4). This PDF is illustrated in Figure 2a.

$$f(x) = \begin{cases} \frac{1}{b-a} & for\ a \leq x \leq b \\ 0 & for\ x \langle a\ or\ x \rangle b \end{cases} \tag{4}$$

where a and b are the minimum and maximum values of the PDF support. In this specific application, a = 0 and b = 0.99, and the index of the selected faulted element ($Element_i$) is determined via Equation (5).

$$\text{Element}_i = \text{floor}(N_{elements} \cdot x) + 1 \tag{5}$$

where x is the random variable between a and b, $N_{elements}$ is the number of elements to be susceptible to failures, $Element_i = 1 \ldots N_{elements}$ is the index of the selected faulted element, and floor(·) represents the rounding toward a negative infinity value. Likewise, it has been considered that all the line lengths present the same probability of being susceptible to short circuits (this fact is not necessarily satisfied in a real system, where the selection of the appropriate PDF will depend on statistics). In this connection, the random short circuit location is also based on the uniform PDF represented via Equation (4). In this case, parameters a and b are equal to 0% and 100% of the line length.

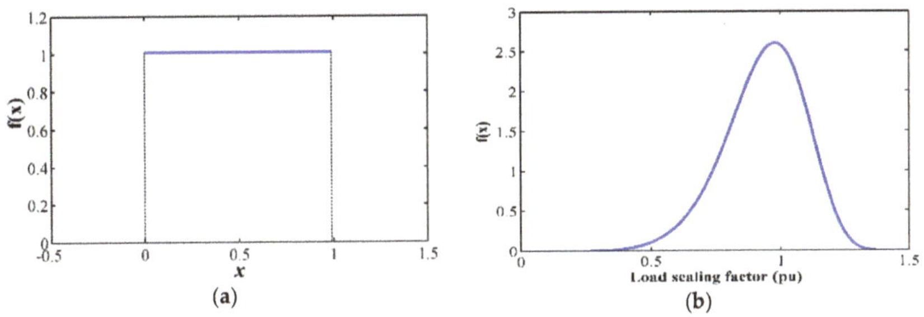

Figure 2. Random N-1 contingencies: (**a**) uniform distribution function for selection of faulted element and (**b**) the Weibull distribution function for load events.

Finally, the Weibull distribution function has been adopted for representing the random load step size to be increased or disconnected in load events. This type of PDF has the property of being able to adapt its shape to the specific physical characteristic which, in this case, is the fact that the load can be completely disconnected (negative step of −100%), but it is practically impossible to be increased in the same percentage (positive step of 100%). Thus, whilst the negative side can vary from 0% to −100%, the positive range might only vary from 0% to a maximum value determined via the statistics in a real system. The Weibull PDF (6) can be easily adapted to represent this characteristic.

$$f(x) = \begin{cases} \frac{k}{\lambda}\left(\frac{x}{\lambda}\right)^{k-1} \cdot e^{-\left(\frac{x}{\lambda}\right)^k} & \text{for } x \geq 0 \\ 0 & \text{for } x < 0 \end{cases} \tag{6}$$

where k > 0 is the shape parameter and λ > 0 is the scale parameter of the distribution. In the proposed application, x corresponds to the load scaling factor, which can vary from 0 to infinity, and represents a factor to be multiplied by the base case load value to acquire the post-event load value. Then, the load step size (in p.u.) will be equal to x − 1. Figure 2b depicts the Weibull PDF employed for representing the load scaling factor, where λ = 1, and k = 7. Once adequate PDFs have been defined, a sampling procedure must be selected to draw the contingency random samples in consistency with their PDFs. Based on the analysis performed in [2], the random sampling technique is selected for the proposed Monte Carlo simulation due to it having shown better results in terms of the mean and variance in a power system probabilistic dynamic analysis.

3.4. General Structure of the Monte Carlo Process

Based on the previously settled system operating policies (i.e., a pre-defined short-term system topology and unit commitment dispatch rules) and the probabilistic models of the input parameters (i.e., the load variation and occurrence of contingencies), a Monte Carlo-type simulation is performed to iteratively evaluate the system responses, which would resemble those signals recorded by the PMUs in real time, with the goal of structuring a performance database.

In connection to this, and depending on the specific application, static or dynamic N-1 contingency simulations, as well as a modal analysis, are performed via the MC process, as depicted in Figure 3. First, the PDFs randomly generate the input variables to be considered in the Monte Carlo simulation, i.e., the load in each bus, the type of contingency (e.g., short circuit, generation outage or load event), the faulted element (line, generator or load), the short circuit location, the amount of load to be changed or the branch outage, in the case of static contingency analysis. Then, optimal power flow (OPF) is performed for every trial set of input variables to define feasible pre-contingency steady-state scenarios. Afterwards, an N-1 static contingency analysis, dynamic time domain simulations, or a modal analysis are performed to obtain the pre- and post-contingency power flow results, the post-contingency dynamic data or the eigenvalues and eigenvectors, respectively. These data will then be used as inputs for machine learning tools.

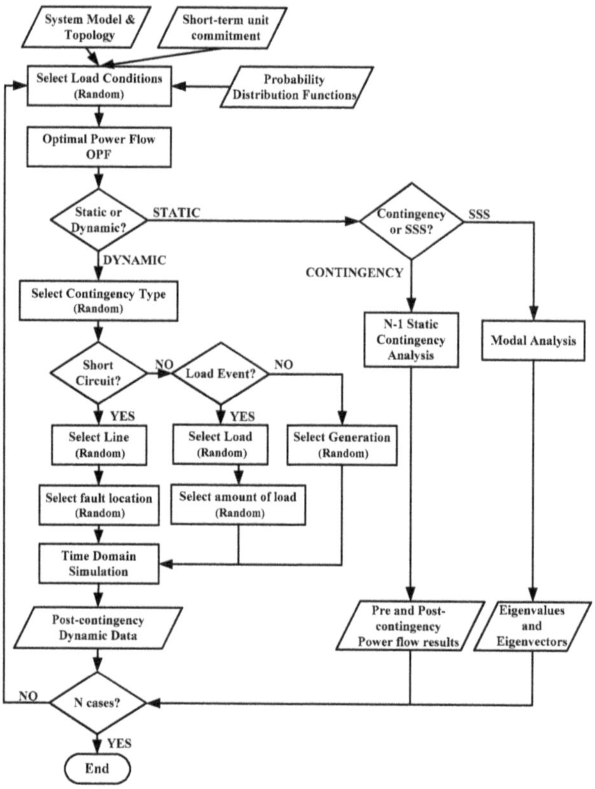

Figure 3. Methodological framework of the proposed MC-based simulations.

4. Results

The proposed MC-based simulation has been applied to the IEEE New England 39-Bus test system [3] to obtain several post-contingency dynamic data that can later be used to train different machine learning tools for performing DVAs. In connection to this, [3,4] the data can be used to perform a DVA for five different symptoms of system stress: transient stability, oscillatory stability, short-term voltage stability, short-term frequency stability, and post-contingency overloads initiated by topological modifications and power injection changes which follow an N-1 contingency. Likewise, [5,6] show the application for generating data for transient stability assessments using machine learning.

Figure 4 illustrates the bus signals (i.e., bus voltage magnitudes and frequencies) for two different Monte Carlo iterations obtained with the application of the proposed methodology.

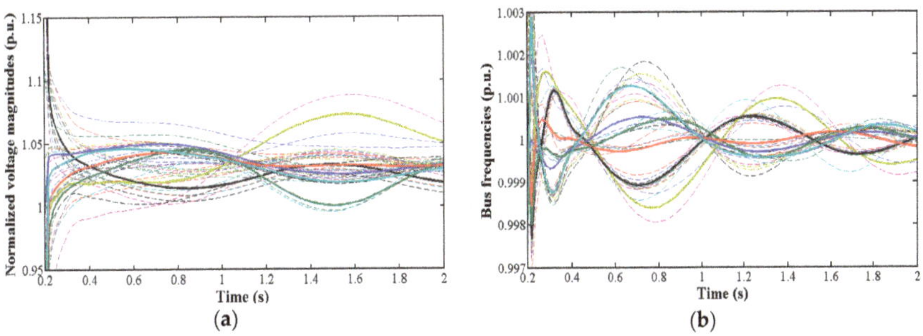

Figure 4. Dynamic data obtained from MC-based simulation: (**a**) normalized voltage magnitudes: three-phase short circuit applied on lines 13–14 at 51.3% of the line length, high-load scenario; (**b**) bus frequencies: three-phase short circuit applied on lines 6–11 at 96.0% of the line length, medium-load scenario. Colorful dotted lines represent the different coherent areas determined in the system and solid lines show the representing centroids for each area.

On the other hand, Table 1 presents the dynamic responses of the single machine equivalent (SIME) method used in [5] for a transient stability assessment, corresponding to three cases of the Monte Carlo simulation: "unstable", "stable" and "very unstable", respectively. Likewise, Figures 5–7 present the rotor angles of synchronous generators for each one of these cases. It is possible to observe how the proposed MC simulation can determine possible cases with different stability statuses, which is important for training machine learning tools for system security assessments.

Table 1. MC-based SIME simulation results.

Case	*Contingency*		*SIME Method*		
	Contingency and Location Resulting from MC	t_{cl} **(ms)**	*Critical Machines*	t_u/t_r **(ms)**	η
Unstable	L/T 2-25, (90%)	114	{G8, G9}	565.7	−1.91
Stable	L/T 13-14, (5%)	80	{G2, G3}	386.7	12.2
Very unstable	L/T 21-22, (15%)	200	{G6, G7}	-	-

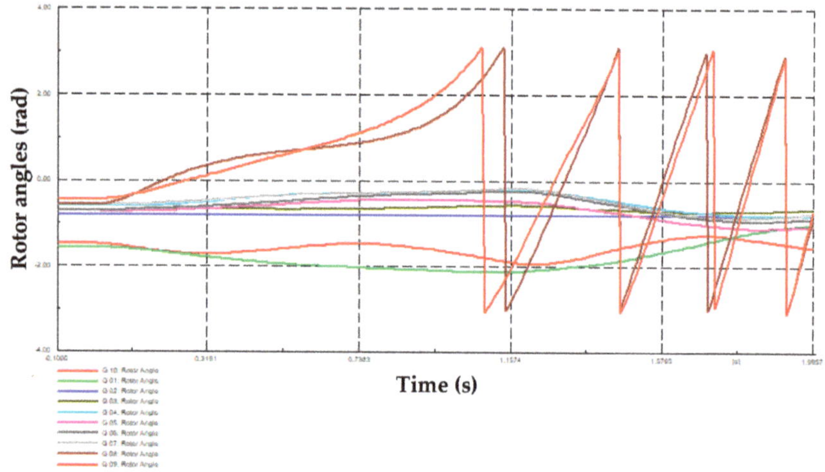

Figure 5. Synchronous machines' rotor angles obtained from the MC-based simulation—unstable case.

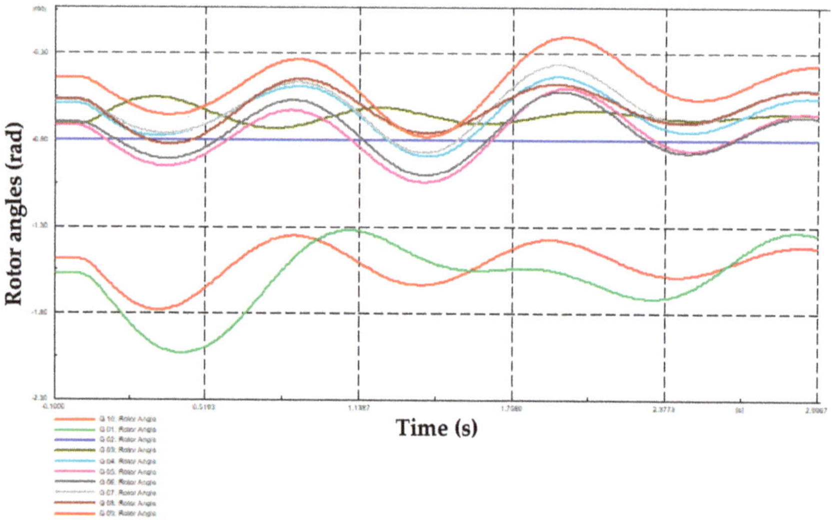

Figure 6. Synchronous machines' rotor angles obtained from the MC-based simulation—stable case.

Finally, with the aim of appreciating how the MC-based dynamic data are capable of being used as the inputs of machine learning tools, the results of applying empirical orthogonal functions (EOFs) [14] to the largest Lyapunov exponent (LLE) [15], as proposed in [6], is presented in Figure 8. This figure presents the parallel coordinates plot of the 15 EOF scores of the LLE, corresponding to the MC-based simulations of G2. In this plot, it is possible to observe a well-defined region of unstable cases, which can later be used as the inputs of a classifier (supervised machine learning).

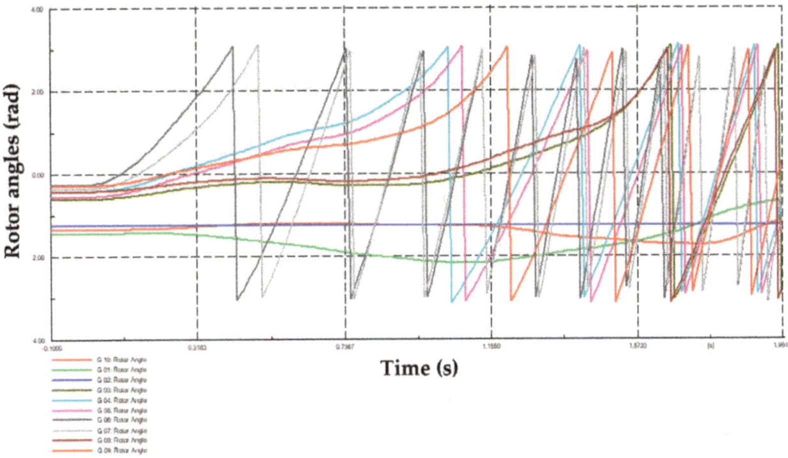

Figure 7. Synchronous machines' rotor angles obtained from the MC-based simulation—very unstable case.

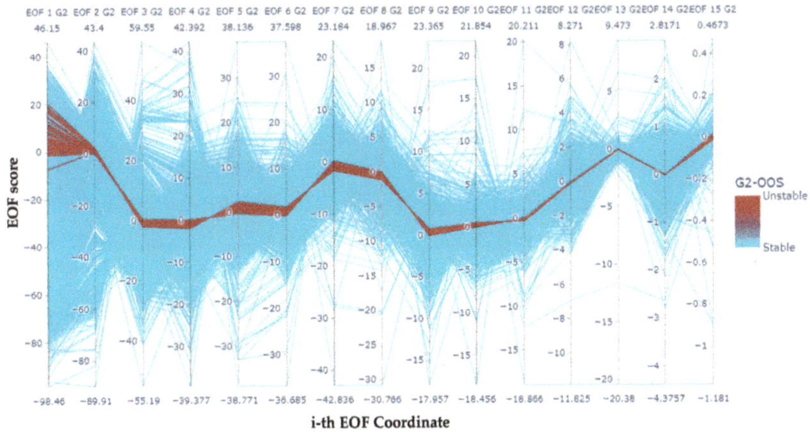

Figure 8. Parallel coordinates of the MC-based 15 EOF scores obtained for the G2 generator.

5. Discussion and Conclusions

The problem of a lack of specific datasets for applying machine-learning-based DVAs has been tackled in this paper; a comprehensive methodology has been presented to perform MC-based simulations to test power systems to determine post-contingency dynamic data that can be used to train machine learning tools.

For this, a well-defined stochastic methodology that can be used to generate dynamic data has been presented, showing good performance. The proposed method is based on a Monte Carlo simulation whose comprehensive structure has been expanded upon in this paper. Further research is warranted to apply this method for the generation of dynamic data from the Ecuadorian national interconnected system for performing security assessments based on machine learning.

This methodology is applied as part of the project PVI-DEE-2023-01, "Dynamic security assessment of the national interconnected system using numerical and artificial intelligence methods". Further research is under development for applying different machine learning tools to the data obtained from the proposed method to accomplish DVAs.

Funding: This research received no external funding.

Institutional Review Board Statement: Not applicable.

Informed Consent Statement: Not applicable.

Data Availability Statement: The data presented in this study are available on request from the corresponding author. The data are not publicly available due to intellectual property of universities.

Conflicts of Interest: The authors declare no conflict of interest.

Nomenclature

AI	Artificial intelligence
DVA	Dynamic vulnerability assessment
EOF	Empirical orthogonal function
k	Weibull shape parameter
LLE	Largest Lyapunov exponent
MC	Monte Carlo
OPF	Optimal power flow
PDF	Probability distribution function
PMU	Phasor measurement unit
SIME	Single machine equivalent
SIPS	System integrity protection scheme
SSS	Small signal stability
t_{cl}	Fault clearing time
t_r	Recovery time
t_u	Unstable time
WAMS	Wide area monitoring system
η	Transient stability margin ($\eta < 0$: unstable, $\eta > 0$: stable)
λ	Weibull scale parameter
μ	Mean or expectation
σ	Standard deviation

References

1. Dong, Z.; Zhang, P. *Emerging Techniques in Power System Analysis*; Springer: Berlin/Heidelberg, Germany, 2010.
2. Rueda, J.L. Evaluación y Mejora de la Estabilidad de Pequeña Señal de Sistemas Eléctricos de Potencia Considerando Incertidumbres. Ph.D. Thesis, Universidad Nacional de San Juan, San Juan, Argentina, 2009.
3. Cepeda, J. Evaluación de la Vulnerabilidad del Sistema Eléctrico de Potencia en Tiempo Real usando Tecnología de Medición Sincrofasorial. Ph.D. Thesis, Universidad Nacional de San Juan, San Juan, Argentina, 2013.
4. Rueda-Torres, J.L.; González-Longatt, F.; Cepeda, J. *Dynamic Vulnerability Assessment and Intelligent Control: For Sustainable Power Systems*; John Wiley & Sons: Hoboken, NJ, USA, 2018.
5. Echeverria, D. Evaluación y Mejora de la Estabilidad Transitoria de Sistemas Eléctricos en Tiempo Real Utilizando PMUs. Ph.D. Thesis, Universidad Nacional de San Juan, San Juan, Argentina, 2021.
6. Vaca, S. Evaluación Predictiva de Estabilidad Transitoria en Tiempo Real Mediante Aprendizaje Automático y la Identificación de Patrones del Máximo Exponente de Lyapunov. Master's Thesis, Escuela Politecnica Nacional, Quito, Ecuador, 2023.
7. Kumar, C.S.; Karuppiah, N.; Kumar, B.P.; Shitharth, S.; Dasu, B. Improvement of the Resilience of a Microgrid Using Fragility Modeling and Simulation. *J. Electr. Comput. Eng.* **2022**, 3074298.
8. Hashish, M.S.; Hasanien, H.M.; Ji, H.; Alkuhayli, A.; Alharbi, M.; Akmaral, T.; Turky, R.A.; Jurado, F.; Badr, A.O. Monte Carlo simulation and a clustering technique for solving the probabilistic optimal power flow problem for hybrid renewable energy systems. *Sustainability* **2023**, *15*, 783. [CrossRef]
9. Alsharif, A.; Tan, C.W.; Ayop, R.; Al Smin, A.; Ahmed, A.A.; Kuwil, F.H.; Khaleel, M.M. Impact of electric Vehicle on residential power distribution considering energy management strategy and stochastic Monte Carlo algorithm. *Energies* **2023**, *16*, 1358. [CrossRef]
10. Hinojosa, V.H. Pronóstico de Demanda de Corto Plazo en Sistemas de Suministro de Energía Eléctrica Utilizando Inteligencia Artificial. Ph.D. Thesis, Universidad Nacional de San Juan, San Juan, Argentina, 2007.
11. Chen, C.L.; Chen, S.L. Short-term Unit Commitment with Simplified Economic Dispatch. *Electr. Power Syst. Res.* **1991**, *21*, 115–120. [CrossRef]
12. Hedman, K.W.; Oren, S.S.; O'Neill, R.P. A review of transmission switching and network topology optimization. In Proceedings of the 2011 IEEE Power and Energy Society General Meeting, Detroit, MI, USA, 24–28 July 2011; pp. 1–7.

13. Li, W. *Risk Assessment of Power Systems: Models, Methods, and Applications*; IEEE Press Series on Power Engineering; IEEE: Piscataway, NJ, USA, 2005.
14. Cepeda, J.C.; Colomé, D.G. Benefits of empirical orthogonal functions in pattern recognition applied to vulnerability assessment. In Proceedings of the 2014 IEEE PES Transmission & Distribution Conference and Exposition—Latin America (PES T&D-LA), Medellin, Colombia, 10–13 September 2014; pp. 1–6.
15. Rosenstein, M.; Collins, J.; Luca, C. *A Practical Method for Calculating Largest Lyapunov Exponents from Small Data Sets*; NeuroMuscular Research Center and Department of Biomedical Engineering: Boston, UK, 1992.

engineering proceedings

Proceeding Paper

Wireless Communication Technologies for Smart Grid Distribution Networks †

Juan Carlos Rodriguez [1,*,‡,§], Felipe Grijalva [2,‡], Marcelo García [3,‡], Diana Estefanía Chérrez Barragán [4,‡], Byron Alejandro Acuña Acurio [4,‡] and Henry Carvajal [5,‡]

1 Analog Devices, Inc., Wilmington, MA 01887, USA
2 Colegio de Ciencias e Ingenierías "El Politécnico", Universidad San Francisco de Quito USFQ, Quito 170157, Ecuador; fgrijalva@usfq.edu.ec
3 Facultad de Ciencias Técnicas, Ingeniería Mecatrónica, Universidad Internacional del Ecuador UIDE, Quito 170411, Ecuador; margarciasaq@uide.edu.ec
4 School of Electrical and Computer Engineering (FEEC), University of Campinas (UNICAMP), Campinas 13083-852, SP, Brazil; diana.cherrez@gmail.com (D.E.C.B.); byronacunia@gmail.com (B.A.A.A.)
5 Faculty of Engineering and Applied Sciences (FICA), Telecommunications Engineering, Universidad de Las Américas (UDLA), Quito 170503, Ecuador; henry.carvajal@udla.edu.ec
* Correspondence: juancarlos.rodriguez@analog.com
† Presented at the XXXI Conference on Electrical and Electronic Engineering, Quito, Ecuador, 29 November–1 December 2023.
‡ These authors contributed equally to this work.
§ The opinions expressed in this publication are those of the authors. They do not purport to reflect the opinions or views of the Analog Devices Inc. (ADI), its subsidiaries or employees.

Abstract: The modernization of the current electric power grid into a smart grid requires the integration of advanced instrumentation, automation, and communication technologies to optimize efficiency, safety, and reliability. In traditional power grids, communication and control tasks are concentrated in substations, limiting their coverage to high-power equipment. As distributed energy resources increase in different sections of the grid, power flow becomes bi-directional. This requires monitoring and control at the Transmission and Distribution (T&D) level, which forms the largest portion of the power grid. To achieve efficient energy flow management and enable consumer participation in demand management, the integration of information and communication technologies (ICTs) is essential. Wireless sensor networks (WSNs) have been identified as a suitable solution for communications within the distribution network. An ongoing challenge, however, is the definition of the best candidates to solve this problem, among the currently available wireless technologies. This paper reviews different wireless communication technologies that provide robustness, reliability, speed, scalability, and cost-effectiveness for monitoring distribution lines. An outline of the architecture for smart grid communications, the definition of sensor network requirements for power line environments, and an overview of specific studies focusing on technology comparisons are the main contributions of this paper. The purpose of this review is to delineate current technologies in order to establish potential future research directions within the field.

Keywords: smart grid; wireless communication; distribution networks; energy management; wireless sensor networks; IoT

Citation: Rodriguez, J.C.; Grijalva, F.; García, M.; Chérrez Barragán, D.E.; Acuña Acurio, B.A.; Carvajal, H. Wireless Communication Technologies for Smart Grid Distribution Networks. *Eng. Proc.* **2023**, *47*, 7. https://doi.org/10.3390/engproc2023047007

Academic Editor: Soraya Lucia Sinche Maita

Published: 4 December 2023

1. Introduction

The smart grid can be defined as the modernization of the existing electric power grid infrastructure, with the aim of optimizing efficiency, safety, and reliability. This modernization facilitates the gradual integration of renewable energy resources through the use of advanced instrumentation, automation, and communication technologies [1].

In the current paradigm of the electricity network, control and communication operations are mainly limited to substations, where transmission lines and distribution feeders are connected via busbars and transformers. Measurement and control equipment in these sections of the grid is usually bulky and has high investment costs. Reliable communication links are always readily available between these high-power appliances, such as power cabinets, transformers, and power stations [2].

The gradual introduction of different distributed generation resources in the future grid will cause the direction of the power flow (which has been historically conceived as "one-way") to be considered "bi-directional" instead. In that situation, to keep security and reliability, the monitoring and control tasks must focus on the lowest sections of the grid. In the new paradigm of the smart grid, generation, Transmission and Distribution (T&D), and customer levels are considered. While the deployment of technologies at the generation and customer levels has already started, at the T&D level, there is still a requirement to develop the infrastructure for power management tasks. The key to accurate power management is monitoring the conditions of the grid over a broad range. Only real-time sensing and data communication technologies at this level will make this possible. As T&D infrastructures are the largest within the power grid, the range to cover is vast, and it has become an urgent issue to identify reliable and cost-effective solutions for these tasks.

To enable the management of energy flow at the T&D level and provide consumers at the customer level with a voice in demand management tasks, the integration of information and communication technologies (ICTs) is going to play a fundamental role in the progress of the smart grid. Currently, research efforts in the field of ICTs for the smart grid are extensive, and there is an ongoing discussion about the best technologies at different levels. Wireless sensor networks (WSNs) are widely considered the most suitable solution for communication nodes that perform sensing functions at different sections of the distribution network. This review paper summarizes the investigations on wireless communication technologies that may offer adequate robustness, reliability, speed, and the ability to scale into large-scale and low-cost networks, to monitor distribution lines (typically of 22 kV). Section 2 presents a brief introduction to the communications architecture in the smart grid, including an overview of the most popular technologies available for the task of smart grid monitoring. Section 3 explores the requirements of sensor networks in the power line environment. Section 4 outlines past studies on the performance of the different available technologies in practical applications and their comparison. Section 5 presents a summary and conclusions, along with recommended potential future research work in this area.

2. Available Wireless Technologies for Smart Grid Communications

The communication system is the key component for monitoring and control tasks in the smart grid infrastructure [3]. Different communication technologies, either wired or wireless, are available for the electric utilities that comprise the different sections of the grid. The smart grid paradigm proposes three main sections: generation, T&D, and customer levels. The development of standardized communication structures has begun at the customer level, where developed countries are gradually modernizing energy metering equipment. The communication between smart meters and the backhaul utility is known as the Advanced Metering Infrastructure (AMI).

The AMI is an advanced instrumentation technology enabled by real-time sensing and data communication to gather and convey raw measurements. Communication technologies must be chosen to address various requirements in the different parts of the AMI. The AMI's communication architecture is depicted in Figure 1.

The Neighborhood Area Networks (NANs) and Home Area Networks (HANs) of the AMI infrastructure are suitable for wireless deployment, as distances are relatively short. Information is concentrated in data aggregation points (DAPs). The backhaul network connecting the AMI head end and the DAPs can either be wireless or wired. The link between the DAPs and consumers requires NANs with coverage of a larger distance. Each DAP can connect to hundreds of smart meters (SMs).

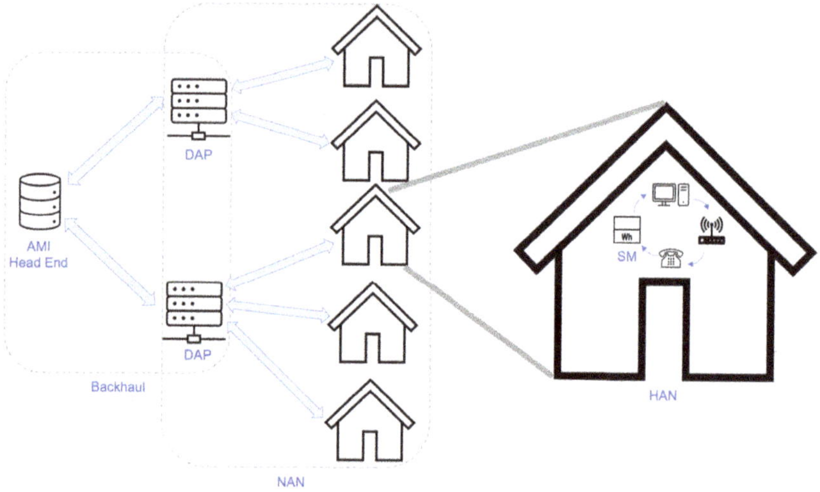

Figure 1. Architecture of the AMI (re-drawn from Ziming et al. [4]).

Elements responsible for sensing the parameters of power lines are usually referred to as sensor nodes. The projected locations of these sensor nodes (i.e., on utility poles in overhead distributed lines and on outer conductor surfaces in underground lines) suggest that the standards of both HANs and NANs may be adequate for defining their communication technologies roadmap. Many communication protocols have been proposed for the composition of NANs and HANs, as well as for other sections of the smart grid (such as the backhaul). A list of the most popular communication technologies considered adequate for different smart grid sections is shown in Table 1.

Table 1. Overview of communication technologies for the smart grid [5].

Technology	Spectrum	Data Rate	Coverage Rate	Limitations
GSM	900–1800 MHz	Up to 14.4 kbps	1–10 km	Low data rates
GPRS	900–1800 MHz	Up to 170 kbps	1–10 km	Low data rates
3G	1.92–1.98 GHz 2.11–2.17 GHz (licensed)	384 Kbps–2 Mbps	1–10 km	Costly spectrum fees
5G	3 to 90 GHz	5–10 Gbps	20–40 km	Costly spectrum fees
WiMAX	2.5 GHz, 3.5 GHz, 5.8 GHz	Up to 75 kbps	10–50 km 1–5 km	Not widespread
PLC	1–30 MHz	2–3 Mbps	1–3 km	Harsh, noisy channel environment
ZigBee	2.4 GHz–868–915 MHz	250 kbps	30–50 m	Low data rates, short range

Technologies for NANs have to provide a radius of coverage in the range of thousands of meters. The reliability of the communication channels between DAPs and smart meters (SMs) dictates that the spectrum used will have to be exclusive or interference-free. Consequently, the most suitable candidates need to be either licensed or leased wireless solutions. HANs have requirements that are not as stringent as those of NANs. In general, the message arrival rate within a customer premise is not as high as that between SMs and DAPs. They have been recognized with the generic name of "last-inch access". A comparison of the desired characteristics of different NAN and HAN communication technologies can be found in Table 2.

Table 2. Comparison of wireless communication technology candidates from the perspective of their suitability for NANs and HANs [4].

Coverage	Technology	Range	Latency	Reliability	Cost and Ease of Deployment
HAN	WiMAX	30 km	Low	High	Medium/Medium
	UMTS/LTE	30 km	Low	High	Medium/Low
	802.22	30 km	Medium	Medium/Low	High/Medium
NAN	WiFi	200 m	Medium–High	Low–Medium	Low
	ZigBee	100 m	Low–Medium	Medium	Low
	Bluetooth	100 m	Low	Medium	Low

In the literature, the preferred topology for smart grid applications at the level that has been discussed is wireless mesh. This technology enables any node in a network with routing capabilities to perform self-healing functions in a smart grid. It is suitable for home energy management and advanced metering infrastructures. Its disadvantages include coverage, the need for encryption techniques due to the information passing through every access point, and loop problems causing a reduction in bandwidth. A summary of the features, strengths, and challenges of each of the communication technologies is presented in Tables 1 and 2, including whether they incorporate deployment in a wireless mesh technology [5].

2.1. ZigBee

ZigBee is recognized as the most suitable technology for smart grid Home Area Networks (HANs) and is preferred by most AMI vendors and metering and energy management systems. It operates in the 2.4 GHz band, with a maximum radio output power of 1 mW, covering up to 100 m at 240 Kbps, using OQPSK modulation. It offers low development and operating costs within an unlicensed spectrum. It is based on the IEEE 802.15.4 standard [6]. Its disadvantages include concerns about low processing and memory capabilities and, mainly, susceptibility to interference from other 802.11 LAN [7] appliances. Interference avoidance schemes and energy-efficient routing protocols are challenges to be overcome.

2.2. WiFi

The IEEE suite of standards for wireless LANs, IEEE 802.11, is the most commonly deployed wireless standard within homes. As such, the devices and integrated circuits (ICs) are relatively cheap, making it an attractive solution. Amendments to the standard incorporate mesh networking capability, which is used in HANs. The data rate of WiFi ranges from 11 Mbps to 54 Mbps. It operates in the 2.4 GHz band, and it has a range of 30–46 m.

2.3. Bluetooth

The Bluetooth specification was designed for personal area networks (PANs) and is, therefore, suitable for HANs. The specification supports functions such as mesh networking. Furthermore, the specification ensures less latency compared to the two previously mentioned standards through the use of a time division multiple access (TDMA) scheme. Similar to ZigBee, it uses CSMA, which can result in large latency if many devices are in operation. It has a very short range of approximately 10 m and a low data rate of 1.5 Mbps. It operates in the 2.4 GHz band.

2.4. IEEE 802.22

The wireless regional area network is an alternative to mainstream broadband wireless that uses the white spaces in the television spectrum. It proposes to use cognitive radio technologies to exploit the unused spectrum in the frequency range allocated to television broadcasting. As the spectrum used is not dedicated, the latency in data transmission could be higher than other solutions for HANs.

Cellular networks are a good option for communication between smart meter nodes and utility far nodes, i.e., in NANs. The cellular communication technologies available to utilities for smart metering deployments include 2G, 2.5G, 3G, WiMAX, and LTE. A data transfer interval of 15 min between the meter and the utility produces a large amount of information, which requires a high data rate connection. GSM and GPRS are currently being used to enable communication between smart meters and the backhaul utility. GPRS is used for data, with a typical rate of 30–80 kbps, whereas EDGE provides 160 to 236 kbps. These ranges are those of cell phones, so coverage is not a concern with these technologies. Other wireless technologies being used for smart grid projects include CDMA, WCDMA, and UMTS, even in the backbone of smart grid communications, as in the case of Verizon's 3G CDMA. SP AusNet in Australia chose WiMAX as the technology for dedicated communication between smart meters and the central system of SP AusNet. Another technology that is being used for this kind of task is 4G. Relaying functionality has also been incorporated into 3GPP Release 10 (commonly known as LTE Advanced), which will allow extended coverage using relay/repeater stations. Since cellular network infrastructure is already built, it presents a cost-effective feature. Cellular networks provide sufficient bandwidth for intermittent applications, security control, and wide coverage. The big drawback of these technologies is their reliability because the services are shared by customers, resulting in network congestion or a decrease in performance under emergency or abnormal situations. This is not acceptable in power utilities. One option for addressing these issues is the implementation of private cellular networks.

From these solutions, WiMAX, which implements IEEE's 802.16 standard for metropolitan networks, is a leading candidate for providing connectivity between DAPs and SMs. It is based on orthogonal division multiplexing access (OFDMA), which assigns slices of the frequency spectrum to different users, avoiding interference among the users and increasing the spectral efficiency of the system. WiMAX is an attractive solution in the sense that its structure is much less sophisticated compared to rival cellular standards such as Long-Term Evolution (LTE). It has a minimum range of 8 km and data rates ranging from 45 to 75 Mbps.

Up to this point, wireless technologies for smart grids have been outlined. They share some common disadvantages, which can be summarized as follows:

- Poor performance in electrically noisy and harsh environments.
- Concerns about security.
- Limited transmission range (except for cell phone technologies) and low data rates.
- Non-industrial-level reliability.
- Standards that are not robust.

Even though wireless technologies are the principal candidates for the deployment of communication among sensor nodes at the distribution level, it is worth mentioning the features that wired technologies can offer in the same context. The two options described below are considered in this summary.

2.5. Power Line Communication (PLC)

PLC uses existing power lines to transmit data at rates of up to 2–3 Mbps. It was the first choice in the implementation of AMIs, where it was used for data communication between smart meters and the data concentrator, whereas GPRS technology was used for transferring the data from the data concentrator to the utility's data center. The utilization of existing infrastructure reduces installation costs, making PLC an attractive solution for urban area applications. However, due to the harsh and noisy nature of power line networks, there are technical challenges. The channel is difficult to model. Also, the low-bandwidth characteristic of 20 kbps and sensitivity to disturbances make PLC unsuitable for data transmission. Medium-Voltage Broadband Power Line Communication (MV-BPLC) technology is considered an important candidate for communication with primary and secondary substations, and [8] proved that MV-PLC is effective in supporting grid management. However, it also suffers from deficiencies like cable length, PV DGs, industrial loads introducing high-frequency noise, and a lack of reliability with faults. Future work includes designing a hybrid model (wireless with MV-BPLC).

2.6. Digital Subscriber Lines (DSLs)

DSL offers an already-built infrastructure with high-speed data transmission that has previously been exploited in smart grid projects. Its disadvantages are reliability, potential downtime, and the expense of installing and constantly maintaining communication cables.

In general, wired technologies, such as DSL, PLC, and even optical fiber, are costly for wide-area development, but they offer the most robust communication capabilities. On the other hand, wireless technologies reduce installation costs but have constraints in terms of bandwidth and security options.

2.7. Fifth-Generation Wireless Networks

The amalgamation of 5G architecture and smart grids presents significant opportunities in Transmission and Distribution. The introduction of 5G slicing adds intricate layers across user and utility domains, impacting energy dynamics and load balancing. The dynamic deployment paradigm aligns network functions with service needs. With end-to-end agreements covering communication service administration and network management for efficiency [9], 5G slicing ensures minimal latency.

The 5G grid segregates networks, which are shared with telecom systems. Industrial control achieves precision through 5G. Adaptable slicing enhances agility, reinforcing security. This approach reduces expenses and manages network velocity [9].

Technical 5G network services unfold through multi-layered slicing. Feeder automation requires intelligent orchestration for latency and bandwidth. Grid integration involves virtualized infrastructure. Wireless communication and power transmission promise progress, but costs and weather vulnerabilities pose challenges. Mobile radio and microwave transmission are alternatives. Cognitive radio supports smart grid wireless communication. The integration of 5G meets energy needs, aiding in forecasting and maintaining balance. Tailored strategies empower cities, and observability ensures distribution stability.

3. Requirements of Sensor Networks Implemented in the Power Line

The basic unit performing measurement tasks at a specific point on a distribution line was previously defined as a sensor node. The concept of the Sensor Network (SN) refers to the gathering of sensor nodes located in different sections of a power line through an appropriate communication link. This concept arises from the necessity of coordinating a large number of sensors for real-time monitoring in different parts of the power grid. It is motivated by the availability of modern low-cost sensing and communication technologies.

Among the challenges that the implementation of reliable SNs presents are the harsh, uncertain, and dynamic operating environments, as well as energy concerns. There are three main technologies that currently need development within the topic of SNs: sensing, networking, and information processing. The challenges, difficulties, and available technologies are discussed by Chen et al. [10,11], from which the following main requirements can be highlighted:

- The need for simple network protocols and algorithms for sensor nodes with limited computational and memory capacities.
- The self-healing capacity of the networking strategy, as sensor nodes are prone to failure in hostile environments.

Wireless communications offer the most flexible and straightforward interconnection between nodes. However, their use in power system environments presents security and reliability concerns, such as EMI, fading signals, bandwidth overloading, and security.

Despite this, wireless networks are still considered the most attractive technology for this development. These technologies are known as wireless sensor networks (WSNs).

The Institute of Electrical and Electronics Engineers (IEEE) has produced several standards related to communication in power systems. Notable standards include IEEE 802.16 (WiMAX), IEEE 802.11 (WiFi), and IEEE 802.15 (wireless personal area networks). These standards may not be specific to power systems but are often used in various applications, including power systems [12].

In [13,14], the authors recognized the four major challenges of WSNs in smart grid applications:

- Harsh environmental conditions: The connectivity of networks can vary due to link failures, and sensors are susceptible to RF interference.
- Reliability and latency: Due to the time sensitivity of sensors, the controller has to receive information in a timely manner.
- Link capacity: The bandwidth of each wireless link depends on the interference level at the receiver, leading to high bit-error rates (BER) ranging from 1×10^{-2} to 1×10^{-6}. The latency at each link varies continuously, so it is hard to meet QoS requirements.
- Design and implementation resources: Energy, memory, and processing resources are limited, so the protocols for WSNs are tailored to ensure high energy efficiency.

Finally, among the most important capabilities that any communication technology developed for a smart grid application should provide are the following:

- Security, as initial experiences in smart meter installations in households have already shown that public concerns about new smart grid technologies should not be ignored.
- Reliability, robustness, and availability, which are provided by most wired technologies. However, hybrid solutions using wireless and wired technologies might be used for the whole infrastructure. For wireless solutions in NANs, wide coverage is needed. Therefore, reliability is important, with latency low enough not only to satisfy demand-side management (DSM) requirements but also to serve all other AMI applications. This translates to a minimum reliability of 99.5% and a latency requirement of less than 1 s, which is a relatively relaxed figure compared to commercial broadband requirements [15]. HANs require a minimum reliability of 99.5% and a latency of less than 5 s, as they are shorter in reach and easier to access within the AMI.
- Scalability through the integration of web services, protocols, instruments, and configurations of the grid.
- Quality of Service (QoS). Quality, in this context, is defined in terms of performance degradation, like delays or outages. Requirements can be defined based on the power price. The impact of delays or outages on a reward system for a house appliance, based on the price, makes it possible to optimize the reward as a means to measure the QoS. Routing methodologies are used to meet the previously defined QoS requirements. This leads to concepts such as dynamic pricing and distributed energy resources management, which are presented in [15].

4. Previous Studies on Selecting Suitable Technologies for Power Line Monitoring

Although the field of WNSs is currently highly developed, there are few works that address the issue of wireless link quality or the experimental performance of different WSN technologies at the Transmission and Distribution (T&D) level [13]. The most significant efforts in applying previous theoretical frameworks of WSNs to practical scenarios are summarized in this section.

Toma et al. [16] introduced a self-powered WSN for underground HV power lines. The authors addressed one of the challenges in developing WSNs for power lines and presented the design of a protocol and network. The main objective of this work involved measuring temperatures and currents in underground conductors, as well as monitoring and controlling the capacity of transmissions (overall ampacity). Wireless sensors were deployed due to the large gaps that the information must travel to reach the gateway. The second challenge of the project was the necessity of finding a way to encapsulate the data from the sensor to transmit them in a minimal number of packages.

Each sensor node consisted of an XBee-PRO 802.15.4 module, including an MC9S08GT60 MCU and an MC13193 RF chip. The backend server was connected to the last node through a GPRS module. The energy harvesting device was magnetic, but the sensor also incorporated a battery. The proposed protocol for communications was based on the Freescale Simple Media Access Controller (SMAC). The WSN used a 10 ms time slot for communication between two consecutive nodes, as well as Time Division Multiple Access (TDMA). The limit of transmission was 123 B, encapsulated according to the HDLC standard protocol, to be added to the 802.15.4 package. Each node's frame of data had a length of 12 B. To maintain accurate clock synchronization and minimize power consumption, the protocol was based on the IEEE 1588 standard. As network reliability was a concern, the protocol provided redundancy, where each node was required to communicate with at least two consecutive nodes. Under normal conditions and at a sampling rate of 10 s, the WSN achieved synchronization after 5 min; after 10 min, all odd nodes disconnected. The power consumption of the battery, in case the line was off, allowed for a duration of 2 weeks for each node.

Yang et al. [17] designed a sensor net for overhead transmission lines to transmit the local information of sensor modules peer-to-peer back to a master node, far away from local sensors, which is powered via a magnetic harvesting device clamped onto the power line. Digi's ZigBee-PRO RF module is used, implementing 802.15.4. It was noted that the protocol specifies Direct Sequence Spread Spectrum(DSSS) and Offset Quadrature Phase-Shift Keying (O-QPSK) to modulate the RF carrier, which helps enhance communication immunity to ambient noises, especially those with wide frequency bandwidth, such as impulse noises commonly observed in high-voltage power line environments. The topology is a cluster tree, along with a central coordinator RF module to provide synchronization services and end routers to connect to the higher subnet. It uses 128-bit encryption. The module is complemented with a TMS320F2812 DSP. The results show an outdoor range of 1.6 km, a data rate of 250 kbps, and a packet transmission rate of 50 packets per second (20 ms per packet), each containing 100 bytes of data.

The device was tested on a power line conductor. It was shown to be autonomously powered by a primary-side current ranging from 100 to 1000 A. To test the communication performance, a host PC was used to continuously transmit a data stream (32 bytes/packet) to the module, and the communication performance was evaluated by receiving the same data stream looped back from the remote PLS module. For the communication performance, two criteria were evaluated: the Received Signal-Strength Indicator (RSSI) and the Percentage of Successful Reception (PSR). In a second experiment, the communication performance was evaluated at a distance of up to 700 m between two nodes in outdoor conditions. The results of both experiments are shown in Tables 3 and 4.

Table 3. Performance of IEEE 802.15.4 in indoor environments from the study by Gupta et al. [17,18].

Current (A)	Range (m)	RSSI (dBm)	PSR
1000	50	−70 to −75	100%
	100	−92	~45%
500	10	−45 to −50	100%
	50	−73	100%
	100	−93	~45%

Table 4. Performance of IEEE 802.15.4 in outdoor environments from the study by Gupta et al. [17,18].

Range (m)	RSSI (dBm)	PSR	Conditions
200	~−76	~95%	Close to line of sight
400	~−83	~80%	Trees
500	−92	~35%	Trees and buildings

The test results showed that the high current did not affect the communication performance. However, the obstacles between the two communication nodes significantly affected the performance. Even though the communication performance was degraded, outdoor transmission at a distance of up to 400 m was achieved with a transmission success rate of up to 80%. The effect of high voltage, mainly due to the corona effect, will be validated in the future.

State-of-the-art sensor communication technology for sensing nodes on power lines, as discussed in [19,20], was implemented usingZigBee technology based on IEEE 802.15.4 in the TI's CC2530 SOC solution. It should be mentioned that this technology is comparable to Bluetooth but with the added advantage of very low latency. The sensor uses an 8051 MCU, 256 KB flash, 8 KB RAM, 12-bit ADC, and eight channels. It takes 400 ms to communicate and go back to sleep. At 100 A, the operational frequency is as high as once every minute. Energy storage ensures 13 cycles of operation after an outage.

As previously mentioned, review studies on smart grid communication technologies, such as [3], have recognized that a big challenge in this area is the difficult estimation of the instantaneous value of the wireless link quality. From all the real-world studies on WSNs where different sensor platforms have been used, the common observations are:

- The recognition of three distinct reception regions in a wireless link: connected, transitional, and disconnected.
- Wireless link quality varies over space and time, unlike standard models often used in simulation tools. The coverage area of sensor radios is neither circular nor convex, and packet losses are common at a wide range of distances, varying over time.
- Link asymmetry (one-way communication) occurs when the transmit power is low.

A measurement of the link quality was performed by Gungor et al. [14], using sensors implementing IEEE 802.15.4. They used a packet length of 30 B and a buffer size of 64 packets. The experiments were conducted in three locations: an outdoor environment near a 500 kV substation, inside a power control room, and in an underground transformer vault. The channel was modeled as a log-normal shadowing path loss, where the Signal-to-Noise Ratio (SNR) γ in dB, was expressed as

$$\gamma(d)_{dB} = P_t - PL(d_0) - 10\eta\left(\frac{d}{d_0}\right) - X_\sigma - P_\eta \tag{1}$$

where P_t is the transmit power in dBm, $PL(d_0)$ is the path loss at the reference distance (d_0), η is the path-loss exponent, X_σ is a zero-mean Gaussian random variable with a standard deviation of σ, and P_η is the noise power in dBm. For different propagation environments, the values of η and σ were calculated from the measured data using linear regression. The results for these parameters are presented in Table 5. Two links for communications were also considered: line of sight (LOS), and non-LOS (NLOS).

Table 5. Mean power loss and shadowing deviation in different electric power environments, as measured by Gungor et al. [14].

Propagation Environment	Path Loss (η)	Shadowing Deviation (σ)
500 kV substation (LOS)	2.42	3.12
500 kV substation (NLOS)	3.51	2.96
Underground transformer vault (LOS)	1.45	2.45
Underground transformer vault (NLOS)	3.15	3.19
Main power room (LOS)	1.64	3.29
Main power room (NLOS)	2.38	2.25

Noise and interference were measured using a TinyOS application that samples RF energy at 62.5 Hz by reading the Received Signal-Strength Indicator (RSSI) on a CC2420 radio in the various aforementioned environments. The noise measurements on an 802.15.4

network indicated an average noise level of around −90 dBm for indoor environments and −105 dBm (of constantly changing) background noise for outdoor environments. The results also showed that a previously existing microwave signal (from a microwave in the study), led to 15 dBm interference in the 2.4 GHz band. Also, interference of 802.11 b was caused by the overlapping band.

The final link-quality measurements were determined using three metrics: the Packet Reception Rate (PRR), RSSI, and Link-Quality Indicator (LQI). The latter is also known as the chip error rate. The distance from receiver to sender varied from 1 to 20 m. The power level of each sensor was −25 dBm and the packet size was 30 B, sending 200 data packets at a rate of 2 packets per second. The results showed that the PRR (ratio of the number of successful packets to the total number of packets transmitted over a certain number of transmissions) strongly correlated with the LQI. Hence, the LQI is a good indicator of the packet reception probability in this kind of experiment. The LQI ranged from 50 to 110.

Comparison of Technologies

Wireless sensor networks are built from nodes that must be low cost and have very low power consumption. At least one of these two requirements is met by the available technologies presented in this summary, such as IEEE 802.15.1 (Bluetooth) and IEEE 802.15.4 (ZigBee). There will be cases where one technology will be preferred over the other, depending on the application type, environment, sensor node technology, and network configuration requirements. Therefore, there is a clear need to analyze different technologies with respect to each other in order to be able to decide which one would be more suitable for a given application, with certain constraints and requirements.

Buratti et al. [21,22] numerically addressed the issue of the lifetime of a WSN to compare Bluetooth and ZigBee in a particular application. This was the first study to compare such technologies in the context of WSNs. The authors used the EMORANS scenario as a framework for the comparison. EMORANS defines the test conditions for wireless communications, as follows:

- The geometry of a square layout of a side is set to 100 m.
- A node density of either 100 or 500 nodes is used with a uniform distribution over the square.
- A sink is required for periodically collecting the measurements performed over the sensed area by nodes.
- The initial battery charge is set to 1 Joule to facilitate shorter simulations. The channel model loss in logarithmic scale should be $k_0 + k_1 ln(d) + s$, where d is the distance, k_0 = 40 dB, k_1 = 13.03 (obtained through experimental measurements made in the field in a rural environment), and s is a Gaussian random variable, with a mean of zero and a standard deviation of σ, modeling channel fluctuations. The capture packet model used states that a packet is correctly received in case the loss is smaller than a given threshold, Lm, which depends on the technology used.
- The three definitions of network lifetime are as follows, where a round is the time elapsed between two consecutive measurements:
  - The interval of time (measured in rounds) from the first transmission in the wireless network to the point when the percentage of nodes with remaining energy drops to zero.
  - The average percentage of nodes that remain reachable within the network over a specified time window.
  - The average percentage of reports sent from nodes to the sink over a defined time window.

If in any of these definitions, the lifetime falls below a specific threshold, which is set according to the type of application, the quality of the lifetime is considered low. The first definition takes into account only energy consumption issues, whereas the second takes into account both energy consumption and connectivity issues. The third definition refers to energy consumption, connectivity, and MAC failures. In Figure 2, a comparison of

Bluetooth and IEEE802.15.4 is provided, considering two networks composed of 500 and 100 nodes.

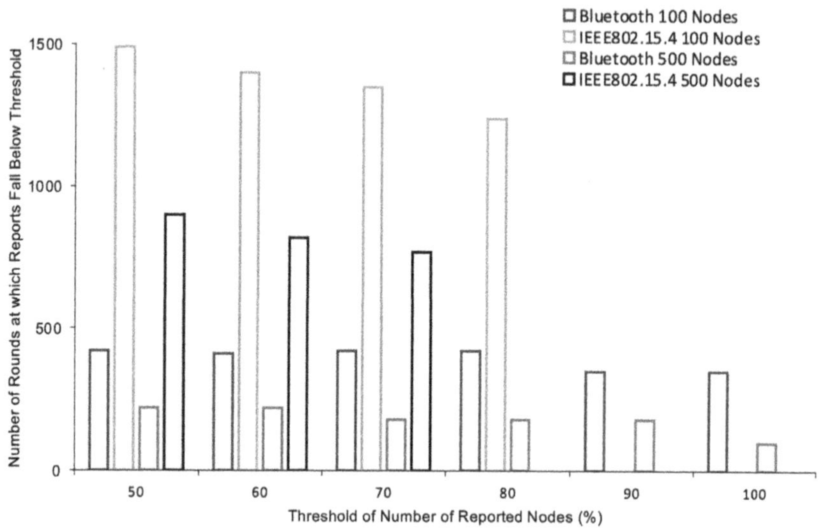

Figure 2. Network lifetime comparison of Bluetooth and ZigBee (taken from Buratti et al. [21]).

Figure 2 shows the rounds at which the number of reports arriving at the sink from the nodes started to fall below a certain percentage (a threshold), indicated on the x-axis. ZigBee exhibited a longer lifetime compared to Bluetooth, supporting up to 80 nodes. However, beyond 80 nodes, the lifetime of ZigBee is not shown because in the first round, more than 10% of packets in the 100-node case were lost. The choice of technology depends on the application requirements. If, in fact, one considers a network composed of 100 nodes and an application that requires that the sink receives 100% or 90% of the packets, one would choose Bluetooth because IEEE802.15.4 cannot attain these percentages. In the case where the application can tolerate a loss of more than 10% of the packets, IEEE802.15.4 is better because it establishes a more energy-efficient network.

Gupta and Malvika [18] presented a comparison of the major characteristics of the two main protocols for short-range terrestrial communications: WiFi and Bluetooth. The study did not focus on WSNs but provided interesting information, which is summarized in Table 6.

As shown in Table 6, the power requirements of Bluetooth devices are significantly lower compared to those of WiFi n/abg devices. However, the data rates of WiFi are higher. Additional considerations about the signal rate and channel characteristics might be useful for considering similar patterns in other WSN technologies.

Finally, Sharman et al. [23,24] briefly evaluated different technologies using standards for synchrophasors. The authors concluded that wired technologies such as ADSL2 and FTTC are the best last-mile possible candidates for smart grid operations. Among wireless technologies, WiMAX is the best witha packet loss of 0.0322, a latency of around 9 ms, and a throughput of 20 Mbps (<2 ms for required protections).

Table 6. A comparison of Bluetooth and WiFi n/abg protocols [21].

	Bluetooth	**WiFi**
Frequency band	2.4 GHz	2.4 GHz, 5 GHz
Coexistence mechanism	Adaptive frequency hopping	Dynamic frequency selection, Adaptive power control
Multiplexing	FH55	DSSS, CCK, OFDM
Future multiplexing	UWB	MIMO
Noise adaptation	Link layer	Physical layer
Typical output power	1–10 mW (1–10 dBm)	30–100 mW (15–20 dBm)
Nominal range	10 m	100 m
Maximum one-way data rate	732 kb/s	31.4 Mb/s
Basic cell	Piconet	BSS
Extension of the basic cell	Scatternet	ESS
Topologies	Various analogies: see Subsection Network Topologies	
Maximum number of devices in the basic cell	8 active devices; 255 in park mode	Unlimited in ad hoc networks (IBSS)
Maximum signal rate	1 Mb/s	54 Mb/S
Channel access method	Centralized: polling	Distributed: CSMA/CA
Channel efficiency	Constant	Decreasing with offered traffic
Spatial capacity	From 0. 1 to 400 Kb/s $\cdot m^2$	About 15 kb/s$\cdot m^2$
Data protection	16-bit CRC (ACL links only)	32-bit CRC
Procedures used for the network setup	Inquiry, Page	Ad hoc networks: Scan, Authentication Infrastructured Scan
Average speed in the network setup without external interference	$5s + n - 1.28s$, where n is the number of Slaves in the piconet, ranging from 1 to 7	n.c. 1.35 ms for an unsaturated network, c probed channels
Authentication	Shared secret, pairing	Shared secret challenge-response
Encryption	Eo stream cipher	RC4 stream cipher RES
QoS mechanism	Link types	Coordination functions
Typical current absorbed	1–35 mA	100–350 mA
Power-save modes	Sniff, hold park, standby	Doze

5. Summary and Conclusions

This report has presented an overview of the candidate technologies for the task of communication for sensing devices located in different sections of the future electricity distribution network, also known as the smart grid. The new paradigm of smart grid communications defines, at the customer level, a communication architecture known as the Advanced Metering Infrastructure (AMI), which includes Home Area Networks (HANs) to facilitate communication among different elements within customer households, including smart meters (SMs). It also includes Neighborhood Area Networks (NANs) to facilitate communication between SMs and concentration points. On the other hand, devices in charge of measuring tasks in distribution lines are known as sensor nodes (SNs). The concept of the AMI is adequate for SNs due to their location. Hence, technologies that are used in NANs and HANs are suitable for use in SNs. Wireless mesh topologies are the most suitable within that paradigm.

Among the suitable wireless technologies for this kind of development are 802.15.4 (ZigBee), IEEE 802.11 (WiFi), and Bluetooth, which are common technologies in HANs. Among them, ZigBee provides the longest coverage range (100 m) but at the smallest data rate (240 kbps). Cellular networks and 802.22 are employed in HANs, as the coverage area can extend up to thousands of meters. Cellular networks have the advantage of wide coverage and have been used as the backbone of smart grid deployments. Their disadvantage is their low reliability for real-time tasks in sensing and power management applications. WiMAX (802.16 implementation) is another candidate that does not encounter the problem of interference issues from other technologies. It has a practical range of up to

8 km and speeds in the order of Mbps. Wired technologies that have been considered in smart grid deployments include Power Line Communication (PLC), which provides data rates in the order of several Mbps and has been used in SM applications. However, the difficulty of channel modeling is a big disadvantage. Digital Subscriber Line (DSL) and fiber optic communication have also been considered, but these wired technologies are costly for wide-area deployment. Correspondingly, wireless technologies reduce installation costs but have constraints in terms of bandwidth and security options.

A large number of sensors are installed along the distribution power line in a wireless sensor network (WSN). WSNs have punctual challenges and requirements, which have been identified in the literature. To address the main bandwidth and energy concerns, WSNs should implement protocols and algorithms designed for limited computational and memory capacities. In addition, the self-healing capacity is important in the environment of a power line. The future IEEE P1777 standard will define the precise requirements and challenges of different technologies. The main challenges for WSNs are the harsh environmental conditions and the energy required to power them. More specifically, the main requirements of WSNs are the security of information, robustness in terms of wide coverage, and reliability to keep the information flowing continuously. This latter feature is very important, as time delays and outages are critical in the power utility grid. Scalability, as the possibility of integrating protocols and grid configurations, is also an important feature in the context of the sensor nodes of different instrumentation and communication technologies. Wireless technologies must address issues such as performance in noisy grid environments, security concerns, the limited range of transmission, low data rates, non-industrial-level availability, and the lack of robust standards for WSNs.

The experimental performance of different WSNs in the environment of power lines has not often been addressed in the literature. Some studies have presented the design of protocols for WSNs. ZigBee technology is a very popular technology. The challenge of maintaining accurate clock synchronization between sensors has been clearly identified. In one study, this was accomplished with the IEEE 1588 standard. Apart from measurements of the coverage range, data rate, and latency, the quality of the channel is an important indicator of communication performance. Metrics for achieving this are not yet standardized, but the Received Signal-Strength Indicator (RSSI) in the nodes, Percentage of Successful Reception (PSR), and Packet Reception Rate (PRR) have been used in some studies. Ultimately, a parameter called the Link-Quality Indicator (LQI) (known as the chip error rate) has proven to be a good indicator of the packet reception probability in WSNs. In outdoor environments, 802.15.4 can achieve a transmission success rate of 80% at 400 m, with physical obstacles as the main drawback. Power line currents of 500 A do not affect communication according to the experiments conducted. The complexity of the WSN's communication channel in power system environments has been addressed using mathematical models like the log-normal shadowing path loss. The gathering of data for this model is dependent on the specific communication environment. ZigBee has been tested using this model, showing around -105 dBm of noise in outdoor environments and interference with 802.11 signals in the 2.4 GHz band.

Future Research Work

While a limited number of studies have analyzed the performance of the various communication technologies in high-voltage distribution line environments, there is scope for future research in this domain. Some promising research directions include:

- Simulation-Based Performance Analysis: To address the dearth of studies on high-voltage distribution lines, conducting simulations of different communication protocols under such conditions is a crucial step. Evaluating the performance of the following prominent wireless sensor network (WSN) technologies will provide insights into the suitability and efficiency of these technologies in the context of high-voltage distribution lines: ZigBee (802.15.4 standard), WiMax, Bluetooth, WiFi, and 5G.

- Mathematical Modeling of Communication Channels: The development of mathematical models for the communication channels within the grid infrastructure, specifically in high-voltage (HV) power lines, is a valuable avenue of research. Techniques such as Finite Element Analysis (FEA) using software like ANSYS Maxwell can be employed. Such models can be informed by the findings from the literature review in this paper, enabling a more comprehensive understanding of the communication channels' behavior in HV environments.
- Performance Evaluation Using the Link-Quality Indicator (LQI): To extend the analysis, an evaluation of the technologies reviewed in this paper, particularly in HV distribution line scenarios, using metrics like the Link Quality Indicator (LQI) is necessary. The methodology for these assessments can draw inspiration from the approaches detailed in the referenced articles. Utilizing network simulation tools like OpNET can facilitate these evaluations.
- Defining WSN Performance in Power System Environments: A significant contribution would be the definition of WSN performance in power system environments, a knowledge gap that currently exists in the literature. Providing insights into the strengths and weaknesses of different WSN technologies within this context can aid in informed decision making for smart grid deployments.
- Development of Protocols and Strategies: Based on the insights gained from initial experiments and performance evaluations, there is potential for the development of new protocols and strategies for smart grid communication. These innovations can be shaped by real-world findings and may lead to advancements in the reliability and efficiency of communication systems within the smart grid.

By exploring these research avenues, we can contribute valuable knowledge to the field of smart grid communication, address existing gaps, and pave the way for more robust, reliable, and efficient energy distribution systems.

Author Contributions: The authors contributed equally to this work. All authors have read and agreed to the published version of the manuscript.

Funding: This work was supported by the Universidad San Francisco de Quito through the Poli-Grants Program under Grant 17993.

Institutional Review Board Statement: Not Applicable.

Informed Consent Statement: Not Applicable.

Data Availability Statement: No new data were created in this study.

Conflicts of Interest: The authors declare no conflict of interest.

References

1. Goudarzi, A.; Ghayoor, F.; Waseem, M.; Fahad, S.; Traore, I. A Survey on IoT-Enabled Smart Grids: Emerging, Applications, Challenges, and Outlook. *Energies* **2022**, *15*, 6984. [CrossRef]
2. Çelık, D.; Meral, M.E.; Waseem, M. A new area towards to digitalization of energy systems: Enables, challenges and solutions. In Proceedings of the 2022 14th International Conference on Electronics, Computers and Artificial Intelligence (ECAI), Ploiesti, Romania, 30 June–1 July 2022; pp. 1–6.
3. Butt, O.M.; Zulqarnain, M.; Butt, T.M. Recent advancement in smart grid technology: Future prospects in the electrical power network. *Ain Shams Eng. J.* **2021**, *12*, 687–695. [CrossRef]
4. Zhu, Z.; Lambotharan, S.; Chin, W.H.; Fan, Z. Overview of demand management in smart grid and enabling wireless communication technologies. *IEEE Wirel. Commun.* **2012**, *19*, 48–56.
5. Gungor, V.C.; Sahin, D.; Kocak, T.; Ergut, S.; Buccella, C.; Cecati, C.; Hancke, G.P. Smart grid technologies: Communication technologies and standards. *IEEE Trans. Ind. Inform.* **2011**, *7*, 529–539. [CrossRef]
6. IEEE 802.15 WPAN Task Group 4. Available online: https://www.ieee802.org/15/pub/TG4.html (accessed on 5 November 2023).
7. IEEE 802.11 Wireless Local Area Networks Standard. Available online: https://www.ieee802.org/11/ (accessed on 5 November 2023).
8. Lopez, G.; Matanza, J.; De La Vega, D.; Castro, M.; Arrinda, A.; Moreno, J.I.; Sendin, A. The role of power line communications in the smart grid revisited: Applications, challenges, and research initiatives. *IEEE Access* **2019**, *7*, 117346–117368. [CrossRef]

9. Dragičević, T.; Siano, P.; Prabaharan, S.S. Future generation 5G wireless networks for smart grid: A comprehensive review. *Energies* **2019**, *12*, 2140.
10. Chen, Z.; Amani, A.M.; Yu, X.; Jalili, M. Control and Optimisation of Power Grids Using Smart Meter Data: A Review. *Sensors* **2023**, *23*, 2118. [CrossRef] [PubMed]
11. Yang, Y.; Lambert, F.; Divan, D. A survey on technologies for implementing sensor networks for power delivery systems. In Proceedings of the 2007 IEEE Power Engineering Society General Meeting, Tampa, FL, USA, 24–28 June 2007; pp. 1–8.
12. Chamana, M.; Bhatta, R.; Schmitt, K.; Shrestha, R.; Bayne, S. An Integrated Testbed for Power System Cyber-Physical Operations Training. *Appl. Sci.* **2023**, *13*, 9451. [CrossRef]
13. Abujubbeh, M.; Al-Turjman, F.; Fahrioglu, M. Software-defined wireless sensor networks in smart grids: An overview. *Sustain. Cities Soc.* **2019**, *51*, 101754. [CrossRef]
14. Gungor, V.C.; Lu, B.; Hancke, G.P. Opportunities and challenges of wireless sensor networks in smart grid. *IEEE Trans. Ind. Electron.* **2010**, *57*, 3557–3564. [CrossRef]
15. Abrahamsen, F.E.; Ai, Y.; Cheffena, M. Communication technologies for smart grid: A comprehensive survey. *Sensors* **2021**, *21*, 8087. [CrossRef] [PubMed]
16. Toma, D.M.; del Rio, J.; Mànuel-Làzaro, A. Self-powered high-rate wireless sensor network for underground high voltage power lines. In Proceedings of the 2012 IEEE International Instrumentation and Measurement Technology Conference Proceedings, Graz, Austria, 13–16 May 2012; pp. 1881–1885.
17. Yang, Y.; Divan, D.; Harley, R.G.; Habetler, T.G. Design and implementation of power line sensornet for overhead transmission lines. In Proceedings of the 2009 IEEE Power & Energy Society General Meeting, Calgary, AB, Canada, 26–30 July 2009; pp. 1–8.
18. Gupta, M.; Singh, S. A survey on the zigbee protocol, it's security in internet of things (iot) and comparison of zigbee with bluetooth and wi-fi. In *Proceedings of the Applications of Artificial Intelligence in Engineering: Proceedings of the First Global Conference on Artificial Intelligence and Applications (GCAIA 2020)*; Springer: Berlin/Heidelberg, Germany, 2021; pp. 473–482.
19. Riba, J.R.; Moreno-Eguilaz, M.; Bogarra, S. Energy Harvesting Methods for Transmission Lines: A Comprehensive Review. *Appl. Sci.* **2022**, *12*, 10699. [CrossRef]
20. Moghe, R.; Yang, Y.; Lambert, F.; Divan, D. Design of a low cost self powered "Stick-on" current and temperature wireless sensor for utility assets. In Proceedings of the 2010 IEEE Energy Conversion Congress and Exposition, Atlanta, GA, USA, 12–16 September 2010; pp. 4453–4460.
21. Buratti, C.; Korpeoglu, I.; Karasan, E.; Verdone, R. Bluetooth or 802.15. 4 technologies to optimise lifetime of wireless sensor networks: Numerical comparison under a common framework. In Proceedings of the 2008 6th International Symposium on Modeling and Optimization in Mobile, Ad Hoc, and Wireless Networks and Workshops, Berlin, Germany, 1–3 April 2008; pp. 539–543.
22. Gupta, S.K.; Singh, S. Survey on energy efficient dynamic sink optimum routing for wireless sensor network and communication technologies. *Int. J. Commun. Syst.* **2022**, *35*, e5194. [CrossRef]
23. Sharma, D.K.; Rapaka, G.K.; Pasupulla, A.P.; Jaiswal, S.; Abadar, K.; Kaur, H. A review on smart grid telecommunication system. *Mater. Today Proc.* **2022**, *51*, 470–474. [CrossRef]
24. Laverty, D.M.; O'Raw, J.; Morrow, D.J.; Cregan, M.; Best, R. Practical evaluation of telecoms for smart grid measurements, control and protection. In Proceedings of the 2011 2nd IEEE PES International Conference and Exhibition on Innovative Smart Grid Technologies, Manchester, UK, 5–7 December 2011; pp. 1–5.

engineering proceedings

MDPI

Proceeding Paper

Methodology for Identifying Representative Rates of Change of Frequency (ROCOFs) in an Electric Power System against N-1 Contingencies †

Carlos Xavier Lozada [1,*], Walter Alberto Vargas [2], Nelson Victoriano Granda [2] and Marlon Santiago Chamba [3]

[1] Operador Nacional de Electricidad (CENACE), Quito 17211991, Ecuador
[2] Department of Electrical Engineering, Faculty of Electrical and Electronic Engineering, Campus José Rubén Orellana Ricaurte, Escuela Politécnica Nacional, Quito 170525, Ecuador; walter.vargas@epn.edu.ec (W.A.V.); nelson.granda@epn.edu.ec (N.V.G.)
[3] Corporación Eléctrica del Ecuador (CELEC EP), Unidad de negocio Coca Codo Sinclair, Quito 170517, Ecuador; marlon.chamba@celec.gob.ec
* Correspondence: clozada@cenace.gob.ec
† Presented at the XXXI Conference on Electrical and Electronic Engineering, Quito, Ecuador, 29 November–1 December 2023.

Abstract: An Electric Power System (EPS) is a dynamic system that, due to continuous variations in the load, the presence of disturbances, switching operations, and/or the operation of the protection system, is never in a steady state. A deficit in generation causes a drop in the system's frequency that, if not controlled, could result in the loss of synchronism between generators or areas and, in the worst-case scenario, a total or partial system collapse. This article presents a methodology to identify a subset of representative events that generalizes the N-1 generation contingency space; this subset can later be applied in the development of Under-Frequency Load-Shedding (UFLS) schemes based on the Rate of Change of Frequency (ROCOF).

Keywords: ROCOF; frequency; data mining

Citation: Lozada, C.X.; Vargas, W.A.; Granda, N.V.; Chamba, M.S. Methodology for Identifying Representative Rates of Change of Frequency (ROCOFs) in an Electric Power System against N-1 Contingencies. *Eng. Proc.* **2023**, *47*, 8. https://doi.org/10.3390/engproc2023047008

Academic Editor: Jaime Cepeda

Published: 4 December 2023

1. Introduction

The stability of an EPS is defined as the property of a power system to remain in a state of operational equilibrium under normal operating conditions and evolve towards an acceptable state of equilibrium after a disturbance [1]. When an EPS operates close to its physical limits it is vulnerable to instability problems, which if left uncontrolled can lead to partial or total system collapse. In general, there are three types of stabilities: (i) angle, (ii) voltage, and (iii) frequency.

Regarding frequency stability, for a safe and reliable operation of the EPS it is necessary to maintain the balance between generation and load. An excess or deficit in power generation produces a change in frequency, possibly resulting in values outside the admissible operating ranges. In order to maintain stability, power-frequency controls are designed, which allow one to maintain the generation–demand balance, achieving frequency values within the admissible operating range [2].

Power-frequency control is organized on three levels: primary, secondary, and tertiary. Each of the levels operates in a specific time range and involves a set of variables: The primary control is the fastest, operating in a time range of 2 and 30 s. The purpose is to limit the frequency deviation after a contingency, recovering the balance between load and generation and placing the system in a new operating point with a frequency value unequal to the nominal value. The primary response comes from the inertia of the generators, the damping of the loads, the speed regulators (governors), and other devices that provide immediate response, such as Battery Energy Storage Systems (BESSs). In the initial instants, after a power unbalance occurs, the ROCOF and the lowest point of the frequency reached

(NADIR) are determined mainly using the magnitude of the power unbalance, the total inertia of the system, and the response of fast-acting devices such as the BESS [3].

The secondary control operates in a time range of 30 s to 10 min. It operates within the control area, considering the frequency and power exchange with neighboring areas, and it is implemented by the Automatic Generation Control (AGC). Finally, the tertiary control operates in a time margin greater than 10 min. It acts in the scope of a large electrical system, seeking optimized load sharing to ensure sufficient energy reserves [3]. During the operation of the EPS, situations may arise in which imbalances between generated power and consumed power are significantly pronounced. In these circumstances, the mechanical valves controlled by the governors may be too slow to react in time before the frequency crosses acceptable operating limits. This may violate safe operating parameters, which could result in damage to the generating units [1,4]. In these cases, remedial strategies consisting of under-frequency load shedding or generation tripping are designed to prevent possible damage to the generating machines and the collapse of the system.

Load-shedding strategies can be classified into three categories: conventional, adaptive, and computational approaches [5]. Within these categories, time-dependent frequency level strategies as well as ROCOF-based strategies are used. Different methodologies have been proposed to implement load-shedding strategies using ROCOF measurements.

In [6], the calculation of the ROCOF is proposed using a local estimation of the center of inertia per generator, which involves detecting inflection points to eliminate local frequency oscillations. For the prediction of ROCOF, equations related to the center of inertia frequency are employed rather than the dynamic response of frequency to various possible events.

In [7], a load-shedding scheme is proposed for electrical systems serving oil platforms. This scheme defines operational scenarios based on generators with the highest probability of going out of operation. Subsequently, a dynamic analysis is conducted to assess the ROCOF behavior of the selected events and the loads that need to be disconnected, with the aid of a priority table. This scheme has a limited number of scenarios, which may not function correctly in scenarios that are not considered.

In [8], a sizing of energy for virtual energy contribution based on the ROCOF is proposed. In this article, the identification of ROCOF measurements is carried out locally, and calculations are performed using center of inertia parameters. This technique is applied to specific areas of the system; however, it does not provide system characterization under different contingencies.

Building on this background, this article introduces a methodology for characterizing a power system through ROCOF measurements, considering multiple operational scenarios in response to N-1 contingencies using Monte Carlo Simulation (MCS). This approach aims to obtain ROCOF values that describe the dynamic frequency behavior, leveraging data mining tools. These adjustments can be applied in a low-frequency load-shedding scheme triggered by the ROCOF.

2. Mathematical Modeling

2.1. System Frequency Response (SFR) Models

The SFR model allows one to calculate the dynamic frequency response when a generation–load imbalance occurs, and it usually consists of a turbine, speed regulator, synchronous generator, and load. More comprehensive models have been proposed that include the effect of the AGC, Under-Frequency Load-Shedding Scheme (UFLS), wind farms, photovoltaic plants, FACTS, and even induction motors. Figure 1 shows the reduced first-order model for N generators, which is applied in this methodology.

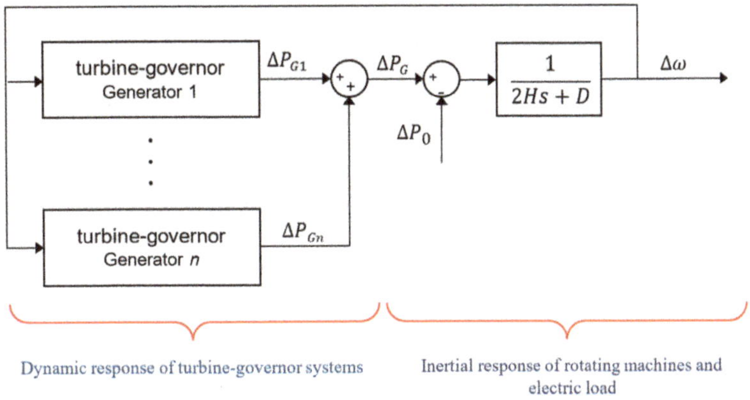

Figure 1. First-order reduced model.

To represent the speed regulator-turbine assembly, a first-order reduced model was proposed in [9], where ΔP_{Gi}, ΔP_o, and Δw, represent electrical power variation, load power variation, and speed variation, respectively, which, when included in the SFR model of the system, yields the average, collective, and coherent response of all generators [10]. This model has a variety of applications, as indicated in [11,12]. However, it has the drawback that all governors of the generators forming the system must be very similar in speed, and, in principle, all generation must be of the steam turbine type [13]. On the other hand, ref. [10] uses a first-order reduced model that accommodates a variety of generation technologies with very different governors. The equivalent SFR model for N generators is shown in Figure 2.

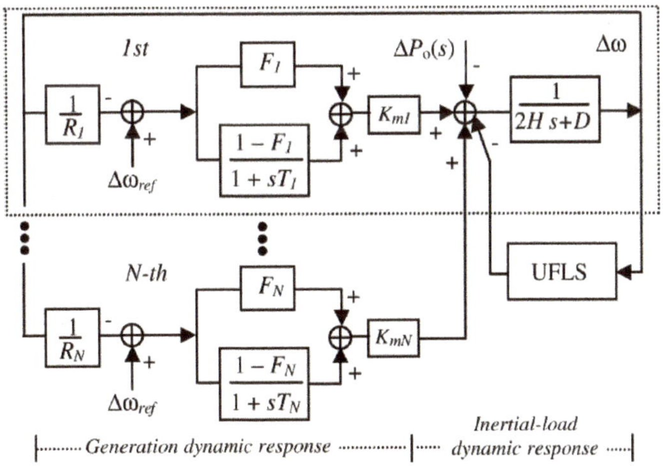

Figure 2. First-order model for dynamic frequency response.

Considering the shown model, the dynamic frequency response $\Delta\omega(s)$ to a generation–load imbalance in the Laplace domain is defined by the following:

$$\frac{\Delta\omega(s)}{\Delta P_o(s)} = \frac{-f_1(s)}{f_2(s)} \tag{1}$$

$$f_1(s) = \prod_{i=1}^{N}(1 + sT_i) \tag{2}$$

$$f_2(s) = \left(2sH_{eq} + D\right)\cdot\prod_{i=1}^{N}(1 + sT_i) + \sum_{j=1}^{N}\left[\frac{Km_i}{R_j}(1 + F_jT_js)\cdot\prod_{i=1, i\neq j}^{N}(1 + sT_i)\right] \tag{3}$$

where Km_i, F_i, T_i, and R_i are the parameters of the first-order reduced model of the i-th generator and are determined according to the procedure described in Section 3.1, while H_{eq} is the equivalent system inertia, and D represents load damping. The equivalent system inertia is calculated as the sum of the product of inertia H_i by the nominal power S_i of each generator, divided by the system base power S_{sys}, as per the equation

$$H_{eq} = \left(\sum_{i=1}^{N} H_i\cdot S_i\right) / \left(S_{sys}\right) \tag{4}$$

The generation–load imbalance is represented by $\Delta P_o(s)$, mathematically modeled using a unit step function, which, in the Laplace domain, results in $\Delta P_o(s) = \Delta P_o/s$. Solving Equation (1), the dynamic frequency response in the Laplace domain or the time domain is given by the following:

$$\Delta\omega(s) = \Delta P_o\cdot\sum_{i=1}^{N}\frac{A_i}{p_i}\left(\frac{1}{s} - \frac{1}{s - p_i}\right) \tag{5}$$

where A_i is the real or complex residue and p_i is the real pole or complex conjugate pair of (1) [10].

2.2. Rate of Change of Frequency

The ROCOF allows for characterizing the robustness of an EPS. The dynamic response of this parameter allows one to estimate the power unbalances occurring in the system, which can be calculated with the following equation.

$$\text{ROCOF} = \frac{\Delta P}{S}\cdot\frac{f}{2H} \tag{6}$$

where ΔP represents the power unbalance due to an event, f is the nominal frequency of the PES, H is the total inertia constant of the system after the event, and S is the nominal power of the system [14]. The ROCOF calculation for the implementation of the proposed methodology is performed considering a window of 0.5 s as recommended in [15–17].

3. Methodology

Figure 3 shows a diagram of the proposed methodology, which consists of the interaction of three main stages: (1) parametric identification; (2) generation of operational scenarios; and (3) data mining.

The methodology uses MATLAB simulation to perform the parametric identification of the SFR model. On the other hand, the MCS generates the operating scenarios, considering demand uncertainty, generator availability, and N-1 generator contingencies. Then, for each operating scenario generated, the dynamic behavior of the frequency is evaluated, which allows for the generation of a database. This database is a multivariate data matrix that allows for descriptive analysis, correlating variables, and identifying patterns by means of data mining tools.

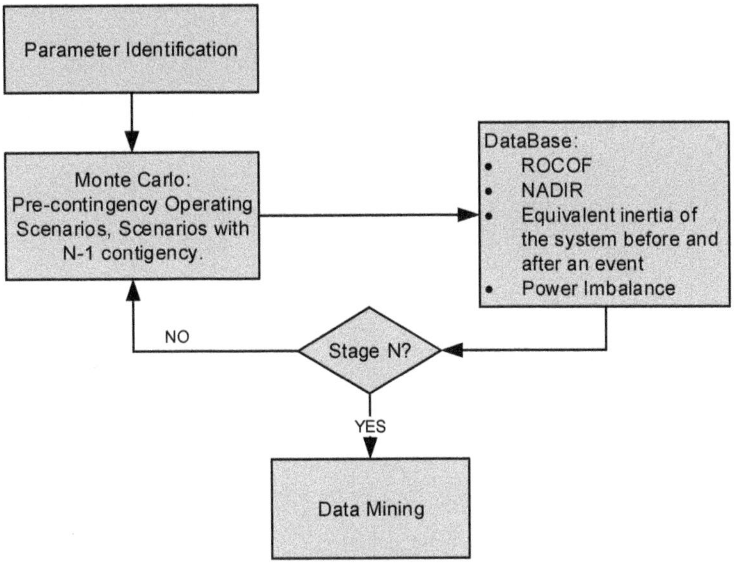

Figure 3. Flowchart of the proposed methodology.

3.1. Parameter Identification

A dynamic equivalent is obtained through a process in which the complexity of a model is reduced while preserving the dynamic response; these equivalents are determined in order to reduce the computational times [18]. At this stage, a validated model of the system behavior is obtained, where a first-order equivalent is implemented that characterizes the dynamic frequency response with the following variables: load unbalance ΔP_o, control system regulation R, and load damping D.

In addition, it is assumed that the dynamic capacity that it can supply is a fraction F of the immediate reserve capacity, the complementary fraction 1-F is a first-order phase shift with a time constant T, and Km is a constant of gain of the spinning reserve of the generators. The proposed model is presented in Figure 2 [10].

In this case, a test system with n generators and complex speed regulators was modeled in the DIgSILENT PowerFactory program to simulate events and emulate real measurements to serve as a reference for the parameter identification of the SFR model. The references extracted from PowerFactory are the mechanical power of each generator and the frequency of the system. Once these measurements are obtained, an optimization tool (lsqcurvefit of Matlab 2022b) is used to identify the constants F, T, and Km of Equation (2).

$$\Delta W = \frac{-\Delta P_o \prod_{i=1}^{N}(1+sT_i)}{(2Hs+D)\prod_{i=1}^{N}(1+sT_i) + \sum_{i=1}^{N}\left(\left(\frac{km_i}{R_i}\right)(1+sT_iF_i)\prod_{j=1,j\neq i}^{N}(1+sT_j)\right)} \quad (7)$$

The objective function of this problem is to solve the nonlinear curve fitting with the least squares error criterion between the measured signals of the system and the signals obtained through the model. Figure 4 shows the behavior of some generators with the identified parameters. The red line is the signal of the proposed model and the blue is the signal obtained from PowerFactory that simulates the real behavior.

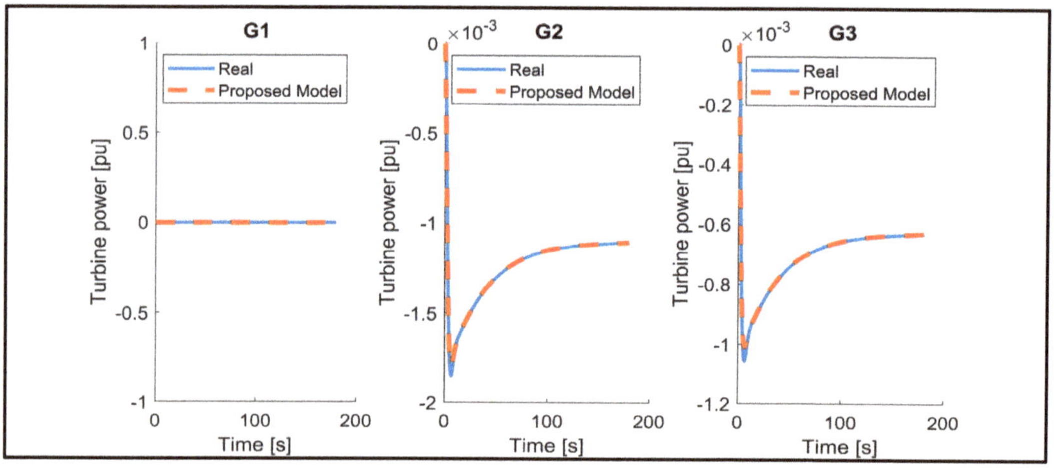

Figure 4. Parameter identification behavior.

3.2. Monte Carlo Simulation

The MCS generates operational scenarios considering uncertainty in demand, generator availability, and N-1 contingencies of the system; then, a dynamic simulation is performed to evaluate the selected variables, which are stored in a database [19].

In the first step, the pre-contingency scenarios are created by using the Matpower tool from Matlab; these scenarios are characterized by operating in a safe region. After obtaining the pre-contingency scenarios under steady-state conditions, the next step involves subjecting each of these scenarios to an N-1 generation contingency. For this stage, it is necessary to create a dynamic model in the Simulink environment with the data acquired through the parametric identification process. For the generation of operating scenarios, the probability distribution functions (FDPs) are used with the data from Table 1.

Table 1. Probability distribution functions.

Component	Variable	PDF	Characteristics
Loads	Active Power	Normal	Mean: [0.5, 0.75, 1] * Variance = 5%
	Power Factor	Uniform	Equal value to base case

* The average of the Normal PDFs is the network condition for a characteristic scenario of the minimum, average, and maximum demand of the system.

3.3. Database

For the elaboration of the database, p numerical variables are identified in a set of n elements that can form a matrix X, of dimension (n × p), necessary to perform the data analysis and identify the existing patterns [16]. The required database consists of the following variables: ROCOF, equivalent inertia before the event (Hpre), equivalent inertia after the contingency (Hpos), lost generation power (Pout), and the lowest frequency value (NADIR), as shown below.

$$X : \begin{pmatrix} \text{ROCOF}_1 & \text{Hpre} & \text{Hpos} & \text{Pout} & \text{NADIR} \\ \cdots & \cdots & \cdots & \cdots & \cdots \\ \text{ROCOF}_n & \text{Hpre}_n & \text{Hpos}_n & \text{Pout}_n & \text{NADIR}_n \end{pmatrix}_{n \times p} \tag{8}$$

The ROCOF calculation is performed using a temporal window of 0.5 s. The database must be restructured; this new matrix must discard scenarios where the system recovers to safe levels after a contingency. The decision variable for this selection is the NADIR;

those scenarios where the 59.4 Hz limit has not been exceeded are considered safe and are discarded from the database, and the value of 59.4 Hz is the first step of the under-frequency load shedding recommended in [20].

3.4. Data Mining

In this stage, data analysis is carried out using data mining tools to find anomalies, patterns, and correlations in the large dataset generated using MCS. Data mining or data exploration (it is the analysis stage of "knowledge discovery in databases", or KDD) is a field of statistics and computer science referring to the process that tries to discover patterns in large volumes of datasets [21]. For data exploration, it is necessary to represent the data resulting from observing several $p > 1$ statistical variables on a sample of n individuals. Each of these p variables is a univariant variable and the set of p variables forms a multivariant variable [22]. Normalization of the data must be performed because the database is formed by variables that have different units; after this result, the desired number of clusters must be selected to apply the "kmeans" algorithm.

4. Application of the Methodology and Analysis of Results

The test system used is an IEEE 39 bus bar system. This system has ten generators, nineteen loads, thirty-five transmission lines, and twelve transformers, whose data can be found in reference [23]. In the present study, a specific modification was made that involves dividing the original generators into eighteen to increase the sensitivity in the analysis of N-1 contingency events.

In the application of the methodology, after performing a selection of data that put the system at risk, three clusters were chosen to represent the frequency dynamics as shown in Figure 5.

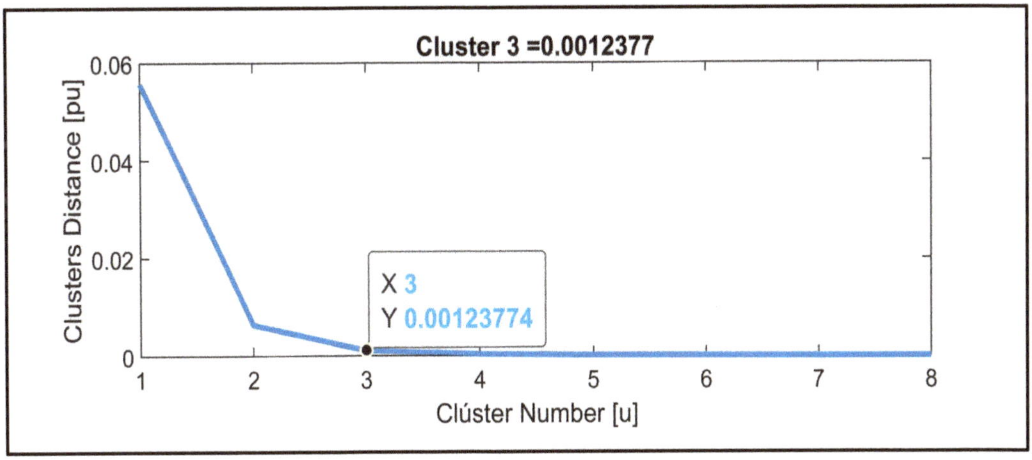

Figure 5. Total cluster distance vs. number of clusters.

For conducting the cluster analysis, the k-means tool from Matlab is used. Figure 6 illustrates the variables displaying strong correlation, showing a linear relationship among them. It can be observed that in cases where an event exhibits substantial strength and the generated output power is higher, it results in increased uncertainty in the NADIR. However, the linear correlation with the ROCOF remains intact. As a result, a UFLS founded on this parameter serves as a reliable predictor of power imbalance magnitude.

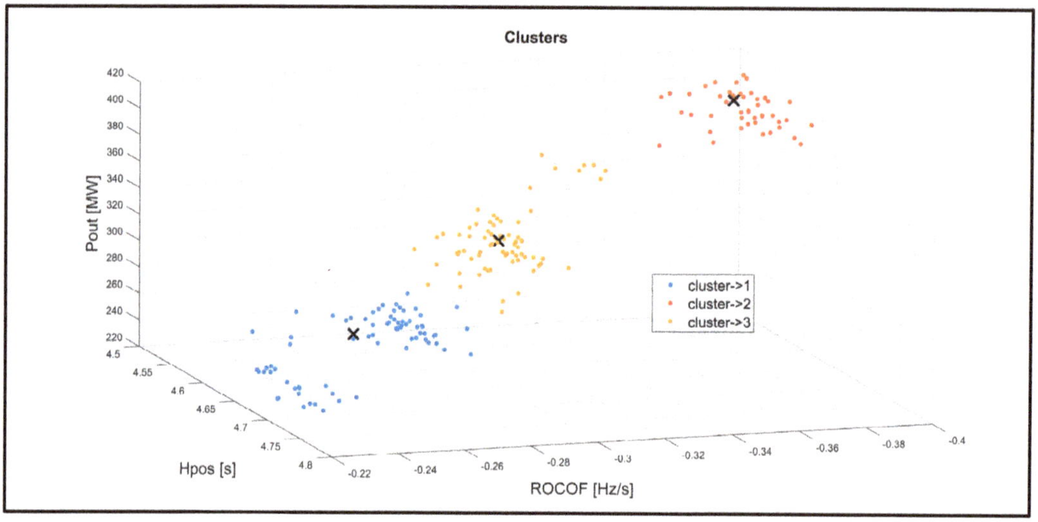

Figure 6. Correlation of variables ROCOF, Hpos, Pout.

Once the groups representing the dynamic frequency response are identified, it is necessary to identify those events which represent the centroid to select them as representative events. Table 2 shows the representative events.

Table 2. Representative events of the contingency space.

Groups	ROCOF [Hz/s]	Hpre	Hpos	Pout	NADIR
Cluster 1	−0.30394497	4.89268039	4.62509346	322.999994	59.0806595
Cluster 2	−0.38361007	4.89527662	4.57441948	403.749999	58.7209959
Cluster 3	−0.25267691	4.88498889	4.66507255	270.696622	59.3051917

5. Discussion of Results

In [6–8], methodologies are presented to obtain a characterization of the ROCOF of an electrical system, of which different strengths and weaknesses can be highlighted as shown in the Table 3.

Table 3. Main advantages and disadvantages of proposed methodologies.

Article	Main Strengths	Main Weaknesses
Underfrequency Load Shedding Using Locally Estimated RoCoF of the Center of Inertia	• Local estimation of the ROCOF in each generator. • Algorithm to eliminate frequency oscillations for ROCOF measurement.	The methodology does not consider the different operational scenarios and contingencies that may occur in the system.
Automatic Load Shedding Scheme for Electrical Systems Serving Oil Extraction Facilities	• System modeling and validation. • Statistical analysis. • Dynamic system simulation.	There are a limited number of contingencies that depend on the designer's experience.
Energy storage sizing for virtual inertia contribution based on ROCOF and local frequency dynamics	• Identification of areas of frequency behavior.	A characterization of the system from the ROCOF point of view is not obtained.

The proposed methodology considers several advantages identified in the table above to obtain ROCOF measurements for the system, including the following:

Developing an equivalent model that allows one to obtain the dynamic frequency response, with the flexibility to adapt to any complete or simplified system.

Conducting an MCS to consider a wide range of operational scenarios and contingencies that may arise in the system.

As a result, groups of ROCOF values characterizing the frequency dynamics in a system are obtained. These results can be implemented to determine the minimum amount of load to be shed in a load-shedding scheme. This procedure helps overcome the weaknesses identified in similar proposals.

6. Conclusions and Recommendations

The proposed methodology stands out for its ability to identify and characterize ROCOF groups in the presence of N-1 contingencies and considering a wide range of possible operating scenarios for the system. This characterization allows for the design of an under-frequency load-shedding scheme through the parameterization of frequency relays using the ROCOF. It is recommended to restructure the database according to the needs of the system since this step eliminates scenarios that are considered safe at the frequency level; in this case, 59.4 Hz was chosen, but it can vary according to the needs. The proposed methodology should be used in conjunction with an optimization algorithm to determine the minimum load to be shed. For the nature of the optimization problem, a heuristic solution method should be considered, and thus the necessary adjustments to guarantee the minimum impact when the system needs a UFLS.

Author Contributions: Methodology, software, investigation: C.X.L.; Conceptualization: W.A.V.; Project administration: N.V.G.; Supervision, review, and editing: M.S.C. All authors have read and agreed to the published version of the manuscript.

Funding: This research received no external funding.

Institutional Review Board Statement: Not applicable.

Informed Consent Statement: Not applicable.

Data Availability Statement: Data are contained within the article.

Acknowledgments: The authors would like to thank the National Polytechnic School for providing support to carry out this research work under project PII-DEE-2023-02 and would like to thank the Research and Development Department of the Ecuadorian ISO.

Conflicts of Interest: The authors declare no conflict of interest.

Abbreviations

ROCOF	Rate of Change of Frequency
EPS	Electric Power System
UFLS	Under-Frequency Load Shedding
Governors	Speed Regulators
BESS	Battery Energy Storage System
NADIR	Lowest Point of the Frequency
AGC	Automatic Generation Control
MCS	Monte Carlo Simulation
SFR	System Frequency Response
FDP	Probability Distribution Function
Hpre	Equivalent Inertia before the Contingency
Hpos	Equivalent Inertia after the Contingency
Pout	Lost Generation Power
KDD	Knowledge Discovery in Databases

ΔP_{Gi}	Electrical Power Variation
ΔP_o	Load Power Variation
Δw	Speed Variation
Km_i, F_i, T_i, and R_i	Meters of the First-Order Reduced Model
H_{eq}	Equivalent System Inertia
D	Load Damping
S_i	Nominal Power of Each Generator
S_{sys}	System Base Power

References

1. Kundur, P. *Power System Stability and Control*; McGraw-Hill Education: New York, NY, USA, 1994.
2. Vargas, W.; Chamba, M.S.; Torre, A.D.L.; Echeverría, D. Testing Framework and Validation for Speed Governing Systems—Practical Application at Delsitanisagua Hydropower Plant. *Rev. Técnica Energía* **2022**, *19*, 22–33. [CrossRef]
3. Chamba, M.S.; Vargas, W.; Riofrio, J.; Echeverría, D. Primary Frequency Regulation with Battery Energy Storage Systems in the Ecuadorian Power System. *Rev. Técnica Energía* **2022**, *19*, 13–21. [CrossRef]
4. Rudez, U.; Mihalic, R. Analysis of Underfrequency Load Shedding Using a Frequency Gradient. *IEEE Trans. Power Deliv.* **2011**, *26*, 565–575. [CrossRef]
5. Laghari, J.A.; Mokhlis, H.; Bakar, A.H.A.; Mohamad, H. Application of computational intelligence techniques for load shedding in power systems: A review. *Energy Convers. Manag.* **2013**, *75*, 130–140. [CrossRef]
6. Sun, M.; Liu, G.; Popov, M.; Terzija, V.; Azizi, S. Underfrequency Load Shedding Using Locally Estimated RoCoF of the Center of Inertia | IEEE Journals & Magazine | IEEE Xplore. Available online: https://ieeexplore.ieee.org/document/9362316 (accessed on 15 October 2023).
7. Jacome, V.N.; Granda, N.V. Automatic Load Shedding Scheme for Electrical Systems Serving Oil Extraction Facilities. *Rev. Técnica Energía* **2023**, *19*, 58–68. [CrossRef]
8. Sørensen, D.A.; Pombo, D.V.; Iglesias, E.T. Energy storage sizing for virtual inertia contribution based on ROCOF and local frequency dynamics. *Energy Strategy Rev.* **2023**, *47*, 101094. [CrossRef]
9. Anderson, P.M.; Mirheydar, M. A low-order system frequency response model. *IEEE Trans. Power Syst.* **1990**, *5*, 720–729. [CrossRef]
10. Aik, D.L.H. A general-order system frequency response model incorporating load shedding: Analytic modeling and applications. *IEEE Trans. Power Syst.* **2006**, *21*, 709–717. [CrossRef]
11. Terzija, V.V. Adaptive underfrequency load shedding based on the magnitude of the disturbance estimation. *IEEE Trans. Power Syst.* **2006**, *21*, 1260–1266. [CrossRef]
12. Anderson, P.M.; Mirheydar, M. An adaptive method for setting underfrequency load shedding relays. *IEEE Trans. Power Syst.* **1992**, *7*, 647–655. [CrossRef]
13. Egido, I.; Fernandez-Bernal, F.; Centeno, P.; Rouco, L. Maximum Frequency Deviation Calculation in Small Isolated Power Systems. *IEEE Trans. Power Syst.* **2009**, *24*, 1731–1738. [CrossRef]
14. Dudurych, O.; Conlon, M. Impact of reduced system inertia as a result of higher penetration levels of wind generation. In Proceedings of the 2014 49th International Universities Power Engineering Conference (UPEC), Cluj-Napoca, Romania, 2–5 September 2014; pp. 1–6. [CrossRef]
15. Wu, L. Power System Frequency Measurement Based Data Analytics and Situational Awareness. Ph.D. Thesis, University of Tennessee, Knoxville, TN, USA, 2018.
16. Wright, P.S.; Davis, P.N.; Johnstone, K.; Rietveld, G.; Roscoe, A.J. Field Measurement of Frequency and ROCOF in the Presence of Phase Steps. *IEEE Trans. Instrum. Meas.* **2019**, *68*, 1688–1695. [CrossRef]
17. European Network of Transmission System Operators for Electricity. *Rate of Change of Frequency (ROCOF) Withstand Capability ENTSO-E Guidance Document for National Implementation for Network Codes on Grid Connection*; ENTSOE: Brussels, Belgium, 2017.
18. Cepeda, J.C.; Rueda, J.L.; Erlich, I. Identification of dynamic equivalents based on heuristic optimization for smart grid applications. In Proceedings of the 2012 IEEE Congress on Evolutionary Computation, Brisbane, QLD, Australia, 10–15 June 2012; pp. 1–8. [CrossRef]
19. Chamba, M.; Vargas, W.; Cepeda, J. Probabilistic assessment of transient stability considering the uncertainty of the demand and risk management. *Energía* **2018**, *15*, 1–10.
20. Centro Nacional de Control de Energía. *Determinación Del Esquema De Alivio De Carga Por Baja Frecuencia Para El Sistema Nacional Interconectado*; CENACE: Quito, Ecuador, 2022.
21. Fayyad, U.; Haussler, D.; Stolorz, P. KDD for Science Data Analysis: Issues and Examples. 2000. Available online: https://cdn.aaai.org/KDD/1996/KDD96-009.pdf (accessed on 15 October 2023).

22. Peña, D. *Análisis de Datos mu Ltivariantes;* Mc Graw Hill Education: Madrid, España, 2010.
23. Athay, T.; Podmore, R.; Virmani, S. A Practical Method for the Direct Analysis of Transient Stability. *IEEE Trans. Power Appar. Syst.* **1979**, *PAS-98*, 573–584. [CrossRef]

engineering proceedings

MDPI

Proceeding Paper

Performance Analysis of Motion Control Algorithms of an Industrial Robot Arm Applied to 3D Concrete Printing Systems †

Carlos Calderon-Cordova *, Leonardo Sarango, Dennis Chamba, Roger Sarango and Raul Castro

Department of Computer Science and Electronics, Universidad Técnica Particular de Loja, Loja 1101608, Ecuador; lmsarango10@utpl.edu.ec (L.S.); dachamba4@utpl.edu.ec (D.C.); rasarango1@utpl.edu.ec (R.S.); jrcastro@utpl.edu.ec (R.C.)

* Correspondence: cacalderon@utpl.edu.ec or cacalderon@ieee.org

† Presented at the XXXI Conference on Electrical and Electronic Engineering, Quito, Ecuador, 29 November–1 December 2023.

Abstract: This paper shows the implementation and performance analysis of motion control algorithms for a 3D Concrete Printing system based on an industrial robotic arm. This work is part of the project of digital fabrication of low-cost housing. Regarding the technological architecture integrated in the present work, the hardware devices used are the EPSON C12 industrial robot arm and the RC700-A controller, and the software tools are Fusion 360 and RC+. To evaluate the point extraction and sequencing algorithms of a 3D structure and the motion control algorithms of the robot arm, three test wall models were designed: a semi-circle with horizontal undulations, a semi-circle with vertical undulations and an orthohedron without undulations. For the performance analysis, 140 trajectory times were extracted for each test model. The extracted values are the trajectory time intervals of the wall model envelope for each layer and of the internal trajectory of the wall model infill. Due to the increasing and decreasing trends of time for specific cases, it was concluded that in the working area of the robot arm, there are parts in which the robot is more efficient and therefore the joints offer less inertia for certain types of movements, for example straight or curved movements and short or long movements.

Keywords: 3D concrete printing; robotic arm; motion control; algorithms; trajectory

Citation: Calderon-Cordova, C.; Sarango, L.; Chamba, D.; Sarango, R.; Castro, R. Performance Analysis of Motion Control Algorithms of an Industrial Robot Arm Applied to 3D Concrete Printing Systems. *Eng. Proc.* **2023**, *47*, 9. https://doi.org/10.3390/engproc2023047009

Academic Editor: Jackeline Abad

Published: 4 December 2023

1. Introduction

In the world, according to the United Nations, the number of people who do not have decent housing is worrying: approximately 1.6 billion people live in inadequate housing and about 900 million live in informal settlements or camps [1]. Similarly, in Ecuador, we find that about 600,000 families do not have their own home, which is why meeting the need for housing has become urgent [2]. Due to the development and introduction of Industry 4.0, digital manufacturing is gaining momentum as it allows a more precise design process and opens up an opportunity to manufacture a great diversity of part geometries compared to traditional manufacturing. Together with 3DCP printing (3D Concrete/Concrete Printing), it allows the materialization of wall blocks or houses, so the printing technology will be a determinant for the heterogeneous geometry to be obtained [3].

In parallel to traditional 3D printing technologies, several companies began to test the creation of 3D ceramic parts, which were the basis for large-scale printing in the construction area. In 2014, the Chinese company Winsun, a pioneer in this area, printed several houses in a period of 24 h, with a method of printing in parts or blocks, which were moved to the place where they would be implemented. In such a way, different techniques and models of 3D printers were tested for different needs [4].

For 3D printing in the construction sector, there are factors to take into account, such as total printing time, costs, materials and printing accuracy. There are two techniques that have been predominant since the inception of 3D Concrete Printing (3DCP) [5,6].

The extrusion-based technique: it works via deposition of material layer by layer through a nozzle that is located on a Cartesian robot or six-axis robotic arm. This technique is used for large-scale construction or printing of complex geometries, depending on the type of robot to be used [7].

Powder-based technique: it works by depositing a binder liquid in a guided manner on the CAD model to be created, then a powder mixture is added for the solidification process. This technique is used for the fabrication of modular components or small-scale constructions [5].

In the academic environment, there are some challenges addressed around 3DCP technology. In [8], a system for mobility limitation for the fabrication of complex structures is presented. In [9], a 3D printing system of a post-tensioned concrete beam is presented. In this project, an industrial robot arm with 6 degrees of freedom is used. In [10], a project for fabrication of structures from a robotic system is presented. This system was implemented at laboratory scale. On the other hand, in [11], the extrusion material is addressed, and the authors conclude that the optimum material should have a balanced behavior of elasticity–viscosity–plasticity properties. The authors performed tests with an industrial robotic arm at a speed of 30 mm/s. Finally, with the aim of analyzing the extraction of key points from a solid intended to be 3D printed, [12] was analyzed. This study presents the use of CAD files for layered 3D printing.

Based on the literature research and to address the technological subsystem (software and hardware) of the digital fabrication project to generate low-cost housing, the objective of this paper is to develop the motion control algorithms and analyze the laboratory-scale performance of the industrial robotic arm applied to 3D concrete printing systems.

The layout of this paper consists of Section 2, Materials and Methods, which addresses the hardware and software architecture of the system and the simulation of the algorithms; Section 3, Results and Discussion, which addresses the physical implementation of the system, the data extraction and the discussion around the trends identified in the data, and finally Section 4, which summarizes the conclusions of the work.

2. Materials and Methods

The present section is composed of the description of the hardware and software architecture of the system and the description of the design and simulation of the algorithms of the experimental system.

2.1. Hardware and Software Architecture of the System

Figure 1 shows a diagram of the system to be designed and implemented composed of software and hardware elements. The workflow starts with the modeling of the test block or wall with the Fusion 360 program. The finished design is exported in the ".step" format compatible with the EPSON RC+ 7.5 R2 software. In the latter, we proceed to elaborate the algorithm, introducing the initial configuration, then we import the designed model; using the CAD To Point 7.5 tool, we extract the points of the wall block.

Once the points of the layers have been obtained, with the "main" function configured, different functions are created to control the movement of the robot arm, generating point-to-point movements for the formation of the wall and the filling. Once the algorithm is completed, from the computer connected via USB to the EPSON RC700-A controller (Seiko Epson Corp., Suwa, Japan), the code is implemented to perform the tests with the EPSON C12 robotic arm (Seiko Epson Corp., Suwa, Japan).

Regarding the hardware devices that are part of the system, these include the EPSON C12 industrial robot arm, which has 6 axes with payload capacity of 12 kg, range up to 1400 mm and accuracy of +/−0.050 mm [13], and the EPSON RC700-A controller, which is a programmable device in SPEL+ language and has digital input/output and Ethernet TCP/IP communication interfaces [14].

The following software tools were used for the development and implementation of the system: Fusion360 was used for the design of the geometric structure intended for 3D printing. EPSON RC+ was used for the implementation of the motion control algorithm of the EPSON C12 robot arm, and the 3D structure point set extraction algorithm.

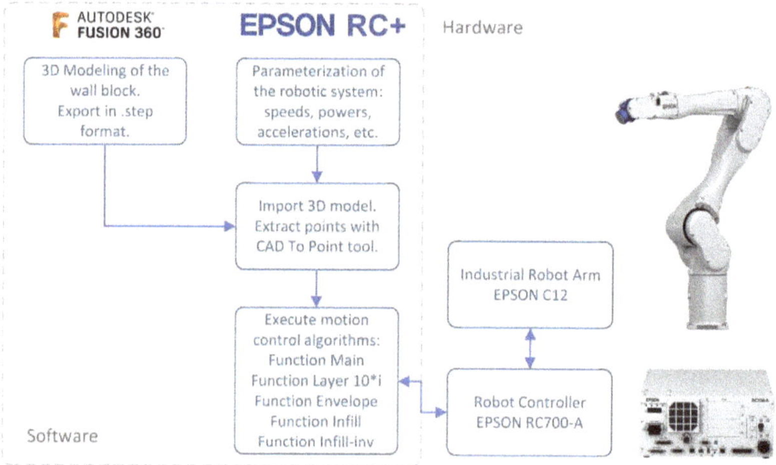

Figure 1. Hardware and software architecture of the system.

2.2. Design of the 3D Models to Be Printed

The model of the wall structures to be printed was designed in Autodesk Fusion 360 2.0 software. Three test models were designed, whose approximate dimensions of the envelope are 1500 × 250 × 490 mm. These designs will be used to evaluate the behavior of the robot in terms of trajectory control and robot movement times. Figure 2 shows the three 3D designs corresponding to the test walls.

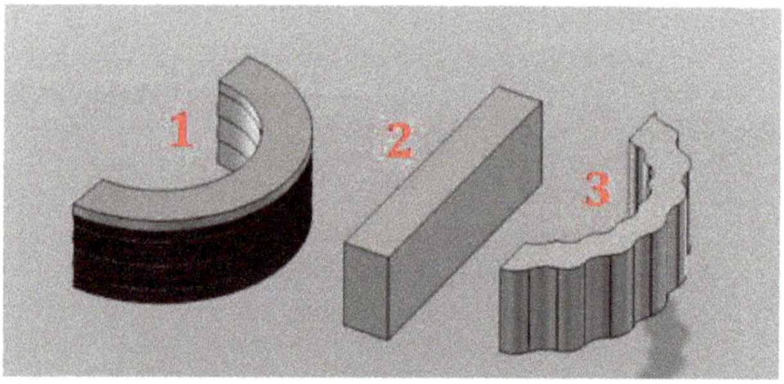

Figure 2. Three-dimensional designs corresponding to the test walls: (**1**) semi-circle with vertical undulations, (**2**) orthohedron without undulations, and (**3**) semi-circle with horizontal undulations.

For the first test design, we took into account geometrical considerations of the 3D-printed housing prototype of the Italian company WASP. The main guidelines of our design are as follows. The design has a 3D semicircle envelope with undulations perpendicular to the base. Each designed layer will have a height of 7 mm, because the material to be deposited has this height. The first 10 layers will have a reduction in radius of 3 mm with respect to the previous one. The next 10 layers will have an increase in radius of 3 mm with respect to the previous one. The second test design is a traditional rectangular wall. The third test design has a geometry of a semicircle with undulations along its perimeter.

2.3. Extraction and Processing of Points from the 3D Model

The CAD To Point tool available in EPSON software is used to extract the points of interest from the 3D design. The design is imported, the extraction is parameterized and finally a table of points is generated. The points of interest of the resulting table are those points that belong to the 3D design envelope and that will be used to generate the movements of the robotic arm.

Regarding point processing, this phase is necessary to link the points in an orderly manner and thus generate the routes that the robot will travel and manufacture. The final part of this stage is the generation of the robot arm movements between the consecutive coordinates of the 3D structure.

2.4. Simulation of Motion Control Algorithms

Once the points of the 3D structure have been extracted and ordered and the control algorithms of the robot movements have been developed, the simulation of the system is carried out. Figure 3 shows the simulation in the EPSON RC+ environment. The red points belong to the design of the 3D structure. These points show 70 layers, and the blue points indicate the path that the robot arm is performing while running the simulation. Figure 3a shows the robot path over the 3D structure envelope and Figure 3b shows the robot path over the 3D structure infill.

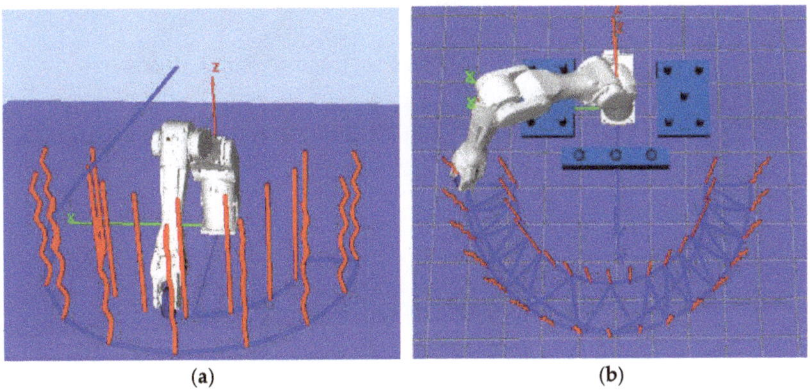

Figure 3. Epson RC+ simulation environment of the robotic arm motion control algorithms: (**a**), robot path over the 3D structure envelope; (**b**), robot path over the 3D structure infill.

3. Results and Discussion

This section presents the results of the implementation of the motion control algorithms in the physical system composed of the EPSON C12 robot arm and the RC700-A controller. In addition, an analysis of the travel times for each of the test blocks will be performed.

3.1. Testing with the EPSON C12 Physical Robot

Figure 4 shows the EPSON C12 industrial robot arm running the path of the 3D designs loaded into the software. It should be noted that at the current stage of the project, the material extruder is not yet integrated to the robot arm. Figure 5 shows the software user interface during program execution on the RC700-A controller, Figure 5a shows the execution of the test wall 1 print program, Figure 5b shows the execution of the test wall 2 print program, Figure 5c shows the execution of the test wall 3 print program, and finally, Figure 5d shows the execution times for each layer. The green points belong to the design of the 3D structure and the blue points indicate the path that the robot arm is performing.

3.2. Analysis and Discussion of Results

This section shows the record of the travel times obtained by the robot arm at the moment of executing the movements to locate the end effector in the coordinates belonging to the three test walls. The total travel time of the envelope and the filler is recorded for each of the 70 layers and for each of the three test walls; a total of 420 time values are obtained.

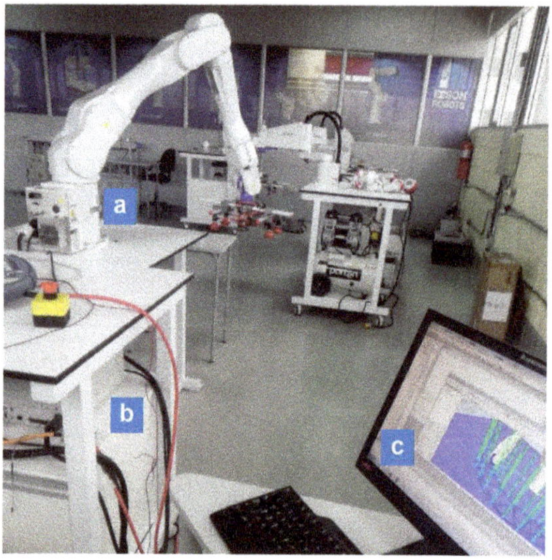

Figure 4. Industrial robot arm running the path of the 3D designs loaded into the software. (**a**) EPSON C12 Robot arm, (**b**) EPSON RC700-A Controller, (**c**) computer to perform real-time visualization of the positions and times of the robot movements.

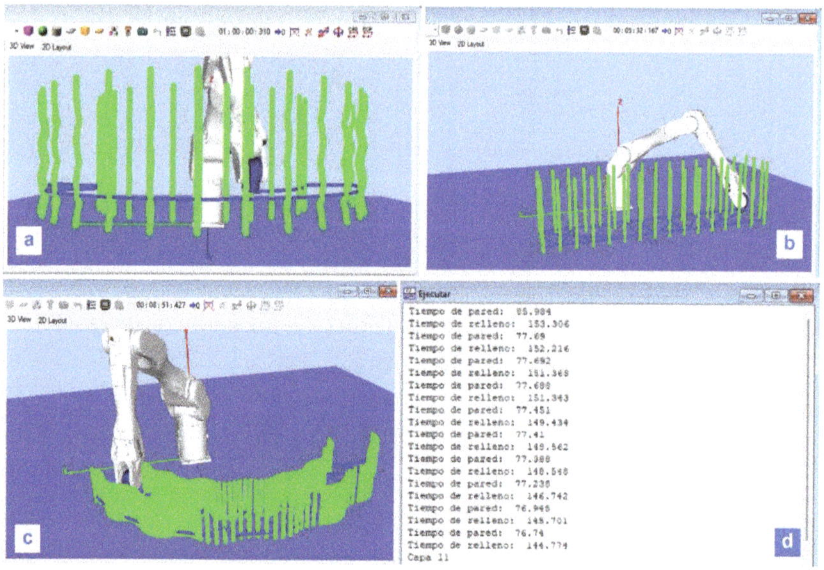

Figure 5. Software user interface during algorithms execution on the RC700-A controller. (**a**) Execution of the test wall 1 motion control program, (**b**) execution of the test wall 2 motion control program, (**c**) execution of the test wall 3 motion control program, and (**d**) execution times for each layer.

Figure 6 shows the fabrication time plot of the 70 layers of the envelope (a) and infill (b) for test wall design 1. The test designs are shown in Figure 2.

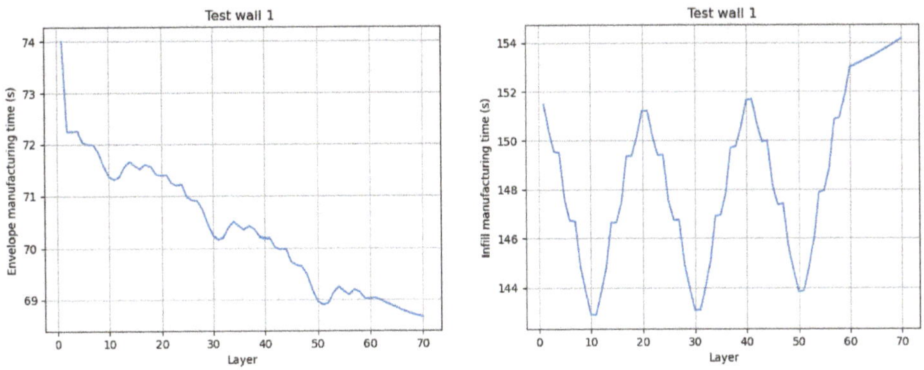

(a) (b)

Figure 6. Manufacturing time plot of the 70 layers of the envelope (**a**) and infill (**b**) for test wall design 1.

The following aspects are observed in Figure 6a:

- The first layer is slower than the others. This is because the robot starts from an idle position.
- The travel time of the layer is decreasing as the wall is being manufactured, with approximately 4 s difference between the printing of the first and last layer. This is because the new positions correspond to segments of the work area where the robot is slightly less efficient.
- Three depressions are identified in the graph due to the design of the wall, since it has three local minima in its vertical path.

The following aspects are observed in Figure 6b:

- It has three local minima due to the geometry of the wall.
- There is a trend of increasing travel time per fill layer; this trend is contrary to the envelope travel.

A preliminary explanation of the above analysis is that the performance of the actual robot arm is altered by the running time, i.e., a longer running time corresponds with a decrease in the task execution times. However, to contradict this hypothesis, it is seen that in Figure 6b, the fill time increases for each layer traversed. Consequently, it is concluded that in the working area of the robot arm, there are parts in which the robot is more efficient and therefore the joints offer less inertia for certain types of movements; for example, straight or curved movements and short or long movements.

Figure 7 shows the fabrication time plot of the 70 layers of the envelope (a) and infill (b) for test wall design 2.

Figure 7a shows the following aspects:

- The first layer is slower than the others. This is due to the fact that the robot starts from a resting position.
- There is a minimum change in the travel time as the vertical displacement of the manufactured layer increases. From layer 2 to layer 52, we can observe a difference of 0.970 s, and from layer 52 to layer 70, the change is 0.229 s.

This reinforces the previous conclusion that the printing plane and the type of movement influences the trajectory travel times of the robot arm.

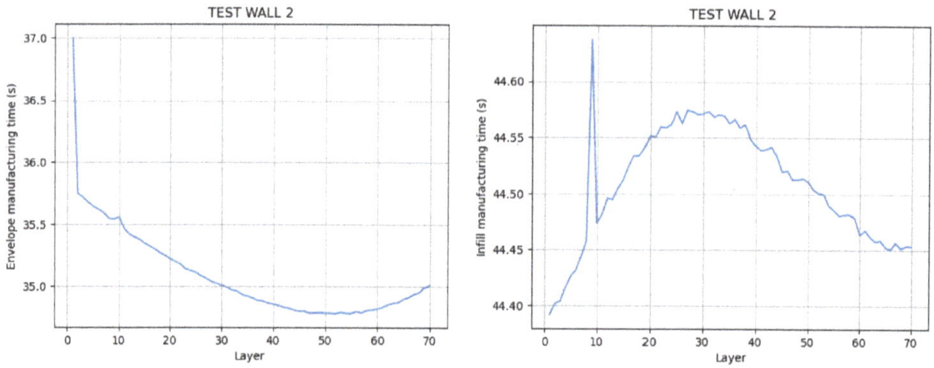

(**a**) (**b**)

Figure 7. Manufacturing time plot of the 70 layers of the envelope (**a**) and infill (**b**) for test wall design 2.

With respect to Figure 7b, two intervals can be seen: one interval of increase and one interval of decrease of the layer travel time. The conclusion that the printing plane and the type of movement influences the travel times of a robot path is more evident in this test design, since here, the design is an Orthohedron; therefore, the path of all the layers is exactly the same size, and the only parameter that changes is the height of the layer to be manufactured.

Figure 8 shows the fabrication time plot of the 70 layers of the envelope (a) and infill (b) for test wall design 3. As in the two previous cases, there is a tendency to decrease the envelope travel time as the robot advances in the fabrication of the upper layers, and, in addition, there is a tendency to increase the filler travel time as the robot advances in the fabrication of the upper layers.

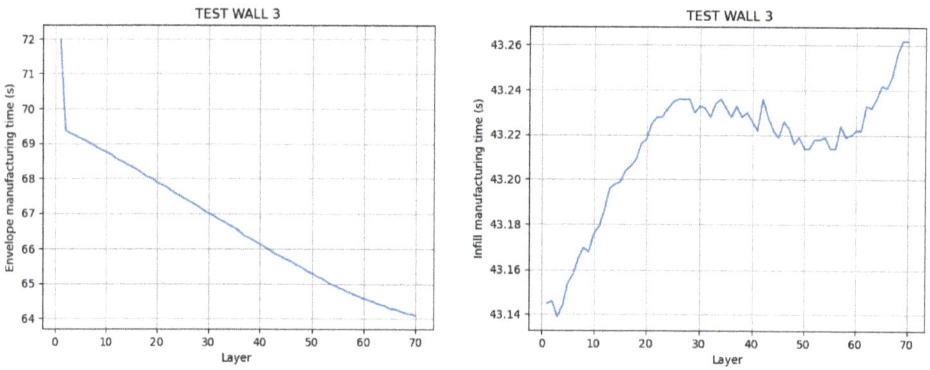

(**a**) (**b**)

Figure 8. Manufacturing time plot of the 70 layers of the envelope (**a**) and infill (**b**) for test wall design 3.

As the system advances to the upper layers, the time decreases, as shown in Figures 6a, 7a and 8a. This characteristic is linked to the simplicity of the path of the envelope with respect to the filling. The points of the envelope are closer to each other, so the robot does not have to accelerate or decelerate abruptly during the entire path. Meanwhile, the infill points are farther apart, so the robot generates velocity changes at the beginning and end of the motion sections. The source data for Figures 6–8 can be found in the supplementary file of this article, Table S1.

4. Conclusions

The performance of motion control algorithms for a 3D Concrete Printing system based on an industrial robotic arm was implemented and analyzed. The integrated technological architecture consisted of the following elements: the EPSON C12 industrial robot arm, the RC700-A robot controller, and the software tools Fusion 360 and RC+.

With respect to the performance analysis of the system's motion control algorithms, trends of increasing and decreasing motion execution times were identified for specific cases. It was concluded that in the working area of the robot arm, there are parts in which the robot is more efficient, and therefore the joints offer less inertia for certain types of movements; for example, straight or curved movements and short or long movements.

Supplementary Materials: The following supporting information can be downloaded at: https://www.mdpi.com/article/10.3390/engproc2023047009/s1, Table S1: Manufacturing times of the 70 layers of the envelope and the infill for Test Wall designs 1, 2 and 3.

Author Contributions: Conceptualization, methodology, C.C.-C. and L.S.; software, L.S. and D.C.; validation, L.S. and D.C.; writing—original draft preparation, C.C.-C. and R.S.; writing—review and editing, C.C.-C., R.S. and R.C.; supervision, C.C.-C. and R.C.; project administration and funding acquisition, C.C.-C. and R.C. All authors have read and agreed to the published version of the manuscript.

Funding: This research was funded by Universidad Tecnica Particular de Loja, grant number PROY_ARTIC_CE_2022_3667.

Institutional Review Board Statement: Not applicable.

Informed Consent Statement: Not applicable.

Data Availability Statement: Data are contained within the article and supplementary materials.

Conflicts of Interest: The authors declare no conflict of interest.

References

1. Millones de Personas Viven Sin Techo o en Casas Inadecuadas, un Asalto a la Dignidad y la Vida | Millions of People Live Homeless or in Inadequate Housing, an Assault on Dignity and Life. Available online: https://news.un.org/es/story/2018/07/1437721 (accessed on 20 August 2023).
2. Casi 600.000 Familias en Ecuador no Tienen Vivienda Propia | Nearly 600,000 Families in Ecuador Do Not Have Their Own Home. Available online: https://www.primicias.ec/noticias/sociedad/familias-ecuador-sin-casa-propia/ (accessed on 21 August 2023).
3. Adam, J.O. *Fabricación Digital: Introducción al Modelado e Impresión 3D*; Ministerio de Educación, Cultura y Deporte: Madrid, Spain, 2016.
4. Wu, P.; Wang, J.; Wang, X. A critical review of the use of 3-D printing in the construction industry. *Autom. Constr.* **2016**, *68*, 21–31. [CrossRef]
5. Sonwalkar, S.P. Roadmap for Adopting 3D Concrete Printing Technology for Production of Affordable Houses. Master's Thesis, University of Twente, Deventer, The Netherlands, 2020. Available online: http://essay.utwente.nl/83162/ (accessed on 10 July 2023).
6. Chen, Y.; Jansen, K.; Zhang, H.; Romero Rodriguez, C.; Gan, Y.; Çopuroğlu, O.; Schlangen, E. Effect of printing parameters on interlayer bond strength of 3D printed limestone-calcined clay-based cementitious materials: An experimental and numerical study. *Constr. Build. Mater.* **2020**, *262*, 120094. [CrossRef]
7. Nematollahi, B.; Xia, M.; Sanjayan, J. Current Progress of 3D Concrete Printing Technologies. In Proceedings of the 34rd ISARC, Taipei, Taiwan, 28 June–1 July 2017. [CrossRef]
8. Cuevas, K.; Chougan, M.; Martin, F.; Ghaffar, S.H.; Stephan, D.; Sikora, P. 3D printable lightweight cementitious composites with incorporated waste glass aggregates and expanded microspheres–Rheological, thermal and mechanical properties. *J. Build. Eng.* **2021**, *44*, 102718. [CrossRef]
9. Vantyghem, G.; De Corte, W.; Shakour, E.; Amir, O. 3D printing of a post-tensioned concrete girder designed by topology optimization. *Autom. Constr.* **2020**, *112*, 103084. [CrossRef]
10. Mechtcherine, V.; Bos, F.P.; Perrot, A.; da Silva, W.R.L.; Nerella, V.N.; Fataei, S.; Wolfs, R.J.M.; Sonebi, M.; Roussel, N. Extrusion-based additive manufacturing with cement-based materials–Production steps, processes, and their underlying physics: A review. *Cem. Concr. Res.* **2020**, *132*, 106037. [CrossRef]
11. Spangenberg, J.; Leal da Silva, W.R.; Comminal, R.; Mollah, M.T.; Andersen, T.J.; Stang, H. Numerical simulation of multi-layer 3D concrete printing. *RILEM Tech. Lett.* **2021**, *6*, 119–123. [CrossRef]

12. Ruffray, N.; Bernhard, M.; Jipa, A.; Aghaei, M.; Montague, N.; Stutz, F.; Wangler, T.; Flatt, R.; Dillenburger, B. Complex architectural elements from HPFRC and 3D-printed sandstone. In Proceedings of the UHPFRC 2017, Montpellier, France, 2–4 October 2017. [CrossRef]
13. Epson C12XL 6-Axis Robot. Available online: https://epson.com/For-Work/Robots/6-Axis/C12XL-6-Axis-Robot/p/RC12XL-A1401ST73SS (accessed on 4 July 2023).
14. Epson RC700A Controller. Available online: https://epson.com/For-Work/Robots/Controllers/Epson-RC700A-Controller/p/R2170011 (accessed on 4 July 2023).

engineering proceedings

MDPI

Proceeding Paper

Physical Downlink Control Channel (PDCCH) Performance Evaluation for 5G/NR Networks at Different Positions of the User Equipment †

Paúl Barona-Castillo, Fabio González-González * and Martha Cecilia Paredes-Paredes

Departamento de Electrónica, Telecomunicaciones y Redes de Información, Escuela Politécnica Nacional, Ladrón de Guevara, Quito 170109, Ecuador; paul.barona@epn.edu.ec (P.B.-C.); cecilia.paredes@epn.edu.ec (M.C.P.-P.)

* Correspondence: fabio.gonzalez@epn.edu.ec

† Presented at the XXXI Conference on Electrical and Electronic Engineering, Quito, Ecuador, 29 November–1 December 2023.

Abstract: This article presents the implementation and the evaluation of a transmission–reception system for Physical Downlink Control Channel (PDCCH) within the context of fifth-generation cellular networks (5G/NR). The research focuses on characterizing the behavior of the PDCCH channel concerning different User Equipment (UE) positions through simulations conducted in MATLAB. Two scenarios are considered: a lossless system and a system with losses, where the outcomes are presented in terms of Bit Error Rate (BER) vs. Signal to Noise Ratio (SNR) and BER vs. Transmit Power, respectively. One contribution of the paper is the explanation transmission–reception system for PDCCH in MATLAB, while the second contribution is the presentation of simulation results. From results, it was deduced that higher UE height corresponds to improved channel performance, greater UE elevation angle leads to enhanced channel performance, and increased carrier frequency results in reduced performance of the physical control PDCCH channel.

Keywords: 5G; NR; PDCCH; UE position; BER; SNR; transmit power

Citation: Barona-Castillo, P.; González-González, F.; Paredes-Paredes, M.C. Physical Downlink Control Channel (PDCCH) Performance Evaluation for 5G/NR Networks at Different Positions of the User Equipment. *Eng. Proc.* **2023**, *47*, 10. https://doi.org/10.3390/engproc2023047010

Academic Editor: Fernando Carrera Suarez

Published: 4 December 2023

1. Introduction

Presently, the widely used fourth-generation (4G) mobile network, known as long term evolution (LTE), is gradually giving way to the anticipated adoption of the fifth-generation (5G) mobile network, known as new radio (NR). The 3GPP technical report 38.913 [1] outlines three primary 5G usage scenarios: enhanced mobile broadband (eMBB), ultra-reliable and low-latency communications (URLLC), and massive machine type communications (mMTC). These scenarios promise significant enhancements, including higher transmission speeds (up to 10 Gbps) and ultra-low latency (as low as 1 ms) [1]. These advancements are set to revolutionize mobile communication services and enable innovative applications [1].

Building on lessons from operational networks like LTE, mobile operators have encountered user grievances related to signal degradation due to coverage issues or channel fading [2]. Thus, a thorough analysis of control channels across various positions of user equipment (UE) within base station coverage areas becomes crucial in devising strategies to mitigate communication losses. Such insights can optimize 5G base station deployment, improving coverage and user services.

The mobility of UEs within coverage areas introduces dynamic changes in channel conditions, influenced by device movement and the stochastic nature of wireless channels [3]. Consequently, assessing the behavior of control channels, crucial for communication control, as UE positions change, becomes pivotal.

With these motivations, this study employs MATLAB simulations to evaluate the performance of the physical downlink control channel (PDCCH) in the new 5G cellular communication network. The paper is structured as follows: Section 2 provides a concise overview of the 5G/NR architecture, followed by Section 3 describing the MATLAB system implementation. Simulation results are presented in Section 4, and finally, Section 5 outlines the conclusions.

2. 5G/NR Network Architecture Overview

The 5G system, along with its predecessors, comprises three main components: user equipment, the 5G core network (5GC), and next generation radio access (NG-RA) [4,5]. User equipment connects to the 5G NR network, while the 5GC manages authentication, configuration loading, and mobility. NG-RA handles radio connections in the telecommunications system [4,5].

In the physical-layer architecture of 5G, there are significant changes compared to 4G. The focus is on data and control channels managed by NG-RA. The transmitter and receiver involve processes like power control, cyclic redundancy check (CRC), 5G polar coding/encoding, modulation/demodulation, and multiple-input multiple-output (MIMO) techniques [4,5].

A channel is the transmission medium through which information travels from a transmitter to a receiver. In 5G/NR there are logical, transport, and physical channels (upstream and downstream) [4,5]. Also, there are channels for user data and control information. The physical layer controls the mapping of transport channels to physical channels [4,5]. One of the most important channels for the control of information is the physical downlink control channel (PDCCH).

The PDCCH channel was designed in Release 15 [4]. The PDCCH finds applications in scheduling both uplink (UL) and downlink (DL) data channels. It has a variety of specialized functions such as uplink power control, slot format indication, and preemptive scheduling [6,7]. Information carried on the PDCCH is called downlink control information (DCI) [6,7]. To receive DCI via the PDCCH, the UE observes a collection of PDCCH candidates during one or more configured monitoring instances within designated control resource sets (CORESETs) according to specified search space set (SS-Set) configurations [6].

The PDCCH employs polar coding with a predefined payload size, rate matching, and QPSK modulation [4,5]. In the context of PDCCH monitoring, NR employs a similar blind decoding approach to the PDCCH, where the UE performs cyclic redundancy checks (CRC). In this process, the CRC bits are concealed using an identifier (RNTI) recognized by the user [4–6].

3. System Implementation in MATLAB

The transmission–reception system of the PDCCH channel is constructed based on the fundamental principles outlined in 3GPP Release 15 [4] and Release 16 [5]. This study comprises two distinct scenarios. Firstly, a transmission–reception system for the PDCCH is implemented to assess the bit error rate (BER) against the signal-to-noise ratio (SNR). This scenario operates within an environment devoid of losses generated by the wireless channel, referred to as the "Lossless System" henceforth in this section. Secondly, for the evaluation of BER against the transmit power, the study transitions to an environment marked by losses in the wireless channel. This setting is situated within a rural macrocell (RMa) scenario, this case is referred to as the "Lossy System".

The implementation and coding of the transmission–reception system for both scenarios are carried out using MATLAB 2021b software [8], with the utilization of functions provided by the 5G Toolbox [8]. In a general sense, the two scenarios consist of the following blocks (Table 1) [4,5]:

Table 1. Block diagram of transmission–reception implemented system [4,5].

Transmitter:	Receiver:
• Source: 140 bits of control information; • DCI encoding—constituted by CRC 24, polar encoding, and rate matching; • PDCCH encoding—constituted by scrambling and QPSK modulation; • Mapping; • OFDM transmission technique. **Channel:** • Wireless channel with and without loss.	• OFDM demodulation; • Unmapping; • PDCCH decoding—constituted by unscrambling and QPSK demodulation; • DCI decoding—constituted by CRC 24 elimination, 5G polar decoding, and rate unmatching; • Calculation of the evaluation parameters.

3.1. Lossless System

The process of transmitting information for the lossless PDCCH initiates on the transmitter side with the transmission of 140 bits of information [4,5]. The following processes are then executed: DCI encoding, PDCCH encoding, mapping, and the OFDM transmission technique [4,5]. The information is transmitted through the additive white Gaussian noise (AWGN) channel. On the receiver side, the inverse processes are carried out: OFDM transmission technique, demapping, PDCCH decoding, and DCI decoding [4,5]. These processes lead to the retrieval of the transmitted 140 bits, which are then used for BER calculation.

3.2. Lossy System

This section provides a brief description of the implementation of the PDCCH with path losses, considering the position of the user equipment (UE) (height and elevation angle) in relation to the 5G NR base station (gNB). The process of transmitting information for the PDCCH also initiates on the transmitter side with the transmission of 140 bits of information [4,5].

The following processes are then executed: DCI encoding, PDCCH encoding, and the OFDM transmission technique with a single-input single-output (SISO) antenna transmission system. The information is transmitted through the channel, incorporating losses generated by the path. On the receiver side, the following processes are carried out: the OFDM transmission technique, PDCCH decoding, and DCI decoding, leading to the retrieval of the transmitted 140 bits. For this scenario, losses generated by the path between the gNB, the UE, and the effect of noise are considered.

Figure 1 depicts the flowcharts of the two implemented scenarios. The main distinction between the two scenarios primarily lies in the channel model and the parameters being evaluated in each scenario. For the first scenario (the lossless system), the outcomes are illustrated in terms of BER vs. SNR. Conversely, for the second scenario (the lossy system), the results are presented in terms of BER vs. transmit power (PTx), which is evaluated at varying UE heights and UE angles.

It is worth noting that the calculations of distances, heights, and angles between the gNB and the UE are conducted using rectangular coordinates. This approach enables the formation of a right triangle that aids us in determining the losses. The angle is determined using the tangent function of the right triangle, as illustrated in Figure 2. The parameters of the right loss triangle are listed below:

- $gNBAlt$: The height of the gNB, which varies between 10 and 150 m;
- $UEAlt$: The height of the UE, can take values from 1 to 10 m. This takes a value of 1.5 m for the system with angle variation losses;
- $DifgNBAlt - UEAlt$: The difference between the height of the gNB and the height of the UE;
- $DisgNB - UE$: The distance that separates the gNB from the UE. The gNB is located at $x = 0$, therefore, the value of the distance $gNB - EU$ represents the position of the UE. Having: 'RMA' scenario and no line of sight. So, 10 m $\leq DisgNB - UE \leq$ 5 km;

- Angle: The angle of elevation (rad)] formed between the UE and the gNB. This value is calculated with the tangent function or can be assumed. The angle value is $\pi/3$ [rad] for the PDCCH channel with height variation losses.

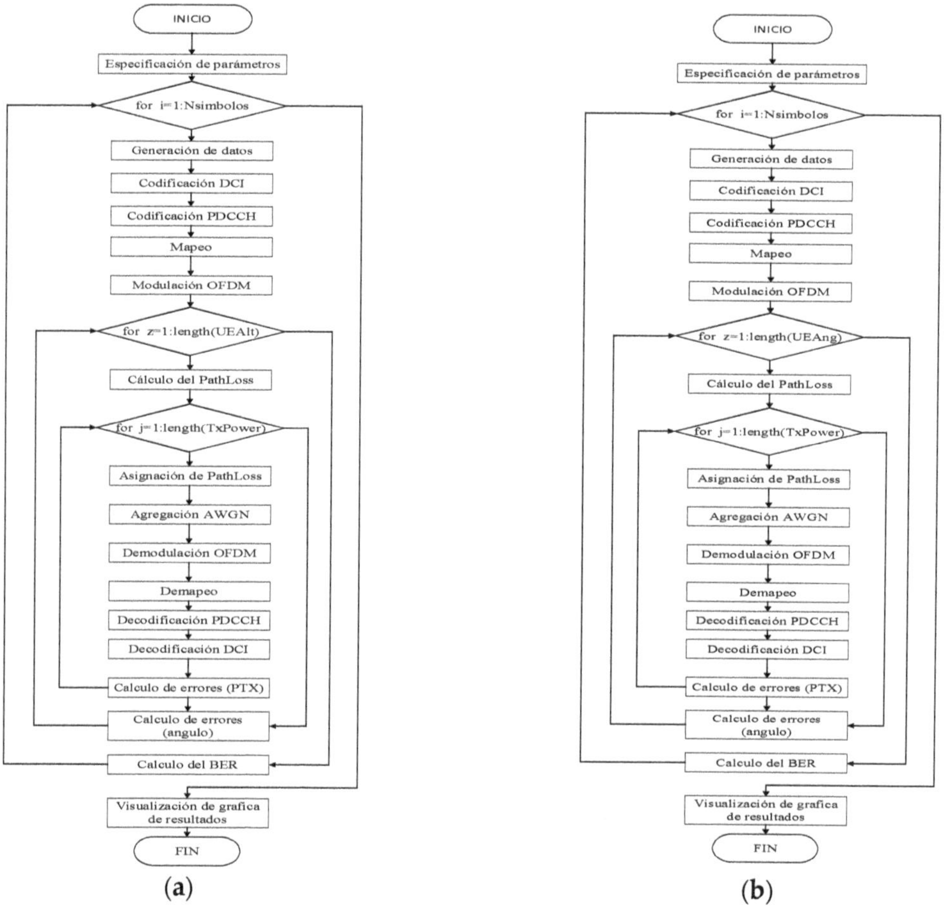

Figure 1. The PDCCH system implemented in MATLAB. (**a**) Lossless System and (**b**) Lossy System.

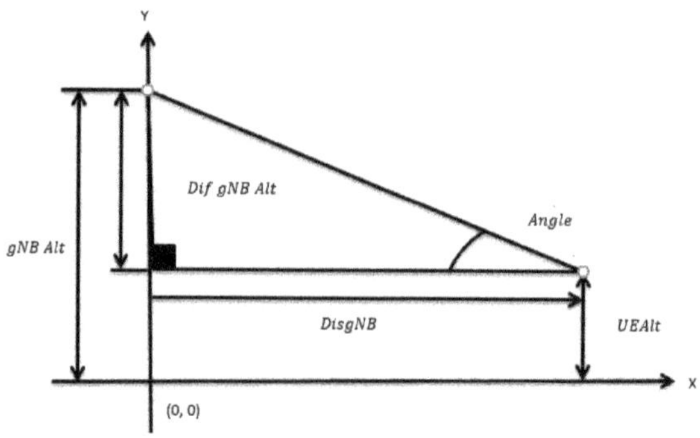

Figure 2. Downlink loss calculation triangle.

All elements relate to the tangent trigonometric function, as shown in Equation (1).

$$Tan(angle) = \frac{DifNBAlt - UEAlt}{DisgNB - UE} = \frac{gNBAlt - UEAtl}{DisgNB - UE} \quad (1)$$

To have the position of the UE on the x-axis of (1), solve for the value of $DisgNB - UE$. The determination of the ranges of allowed angles works with $gNBAlt = 150$ m, while the limits are: 10 m $\leq$ $DisgNB - UE$ $\leq$ 5 km and $UEAtl$.

4. Results and Discussion

Below are the results obtained from simulations of the two scenarios. The lossless systems are presented in terms of BER vs. SNR, and, for the lossy system, in terms of BER vs. transmit power (PTx). The BER vs. SNR behavior is analyzed considering parameters of bandwidth (BW), subcarrier space (SCS), and number of resource blocks (NRB). Then, BER vs. transmit power (PTx) in the lossy system is analyzed, taking into account the variation of parameters such as the angle elevation of the UE (angUE), heights of the UE (hUE), and the frequency range FR1.

4.1. Losslesss System

Figure 3 depicts the BER vs. SNR curves, where various values of SCS, NRB, and BW are considered. In Figure 3a, a BW of 10 MHz, SCS values of {15, 30, 60} kHz, and NRB values of {52, 24, 11} are taken into account. Upon analyzing the outcomes, it is evident that at a probability of error of 10^{-3}, the SNR value for SCS = 15 kHz is −15.79 dB, for SCS = 30 kHz it is −14.62 dB, and for the SCS = 60 kHz curve, it is −10.07 dB. In Figure 3b, the outcomes are presented for a BW of 20 MHz, SCS values of {15, 30, 60} kHz, and NRB values of {106, 51, 24}. For a probability of 10^{-3}, the SNR value for SCS = 15 kHz is −18.85 dB, for SCS = 30 kHz it is −15.76 dB, and for SCS = 60 kHz it is −14.69 dB. Consequently, the curve that exhibits the best performance for a BW of 20 MHz is with SCS = 15 kHz and NRB = 10. It is evident that, with a larger BW there is a greater NRB, which translates to higher performance of the PDCCH channel.

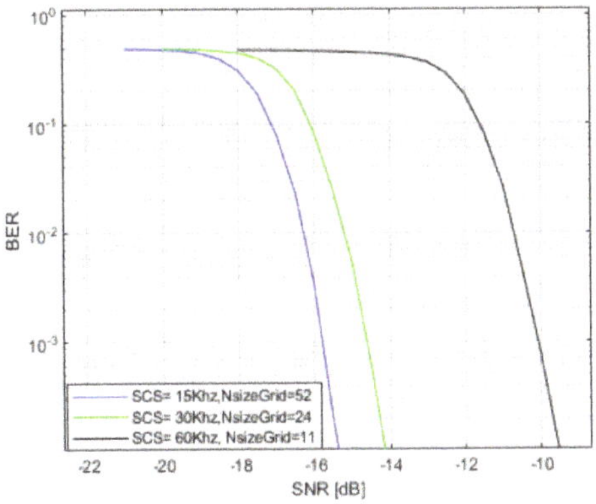

(a)

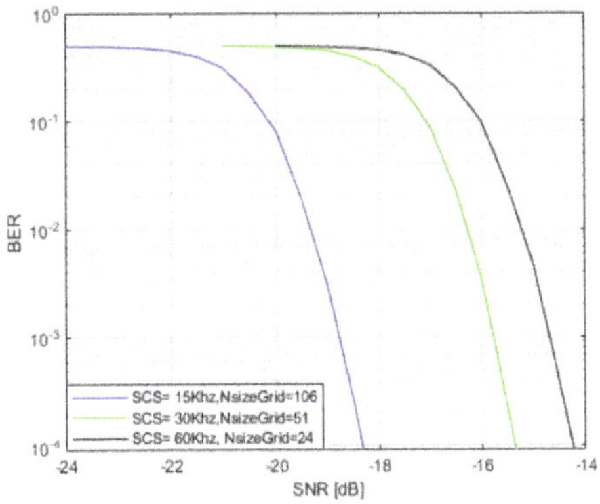

(b)

Figure 3. BER vs. SNR for BW = {10, 20} MHz and several values of SCS and NRB. (**a**) $BW = 10$ MHz and (**b**) $BW = 20$ MHz.

Figure 4 shows the BER vs. SNR curves for various values of BW, SCS, and NRB. In Figure 4a, a fixed SCS of 15 kHz is considered, with BW values of {5, 10, 20, 30} MHz and NRB values of {25, 52, 106, 160}. Upon analyzing the results, it is evident that, for a probability of error of 10^{-3}, the SNR value for BW = 5 MHz is −14.72 dB, for BW = 10 MHz it is −15.79 dB, for BW = 20 MHz it is −18.85 dB, and for BW = 30 MHz it is −21.73 dB. In

Figure 4b, a fixed NRB value of 51 is considered, with BW values of {15, 30, 50} MHz and SCS values of {15, 30, 60} kHz. Upon analyzing the results, it is evident that, at a probability of error of 10^{-3}, the SNR value for the curve with BW = 15 MHz and SCS = 15 kHz is −15.73 dB, for the curve with BW = 30 MHz and SCS = 30 kHz it is −15.76 dB, and for the curve with BW = 50 MHz and SCS = 60 kHz it is −15.78 dB.

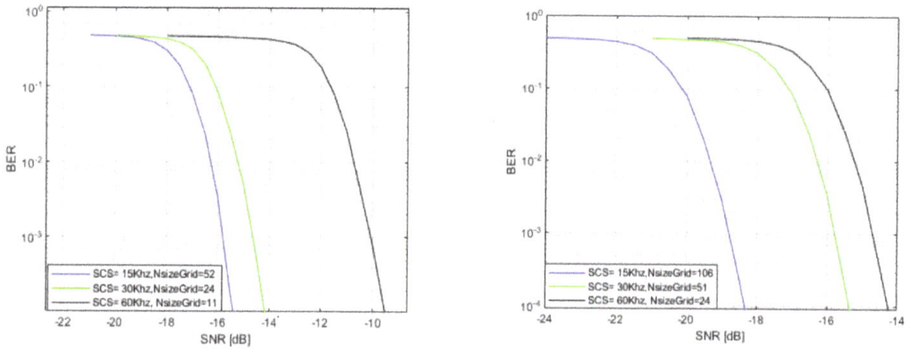

(a) (b)

Figure 4. BER vs. SNR for different values of BW, SCS and NBR. (**a**) SCS of 15 kHz, BW = {5, 10, 20, 30} MHz and NRB = {25, 52, 106, 160}. (**b**) NRB = 51, BW = {15, 30, 50} MHz and SCS = {15, 30, 60} kHz.

4.2. Lossy System

Figure 5 shows the curves of BER as a function of PTx for the lossy channel are shown. It considers an angle of the UE (hUE) of 60°, heights of the UE (hUE) = [1, 1.5, 4, 5, 8, 9, 10] m and a frequency range FR1. Upon analyzing the results, it is evident that at a probability of error of 10^{-3}, the PTx value for the hUE1 curve is 0.4199 dBm, for the hUE2 curve it is 0.3975 dBm, for the hUE3 curve it is 0.1918 dBm, and for the hUE curve it is 0.1322 dBm.

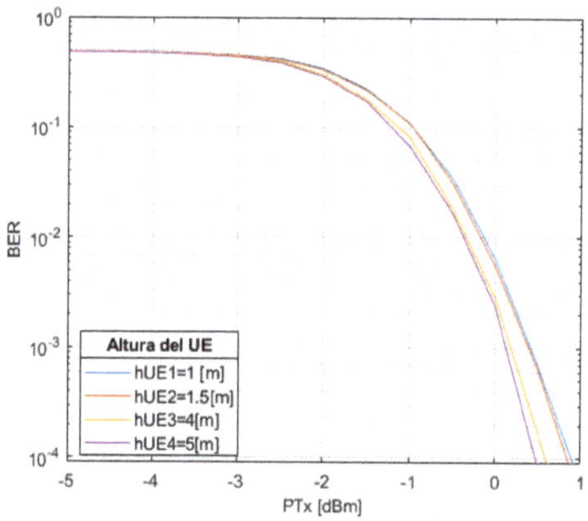

Figure 5. BER vs. PTx curves with angUE = 60°, hUE = [1, 1.5, 4, 5] m.

Figure 6 shows the BER vs. PTx curves for the constant hUE = 1.5 m, angUE = {$\pi/18$, $\pi/12$, $\pi/19$, $\pi/6$} rad for the FR1 frequency range. Upon analyzing the results, it is observed that, at a 10^{-3} error probability, the PTx value for the $AngUE_1$ curve is 24.28 dBm, for $AngUE_2$ it is 17.75 dBm, for $AngUE_3$ it is 13.55 dBm, and for $AngUE_4$ it is 7.57 dBm. At a 10^{-2}, error probability, the PTx value for the $AngUE_1$ curve is 23.67 dBm, for $AngUE_2$ it is 17.28 dBm, for $AngUE_3$ it is 12.97 dBm, and for $AngUE_4$ it is 6.85 dBm.

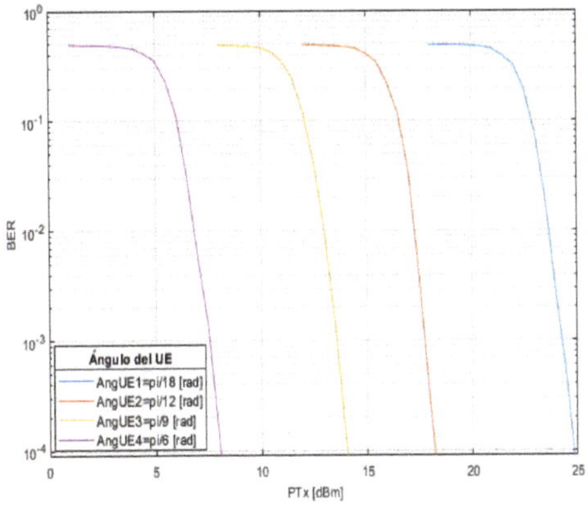

Figure 6. BER vs. PTx for hUE = 1.5 m, angUE = {$\pi/18$, $\pi/12$, $\pi/19$, $\pi/6$} rad.

Figure 7 illustrates the results obtained when considering a hUE = 1.5 m, hUE = 60°, and different frequencies (Fc) = [3.5, 6, 24.5] GHz. It can be observed that, at a 10^{-3} error probability, the PTx value for the Fc1 curve is 0.304 dBm, for the Fc2 curve it is 5.152 dBm, and for the Fc3 curve it is 17.272 dB.

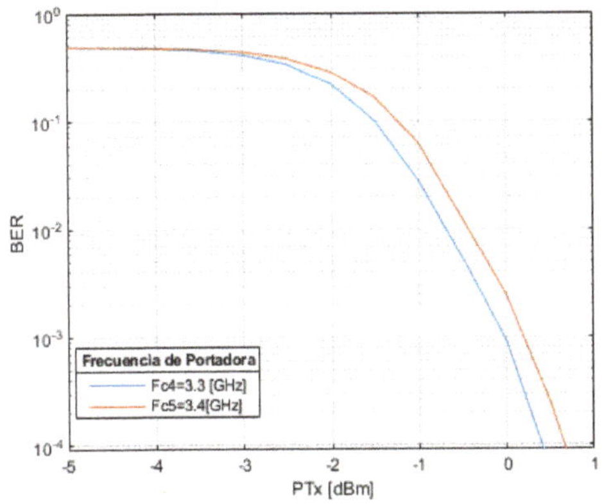

Figure 7. BER vs. PTx for hUE = 1.5 m, hUE = 60°, and Fc = [3.5, 6, 24.5] GHz.

5. Conclusions

In conclusion, this study has addressed the performance evaluation of the physical downlink control channel (PDCCH) in two different scenarios: one without losses and another with losses generated by the path. Through the implementation of a transmission–reception system in MATLAB R2021b and the analysis of BER vs. SNR and BER vs. transmit power curves, valuable insights have been gained into how key parameters impact channel performance. In the first scenario, where a lossless channel was considered, the significance of the number of resource blocks (NRB) in enhancing PDCCH performance was highlighted. As the number of NRB increases, a reduction in the probability of bit errors was observed, indicating a more reliable transmission. The second scenario, incorporating channel losses, revealed the positive influence of user equipment (UE) height and elevation angle on attenuation loss reduction. By increasing the UE's height or elevation angle, the link distance decreased, leading to improved BER performance and ultimately enhancing the system's efficiency. Furthermore, the analysis of carrier frequency variation demonstrated that higher carrier frequencies result in higher bit error rates, underscoring the importance of carefully considering frequency selection in system design.

Author Contributions: Conceptualization, M.C.P.-P.; methodology, P.B.-C., F.G.-G. and M.C.P.-P.; software, P.B.-C.; validation, P.B.-C., F.G.-G. and M.C.P.-P.; formal analysis, P.B.-C., F.G.-G. and M.C.P.-P.; investigation, P.B.-C.; resources, F.G.-G.; data curation, P.B.-C.; writing—original draft preparation, P.B.-C.; writing—review and editing, F.G.-G. and M.C.P.-P.; visualization, P.B.-C. and F.G.-G.; supervision, M.C.P.-P.; project administration, M.C.P.-P.; funding acquisition, F.G.-G. All authors have read and agreed to the published version of the manuscript.

Funding: This research was funded by Escuela Politécnica Nacional.

Institutional Review Board Statement: Not applicable.

Informed Consent Statement: Not applicable.

Data Availability Statement: Data are contained within the article.

Conflicts of Interest: The authors declare no conflict of interest.

References

1. 3GPP. *Study on Scenarios and Requirements for Next Generation Access Technologies*; 3GPP TR 38.913; 3GPP: Sophia Antipolis, France, 2017.
2. Sudhamani, C.; Roslee, M.; Tiang, J.J.; Rehman, A.U. A Survey on 5G Coverage Improvement Techniques: Issues and Future Challenges. *Sensors* **2023**, *23*, 2356. [CrossRef] [PubMed]
3. Audhya, G.K.; Sinha, K.; Ghosh, S.C.; Sinha, B.P. A survey on the channel assignment problem in wireless networks. *Wirel. Commun. Mob. Comput.* **2011**, *11*, 583–609. [CrossRef]
4. 3GPP. *Study on New Radio (NR) Access Technology (Release 15)*; 3GPP TS 39.912 Tech. Rep. V15.2.0; 3GPP: Sophia Antipolis, France, 2018.
5. 3GPP. *Services Provided by the Physical Layer (Release 16)*; 3GPP TS 38. 212 Tech. Rep V16.1.0; 3GPP: Sophia Antipolis, France, 2020.
6. Braun, V.; Schober, K.; Tiirola, E. 5G NR physical downlink control channel: Design, performance and enhancements. In Proceedings of the 2019 IEEE Wireless Communications and Networking Conference (WCNC), Marrakesh, Morocco, 15–18 April 2019; pp. 1–6.
7. Mozaffari, M.; Wang, Y.P.E.; Kittichokechai, K. Blocking Probability Analysis for 5G New Radio (NR) Physical Downlink Control Channel. In Proceedings of the ICC 2021—IEEE International Conference on Communications, Montreal, QC, Canada, 14–23 June 2021; pp. 1–6. [CrossRef]
8. *MATLAB and Statistics Toolbox Release 2021b*; The MathWorks, Inc.: Natick, MA, USA, 2021.

engineering proceedings

MDPI

Proceeding Paper

Dielectric Characterization of PCB Laminate Materials Using Improved Capacitive Coupled Ring Resonators †

Evelyn Quinapaxi Cabrera, Aldrin Reyes Narváez * and Hernan Barba Molina

Departamento de Electrónica, Telecomunicaciones y Redes de Información, Escuela Politécnica Nacional, Quito 170525, Ecuador; evelyn.quinapaxi@epn.edu.ec (E.Q.C.)

* Correspondence: aldrin.reyesn@epn.edu.ec

† Presented at the XXXI Conference on Electrical and Electronic Engineering, Quito, Ecuador, 29 November–1 December 2023.

Abstract: An improved ring resonator structure consisting of a capacitive coupling fixture is presented. Dielectric characterization is carried out using a fitting method that allows for comparison between measured and simulated results by setting a coupling level of about −4 dB and an unloaded quality factor of about 54. The key factor in this approach is the coupled energy, which permits measurements to be made well above the noise threshold. Furthermore, when applied within a frequency range, the dielectric dispersion can be studied as well. As proof-of-concept fixtures, five structures for the L-band at resonance frequencies of 1 GHz, 1.25 GHz, 1.5 GHz, 1.75 GHz, and 2 GHz are designed, manufactured, and measured. The method applied in this study provides good results as verified by error calculations.

Keywords: ring resonator; dielectric characterization; capacitive coupling

Citation: Quinapaxi Cabrera, E.; Reyes Narváez, A.; Barba Molina, H. Dielectric Characterization of PCB Laminate Materials Using Improved Capacitive Coupled Ring Resonators. *Eng. Proc.* **2023**, *47*, 11. https://doi.org/10.3390/engproc2023047011

Academic Editor: Fernando Carrera Suarez

Published: 4 December 2023

1. Introduction

Dielectric characterization of materials has played an important role in the field of the radio frequency (RF) application design. It is well known that different microwave circuit and antenna approaches require knowledge of the dielectric properties of the materials used for design and implementation. In other fields, such as agriculture and medicine, dielectric measurements have been used to determine physical relationships in order to characterize products [1] and human tissues [2].

Planar structures are widely used as circuit elements in RF components due to their ease of fabrication and the simplicity of determining their electric characteristics in their transmission structures, for instance, the impedance. The most common planar transmission structures are stripline, microstrip, slotline, coplanar waveguide, coplanar strips, inverted microstrip, suspended microstrip, microstrip with overlay, strip dielectric waveguide, and inverted strip dielectric waveguide [3]. In all of the above cases, these planar structures are mechanically supported/wrapped by a dielectric laminate material characterized by a determined complex permittivity (lossy), which is frequency dependent, i.e., dispersion is present [4].

Thanks to its ease of acquisition, FR-4 is one of the most widely used dielectric laminate materials. It is especially popular with students and young researchers for proof-of-concept approaches and elementary experiments, as well as in countries with emerging economies. This material is composed by glass fibers, which offer reinforced mechanical support. Therefore, FR-4 laminates become anisotropic, which determines the dielectric behavior of the material as a function of the frequency and the direction of the established electric field [5].

One technique used to determine the dielectric properties and dispersion that occur in planar structures utilizes large ring resonators fed by transmission probes through coupling gaps which do not significantly perturb the field of the ring resonator [6]. To determine the dispersion effect and avoid the impact of mutual inductance, [6] recommended a large ring resonator greater than 5λ. In addition, transmission *Q* measurements are made by varying the coupling gap distance. Although the length restriction is provided here with the aim of studying the dispersion effect, there exist other proposals of compact ring-resonators sensors for specific applications, such as the one reported in [7].

When the ring resonator has been measured, analyses supported by the well-known theory are performed to determine and characterize the relative permittivity, loss tangent, conductor losses, dielectric losses, and even the radiation losses of microstrip or stripline structures. A comparison of different theoretical models was reported in [8], supported by experimental research which established important observations and recommendations to take into account when considering the application of these methods.

Dielectric characterization procedures using ring-resonator methods have been utilized as part of different research fields, such as dielectric RF sensors [9,10] and the design and implementation of wearable antennas in the textile industry [11,12].

In this paper, ew present a microstrip ring-resonator structure excited by an improved capacitive coupler fed by micro-stripline probes for dielectric characterization of PCB laminates. Section 2 shows the design considerations for the measuring structure. The characterization procedure of the FR-4 PCB laminate in the L Band is described in Section 3, the results are detailed in Section 4, and Section 5 provides our conclusions.

2. Dielectric Characterization Measuring Setup

Figure 1 shows the proposed measuring setup for dielectric characterization of planar laminates used with microstrip technology. The structure consists of transmission line probes, capacitive coupling structures, and the ring resonator. The mean radius *r* is designed for the fundamental resonance provided by $n = 1$ (TM_{110} mode) in (1):

$$2\pi r = n\lambda_g, \tag{1}$$

where λ_g is the guided wavelength in the medium. As seen in (1), the operation frequency at the fundamental resonance is reduced by increasing the ring radius *r*.

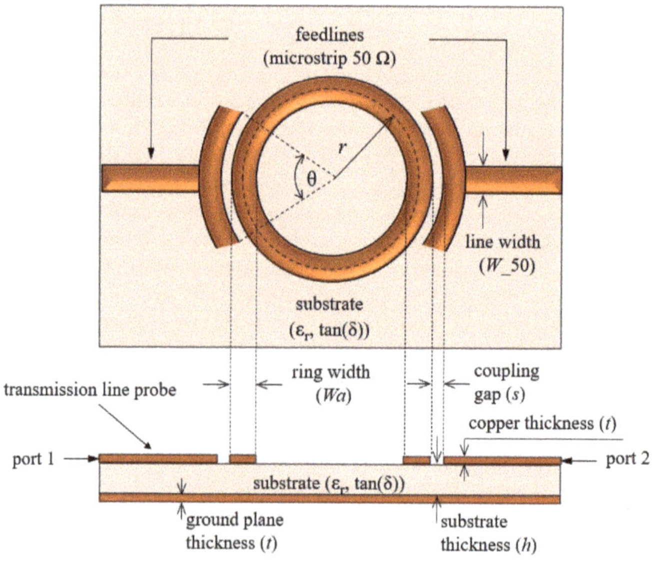

Figure 1. Two-port ring resonator layout fed by a capacitive coupling structure through microstrip transmission line probes: top view (**top**) and side view (**bottom**).

Capacitive coupling is achieved through concentric ring segments at the end of the microstrip transmission line probes. The aim of this structure is to increase the coupling energy in the ring resonator, thereby allowing a better determination of the measured quality factor of the resonator (unloaded $Q \approx 54$) by ensuring a coupling level of about -4 dB, i.e., a much better coupling level than that of -60 to -40 dB recommended in [5]. This capacitive structure holds a constant coupling gap from the outer radius of the ring resonator, which should be as small as possible while taking into account any subsequent manufacturing limitations.

The length of the concentric ring segments is determined by the aperture angle θ. In order to design an adequate aperture angle, the coupling quantity of the capacitive structure is analyzed by simulating the S_{21} parameter at a resonance frequency. Figure 2 shows the variation in the magnitude of the S_{21} parameter at a resonance frequency of the TM_{110} mode versus the aperture angle θ of the capacitive coupling structure. The slope of the curve in Figure 2 is analyzed, establishing a convergence at $\theta = 30^{o}$. Beyond this point, variations in the magnitude of the S_{21} parameter are negligible.

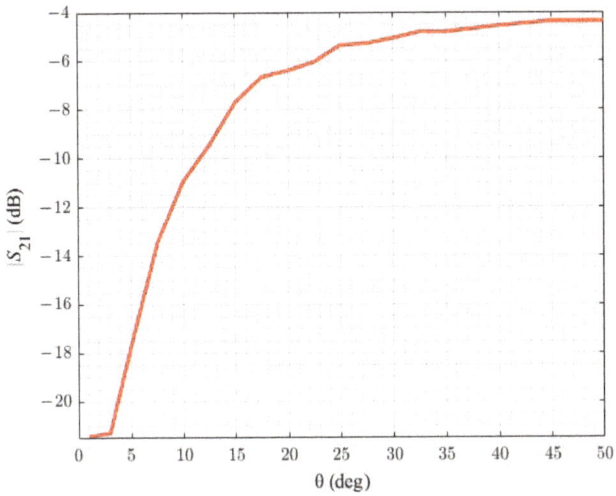

Figure 2. Magnitude of the S_{21} parameter with respect to the opening angle θ.

3. L-Band Dielectric Characterization of an FR-4 Laminate

Due to the high coupled energy realized by the capacitive structure in the measuring setup, the dielectric characterization procedure can be achieved by measuring the S_{21} parameter at a given resonance frequency and then fitting this measured result to those obtained by three-dimensional electromagnetic simulations. Thus, the matching procedure varies ϵ_r and $\tan(\delta)$, which, under tolerance considerations, provide the dielectric parameters.

For this experimentation, an FR-4 laminate with $h = 1.5$ mm, $t = 17$ µm, estimated dielectric relative permittivity $\epsilon_r = 4.3$, and estimated loss tangent $\tan(\delta) = 0.025$ was used as the device under test. Five circular one-wavelength ring resonators ($Wa = 2.77$ mm) with capacitive coupling were implemented for the L-band at the resonant frequencies 1 GHz, 1.25 GHz, 1.5 GHz, 1.75 GHz, and 2 GHz, with respective mean radii of 26.48 mm, 21.18 mm, 17.65 mm, 15.13 mm, and 13.24 mm. The length of the rings was designed utilizing the estimated dielectric values. The transmission lines probes were 50-Ω-microstrip lines ($W\_50 = 2.77$ mm) fed by coaxial lines through SMA connectors attached to the laminate. The coupling gap, as mentioned above, was $s = 0.3$ mm to allow fabrications. Figure 3 shows a photograph of the five manufactured structures.

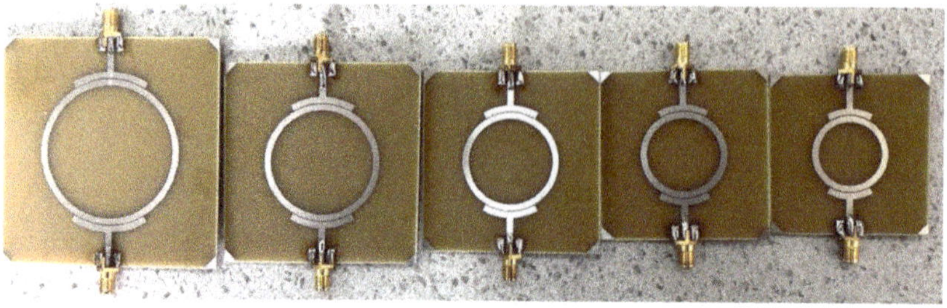

Figure 3. Photograph of the manufactured ring resonator structures for dielectric characterization of an FR-4 laminate in the L band: (**a**) Structure "A", with a designed resonance frequency at 1 GHz; (**b**) Structure "B", with a designed resonance frequency at 1.25 GHz; (**c**) Structure "C", with a designed resonance frequency at 1.5 GHz; (**d**) Structure "D", with a designed resonance frequency at 1.75 GHz; and (**e**) Structure "E", with a designed resonance frequency at 2 GHz.

4. Results and Discussion

Figure 4 shows the measured and simulated S_{21} parameter results after the fitting procedure. The structures were simulated using finite-element simulations with CST Studio Suite, while measurements were realized using a vector network analyzer. The fitting procedure focused on the resonance frequency by varying the permittivity value ϵ_r and on the coupling level by varying tan δ. It can be seen that the resonance frequency in all fixtures is shifted down. This reveals that the real permittivity value ϵ_r is larger than the estimated one of 4.3. In addition, the dielectric losses determine the coupling level, on the one hand due to the conductance effect of the resulting equivalent circuit derived from the transmission line model analysis of the closed-loop ring resonator and on the other due to the capacitive effect of the coupling gap [13]. In the same way, the obtained bandwidth is determined by dielectric losses through the relationship of the former using the unloaded quality factor provided in (2) [14]:

$$|S_{21}| = 20\, log_{10}\left(1 + \frac{Q_L}{Q_U}\right) \tag{2}$$

The differences between the measured and simulated results of the bandwidth are, in our understanding, twofold: first, due to the natural anisotropy of the FR-4 substrate, and second due to dispersion. Table 1 summarizes the adjusted dielectric values.

Table 1. Dielectric characteristic values of the resonators.

Parameter	Structure "A" Designed Resonance Frequency at 1 GHz	Structure "B" Designed Resonance Frequency at 1.25 GHz	Structure "C" Designed Resonance Frequency at 1.5 GHz	Structure "D" Designed Resonance Frequency at 1.75 GHz	Structure "E" Designed Resonance Frequency at 2 GHz
ϵ_r	4.3975	4.4673	4.4601	4.4606	4.4140
tan(δ)	0.0139	0.0216	0.0174	0.0180	0.0208

Table 2 outlines the error calculations for the S_{21} parameter, resonance frequency, and quality factors obtained after fitting the five structures. All of the results show small absolute errors; thus, the aim of the work is completed.

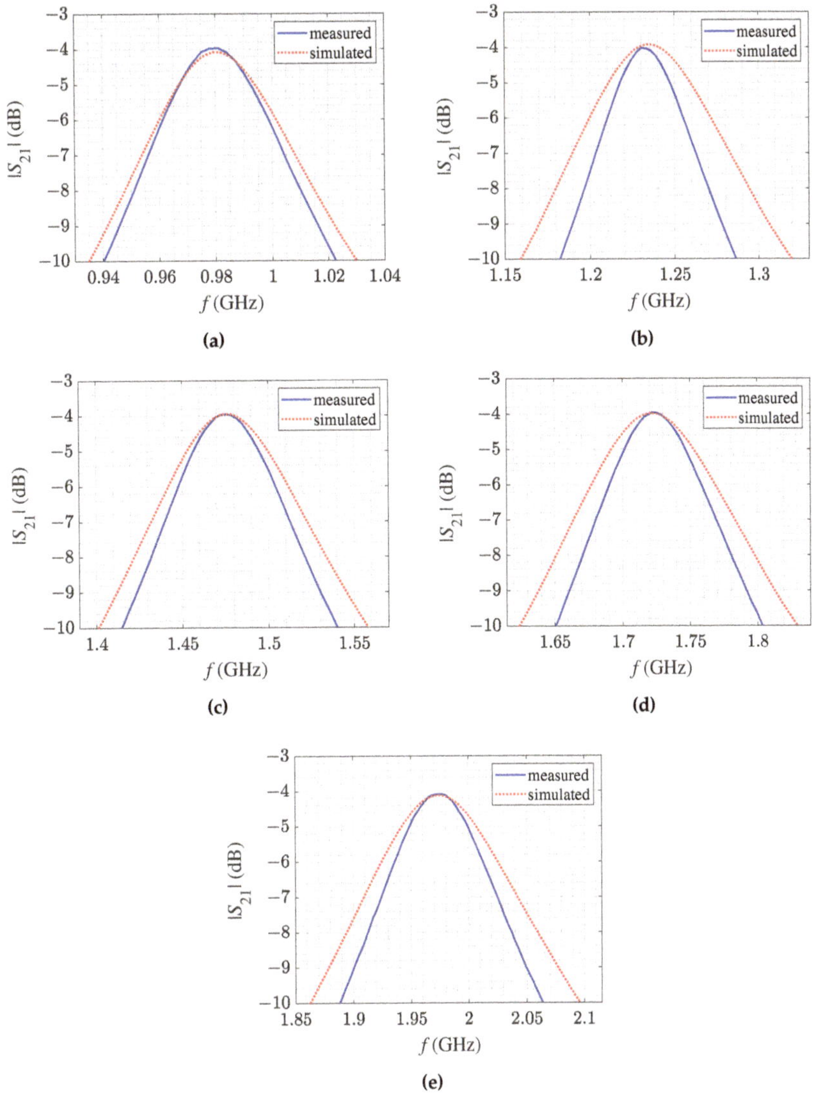

Figure 4. Measured (continuous blue) and simulated (dashed red) S_{21} parameter results for the five ring resonators designed at the resonant frequencies with estimated dielectric values: (**a**) Structure "A", designed with a resonance frequency at 1 GHz; (**b**) Structure "B", designed with a resonance frequency at 1.25 GHz; (**c**) Structure "C", designed with a resonance frequency at 1.5 GHz; (**d**) Structure "D", designed with a resonance frequency at 1.75 GHz; and (**e**) Structure "E", designed with a resonance frequency at 2 GHz.

Table 2. Error calculation of the S_{21} parameter, resonance frequency (f_r), loaded Q, and unloaded Q for the five ring resonator structures.

Structure	Designed Resonance Frequency [GHz]	Parameter	Measured	Simulated	Absolute Error
A	1	S_{21} [dB]	−3.9679	−4.0805	0.1129
		f_r [GHz]	0.9810	0.9804	0.0006
		Loaded Q	20.4918	17.7305	2.7613
		Unloaded Q	55.8807	47.2986	8.5821
B	1.25	S_{21} [dB]	−4.0239	−3.9204	0.1035
		f_r [GHz]	1.2300	1.2346	0.0046
		Loaded Q	20.6186	13.5685	7.0500
		Unloaded Q	55.6091	37.3548	18.2544
C	1.5	S_{21} [dB]	−3.9519	−3.9535	0.0016
		f_r [GHz]	1.4760	1.4760	0.0000
		Loaded Q	20.4918	16.4474	4.0444
		Unloaded Q	56.0592	44.9805	11.0787
D	1.75	S_{21} [dB]	−3.9608	−3.9807	0.0199
		f_r [GHz]	1.7220	1.7218	0.0002
		Loaded Q	19.4175	14.5773	4.8402
		Unloaded Q	53.0259	39.6510	13.3749
E	2	S_{21} [dB]	−4.0899	−4.1222	0.0323
		f_r [GHz]	1.9740	1.9739	0.0001
		Loaded Q	19.1939	14.3885	4.8054
		Unloaded Q	51.1101	38.0792	13.0309

5. Conclusions

An improved structure for the determination of dielectric characteristics in planar laminates has been presented in concept as a fitting method using adjusted simulated parameters and with five proof-of-concept structures. The laminate is of an FR-4 type of material, which has been dielectric characterized within the L band at frequencies of about 1 GHz, 1.25 GHz, 1.5 GHz, 1.75 GHz, and 2 GHz. The structure shows a ring resonator for the fundamental resonance, i.e., the length of the ring is one wavelength. A capacitive coupling structure for the ring resonators is the key approach of this work. Compared with other implementations, where coupling levels of about −45 to −28 dB on an FR-4 substrate [5], −15 to −10 dB on a TLY5 substrate [15], and −15 dB on textile materials [11] have been reported, the coupling level of about −4 dB for the structures presented here allows for the establishment of a reliable fitting method to determine the relative permittivity and dielectric losses. Error calculations between the simulated and measured results of coupling level, resonance frequency, and quality factors confirm the good performance of the proposed method.

In consequence, the method we have proposed in this paper can be widely used in the dielectric characterization of planar laminates. The functionality of this method expands the knowledge around determining the dispersion behavior of substrates.

Author Contributions: Conceptualization, E.Q.C. and Hernan Barba Molina; methodology, E.Q.C.; validation, E.Q.C. and A.R.N.; investigation, E.Q.C. and A.R.N.; writing—original draft preparation, E.Q.C. and A.R.N.; writing—review and editing, A.R.N. and H.B.M. All authors have read and agreed to the published version of the manuscript.

Funding: This research received no external funding.

Institutional Review Board Statement: Not applicable.

Informed Consent Statement: Not applicable.

Data Availability Statement: All data are contained within the article.

Conflicts of Interest: The authors declare no conflict of interest.

Abbreviations

The following abbreviations are used in this manuscript:

FR-4	Flame Retardant-4 (a woven glass-reinforced epoxy resin)
RF	Radio Frequency
PCB	Printed Circuit Board

References

1. Nelson, S.O. Agricultural applications of dielectric measurements. *IEEE Trans. Dielectr. Electr. Insul.* **2006**, *13*, 688–702. [CrossRef]
2. Li, X. *Body Matched Antennas for Microwave Medical Applications*; KIT Scientific Publishing: Karlsruhe, Germany, 2014; Volume 72.
3. Garg, R.; Bahl, I.; Bozzi, M. *Microstrip Lines and Slotlines*; Artech House: Norwood, MA, USA, 2013.
4. Pozar, D.M. *Microwave Engineering*; John Wiley & Sons: Hoboken, NJ, USA, 2011.
5. Heinola, J.M.; Strom, J.P. Evaluation of dielectric properties of printed wiring board materials by using a microstrip-ring and strip-line ring resonator methods. *IEEE Electr. Insul. Mag.* **2007**, *23*, 23–29. [CrossRef]
6. Troughton, P. Measurement techniques in microstrip. *Electron. Lett.* **1969**, *2*, 25–26. [CrossRef]
7. Ali, A.M.; Ramahi, O.M. Microwave imaging of subsurface defects in coated metallic structures using small ring resonators. In NDT in Canada 2015 Conference, Edmonton, AB Canada, 15–17 June 2015; pp. 16–17.
8. Heinola, J.M.; Tolsa, K. Dielectric characterization of printed wiring board materials using ring resonator techniques: A comparison of calculation models. *IEEE Trans. Dielectr. Electr. Insul.* **2006**, *13*, 717–726. [CrossRef]
9. Abd Rahman, N.; Zakaria, Z.; Abd Rahim, R.; Dasril, Y.; Bahar, A.A.M. Planar microwave sensors for accurate measurement of material characterization: A review. *TELKOMNIKA (Telecommun. Comput. Electron. Control)* **2017**, *15*, 1108–1118. [CrossRef]
10. Raveendran, A.; Raman, S. Low cost multifunctional planar RF sensors for dielectric characterization and quality monitoring. *IEEE Sens. J.* **2021**, *21*, 24056–24065. [CrossRef]
11. Ahmed, M.; Ahmed, M.; Shaalan, A. Investigation and comparison of 2.4 GHz wearable antennas on three textile substrates and its performance characteristics. *Open J. Antennas Propag.* **2017**, *5*, 110–120. [CrossRef]
12. Mallavarapu, S.; Lokam, A. Circuit modeling and analysis of wearable antennas on the effect of bending for various feeds. *Eng. Technol. Appl. Sci. Res.* **2022**, *12*, 8180–8187. [CrossRef]
13. Chang, K.; Hsieh, L.H. *Microwave Ring Circuits and Related Structures*; John Wiley & Sons: Hoboken, NJ, USA, 2004; Volume 156.
14. Ragan, G. *Microwave Transmission Circuits*; McGraw-Hill Book Company, Inc.: New York, NY, USA, 1948.
15. Rashidian, A.; Aligodarz, M.T.; Klymyshyn, D.M. Dielectric characterization of materials using a modified microstrip ring resonator technique. *IEEE Trans. Dielectr. Electr. Insul.* **2012**, *19*, 1392–1399. [CrossRef]

engineering proceedings

MDPI

Proceeding Paper

A Deep Reinforcement Learning Algorithm for Robotic Manipulation Tasks in Simulated Environments †

Carlos Calderon-Cordova * and Roger Sarango

Department of Computer Science and Electronics, Universidad Técnica Particular de Loja, Loja 1101608, Ecuador; rasarango1@utpl.edu.ec

* Correspondence: cacalderon@utpl.edu.ec or cacalderon@ieee.org

† Presented at the XXXI Conference on Electrical and Electronic Engineering, Quito, Ecuador, 29 November–1 December 2023.

Abstract: Industrial robots are used in a variety of industrial process tasks, and due to the complexity of the environment in which these systems are deployed, more robust and accurate control methods are required. Deep reinforcement learning emerges as a comprehensive approach that directly allows for the mapping of sensor data and the setting of motion actions to the robot. In this work, we propose a robotic system implemented in a semi-photorealistic simulator whose motion control is based on the A2C algorithm in a DRL agent; the task to be performed is to reach a goal within a work area. The evaluation is executed in a simulation scenario where a fixed position of a target is maintained while the agent (robotic manipulator) tries to reach it with the end-effector from an initial position. Finally, the trained agent fulfills the established task; this is demonstrated by the results obtained in the training and evaluation processes, and the reward value increases when the measured distance decreases between the end-effector and the target.

Keywords: robotics manipulation; DRL algorithms; deep learning; reinforcement learning; robotic simulator; CoppeliaSim

Citation: Calderon-Cordova, C.; Sarango, R. A Deep Reinforcement Learning Algorithm for Robotic Manipulation Tasks in Simulated Environments. *Eng. Proc.* **2023**, *47*, 12. https://doi.org/10.3390/engproc2023047012

Academic Editor: Jackeline Abad

Published: 4 December 2023

1. Introduction

In recent decades, Industry 4.0 (I4.0) has promoted the upgrade of automation and the digitization of industrial processes, using emerging technologies such as cloud computing, artificial intelligence (AI), internet of things (IoT), cyber-physical systems, etc. [1]. Some of the industrial processes such as welding, manufacturing, and object manipulation require robotic systems that enable decision making to execute specific tasks within production lines [2].

The integration of artificial intelligence improves the efficiency of factory automation; in particular, machine learning (ML) is implemented, which gives systems the ability to learn to execute a task and has been demonstrated to improve control in many applications [3]. The learning process is limited by several factors: the volume of data required for the system, the elements of scenarios in industrial areas, deployment time, the cost of equipment, and others. Currently, there are several simulation platforms [4] that allow for the simulation of various scenarios, generating synthetic data and robotic components that allow for the developing of an end-to-end system based on artificial intelligence.

In machine learning, Deep Reinforcement Learning (DRL) has been of great interest to researchers with academic and industrial approaches for several application areas [5]. In the field of robotic manipulation, there are several frameworks and simulation platforms that allow for the implementation and evaluation of various control algorithms for the application of RL in robotics [6]. Despite the success of DRL, there are also methods that allow for the optimization of DRL algorithms by performing CPU and GPU combinations, thus allowing larger data batches to be used without affecting the final performance [7].

In the literature, there is a large number of papers presenting reviews of DRL algorithms for robotic manipulation; however, in these papers, they do not provide details of the implementation process for a specific problem. The contribution of this work is to implement a robotic system in a simulation platform; the end-to-end control approach is based on DRL, and its performance in the classical task of reaching a goal using a robotic manipulator is evaluated.

The structure of the paper is described as follows: Section 2 describes the main fundamentals and structure involved in deep reinforcement learning applied to robotic manipulation. Section 3 describes the components and architecture of the robotic system implemented in simulation. Section 4 provides the details of the training and evaluation processes of the robotic system whose control is based on DRL. Section 5 presents some results obtained with the simulation implementation of the DRL algorithms. Finally, some conclusions obtained from the complete development process of this work are presented.

2. Deep Reinforcement Learning

Deep reinforcement learning algorithms are based on the iterative trial-and-error learning process. The interaction between the agent and the environment in which the task is performed is modeled after a Markov Decision Process (MDP). With this method, the interaction is reduced to three signals: the current state of the environment (observations), a decision made by the agent based on the state (action), and feedback with positive or negative value depending on the action performed by the agent (reward) [8]. The mathematical foundations of DRL approaches are based on the Markov Decision Process (MDP) [9], which consists of five elements:

$$MDP = (S, A, P, R, \gamma), \quad (1)$$

where S is the set of states of the agent and the environment, A is the set of actions executed by the agent, P is the model of the system—in other words, it is the transition probability of a state—R is the reward function, and γ is a discount factor [10]. The DRL objective function has two forms: the first is a value function that defines the expectation of the accumulated reward. $V^{\pi}(s)$ represents the estimates of the state-value function in the policy for the MDP; given a policy π, we have the expected return:

$$V^{\pi}(s) = E_{\pi}\left[r_t + \gamma r_{t+1} + \gamma r_{t+2} + \cdots | s_t = s\right], \quad (2)$$

The second is the action-value function, which is known as the Q function; this function indicates that after performing an action a, based on the policy π on the state s, an accumulative reward (r) is generated. $Q^{\pi}(s,a)$ represents the estimates of the action-value function in the policy for the MDP; given a policy π, we have the following expected return:

$$Q^{\pi}(s,\alpha) = E_{\pi}\left[r_t + \gamma r_{t+1} + \gamma r_{t+2} + \cdots | s_t = s, a_t = a\right], \quad (3)$$

The Advantage Actor-Critic (A2C) algorithm combines two elements of reinforcement learning: a policy gradient (actor) and a learned value function (critic). The Actor network learns a parameterized policy, and the Critic network learns the value function that evaluates pairwise state–action signals. The Critic network provides a reinforcement signal to the Actor network. The Advantage function of this algorithm is that the Actor network chooses an action at each time step, and the Critic network evaluates that action based on the Q value of an input state. As the Critical network learns which states are better, the Actor network uses that information to teach the agent to search for better states.

$$A_t = R_t + V\gamma(r_{t+1}) - V\gamma(r_t), \quad (4)$$

where A_t is the action of the state, R_t is the reward obtained in a given a state S_t, $R_t = R(S_t)$, and R is the reward function. To have a correct advantage function, the TD error is used as a better estimator of the decision-making process when choosing an action in a given state [9]. Agents can be trained using A2C to estimate the advantage function by combining the estimated reward value and the observed reward value.

3. Proposed System

The robotic system proposed in this work was developed using free software tools for robotics research, the agent to be evaluated is a robotic manipulator of 7 degrees of freedom, and the task to be performed is to reach a target within its working area in a semi-photorealistic scene. The training data are obtained directly from the simulator, and by means of the API integrated in this platform, the data flow control and control of the robot is performed.

3.1. System Architecture

The robotic manipulator control approach is based on the general DRL scheme; the software system architecture is a client–server type. The client contains the components to train and evaluate the DRL agent located on the server side, which consists of the simulation platform. Since this is an end-to-end control approach, the input data to the system consists of the position of the robot joints, the position (x, y, z) of a tip located at the end of the end-effector, and the measured distance between the end-effector and the target sphere. The control actions are angles for each of the joints of the 7-DoF robot.

This configuration is shown in Figure 1 and describes the process of learning an agent using DRL for robotic manipulator in CoppeliaSim.

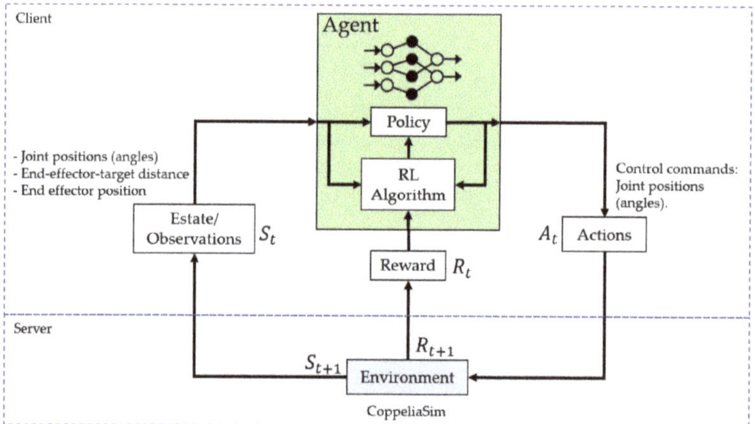

Figure 1. DRL process architecture for the robotic manipulator.

When using an A2C algorithm with an Actor-Critic policy, the model structure consists of two networks: (1) the Actor network recommends the action to be taken by the agent and gives the probability distribution of the state, and (2) the Critical network will give an estimate of the reward value for the executed actions. It yields the estimated total rewards in the given state in the future. The neural network architecture consists of the structure shown in Figure 2, which consists of the two networks Actor and Critic; each network layer has a size of 256, 512, and 256, respectively, with an ReLU activation function.

3.2. Reward Function

The reward function should make the robotic manipulator reach the target position, whose end-effector should approach a sphere located in the working area. After executing an action, an observation is obtained, and the total reward is calculated, which is achieved with the following expression:

$$R_t = r_{dist} + r_{ctrl} + r_{pa} + r_l + r_c, \tag{5}$$

where r_{dist} represents the distance reward, which quantifies the Euclidean distance between the end of the robotic manipulator and the target sphere. This reward has a more negative value when the end of the manipulator is farther away from the target. On the other hand, r_{ctrl} corresponds to the control reward, which is calculated as the squared Euclidean norm of the action taken by the agent. This reward penalizes the agent when it makes excessively large movements. In addition, the reward r_{pa} is obtained by comparing the previous distance with the current distance after performing an action. If the previous distance is less than the current distance, the agent receives a negative reward, as this indicates that the end-effector of the manipulator is moving away from the target position. The difference reward is calculated by the following function:

$$r_{pa} = d_p - d_a, \tag{6}$$

where d_p is the previous distance between the end-effector and the target to be reached, this distance value is calculated before executing an action of the robot, and d_a, indicates the current distance between the end-effector position and the sphere position. The arrival reward r_l of the manipulator is defined as a positive value that is within a threshold distance between the end-effector and the target. The collision reward r_c is a positive value given to the agent when the effector touches the target object. The total reward is composed of the five elements already mentioned and is defined by the following function:

$$R_t = \begin{cases} r_{dist} : -\sqrt{\sum_{i=1}^{n}(y_i - x_i)^2} \\ r_{ctrl} : -\sum_{i=1}^{n}(Action)^2 \\ r_{pa} : d_p - d_a \\ r_l : if|d_a| < d_u \\ r_c : if\ collision = True \end{cases} \tag{7}$$

where *Action* is a vector composed of the angular positions of the seven joints of the robot. The term d_u is the threshold distance that determines that the end-effector is approaching the target, and the term "collision" has a logical value that represents the collision between these two objects in the simulation; once this condition is fulfilled, the agent is given a positive value for reaching the target. These rewards are essential components in guiding the agent's on-task behavior, encouraging precise movements toward the target sphere and discouraging abrupt or distant actions.

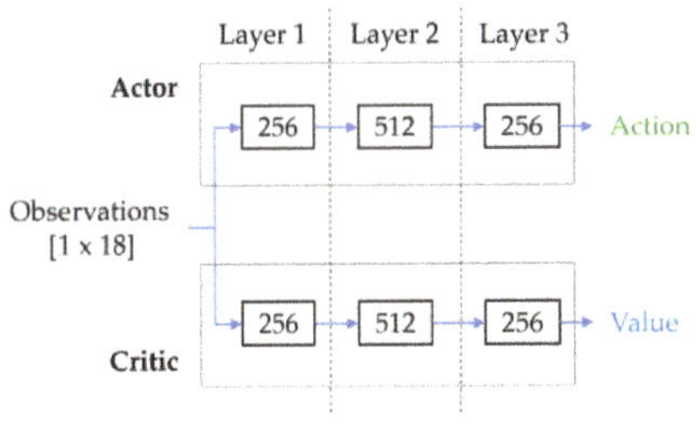

Figure 2. Neural network structure for A2C.

3.3. *Action and Observation Space*

The robotic manipulator consists of seven joints (7 DoF); as a result, we have the action space of a vector containing the angles of each joint of the robot. The observation space is a vector of 1×18 elements containing the position of the end-effector, the angles of each joint, and the distance between the end-effector and the target sphere.

3.4. *Simulation*

The implementation of the robotic manipulator occurs in the CoppeliaSim simulator [11]; this platform provides an API in Python that allows for communication with external software, data acquisition, and control functions. The base architecture for this system consists of two elements: the first is the server in which the robotic manipulator (Franka Emika Panda) and a static element (red sphere) that will be the target to reach are deployed. The second is the client, which consists of Python scripts for the creation of a customized environment with Gymnasium [12] and the training and evaluation of the agent with Stable Baselines 3 [13].

The simulation environment is presented in Figure 3; the objective of the agent is to move the end-effector to a target position, which is marked by a red sphere. At the end of the end-effector, a mark is placed in order to set a parameter for the observation space; the working area is inside the white colored area.

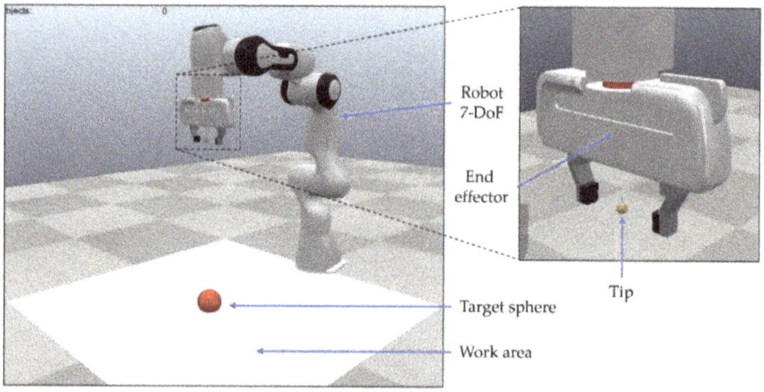

Figure 3. Simulation scene for the robotic manipulator.

4. Training and Evaluation Details

The robotic system was trained and evaluated in a simulation environment, and the following software components were used: Python as the base programming language, CoppeliaSim as the simulation platform, Stable Baselines 3 as the DRL framework that allowed for the integration of a deep neural network (DNN), and the A2C algorithm in the test agent. The hardware used for the implementation of this system has the following features: Ryzen 7 CPU, NVIDIA GeForce RTX 3070 8 GB GPU, and 16 GB RAM.

During the training phase, the robotic manipulator searches to approximate the position of the target located at a point in the scene; the coordinates of the robot joints towards the target are given randomly and updated after several episodes, making the robot perform its task better. This makes the DRL training efficient and robust.

Figure 4 presents the main elements used for this robotic system in software. In (a), the custom environment for CoppeliaSim is presented, and the class is created; Gymnasium methods for RL and control functions and the data acquisition of the simulator are established. (b) is part of the client in which the agent training is performed, and the model is saved. (c) is also part of the client, and the trained agent is evaluated, which involves loading the model and running it on the simulation platform to verify that the assigned task is being performed.

The selected algorithm is the Actor-Critical Advantage (A2C) [14], because it meets the requirements of the action and observation spaces used for Gymnasium that are supported by the DRL framework, and it is one of the algorithms used for robotic manipulation [15]. The input state of the deep neural network consists of a vector with these elements: the position of the tip point of the end-effector, the angles of the joints, and the distance from the end-effector to the target. The action performed by the system is a vector containing seven elements corresponding to each of the robot's joints.

For additional details on the development and implementation of the robotic system in the simulation platform, a link of the repository is attached here: https://github.com/RogerSgo/DRL-Sim-Reach-Target (accessed on 2 October 2023).

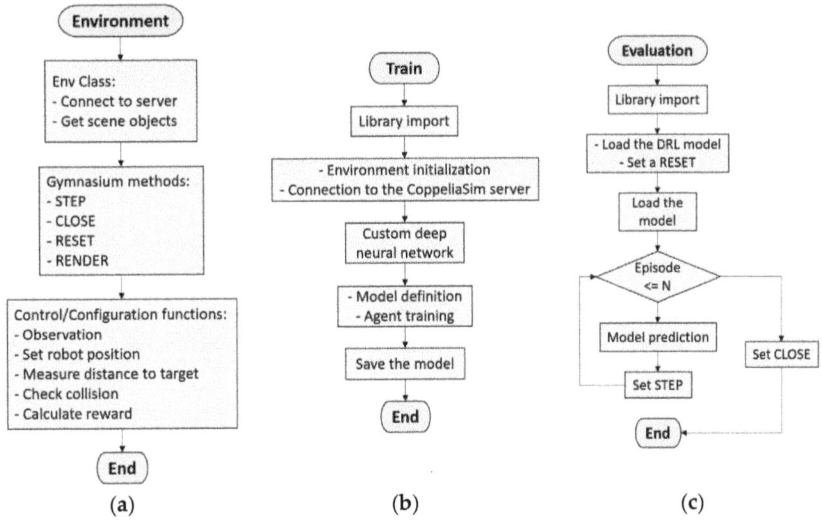

Figure 4. The flowcharts of the software system: (**a**) Customized environment for CoppeliaSim; (**b**) Flowchart for client training; (**c**) Flowchart for the evaluation of the trained model.

5. Results

This section presents the main results obtained from the experimentation carried out in the CoppeliaSim EDU 4.5 simulation environment. The model was trained with 1000 episodes and with a standard Actor-Critic neural network architecture of 3 layers of the sizes 256, 512, and 256 in each network. The resulting test model took about 1 h to train.

Table 1 shows the reward values obtained during the training of the agent with the A2C algorithm during 1000 episodes; it can be seen that the training worked according to the criteria established for the assigned task. In this case, it is observed that the as the distance (d_a) between the end-effector and the target decreases, the reward value (R_t) increases, which indicates the agent learning to perform this movement action.

Table 1. Reward values.

Episode	d_p	d_a	r_{dist}	r_{ctrl}	r_{pa}	r_l	r_c	R_t
1	0.467	0.465	−0.465	−0.350	0.002	0.000	0.000	−0.813
500	0.158	0.145	−0.145	0.552	0.013	0.145	0.00	−0.538
1000	0.092	0.090	−0.090	−0.0333	0.001	0.090	0.00	−0.332

The learning curve of the A2C agent is based on the training reward. To obtain the learning curve, the sum of the rewards that the agent obtains during the set of episodes determined for training is measured. A high cumulative reward indicates that the agent is successful in the task. The learning curve is presented in Figure 5.

To verify the performance of the reach-to-target task using the A2C algorithm, the trained model is evaluated by performing multiple simulation experiments in CoppeliaSim. In Figure 6, which robotic manipulator can execute the task of reaching a target using the DRL control method is demonstrated. The robot starts from an initial position, and with the use of the trained model, it performs motion predictions until it reaches the target located at a position in the working area of the environment.

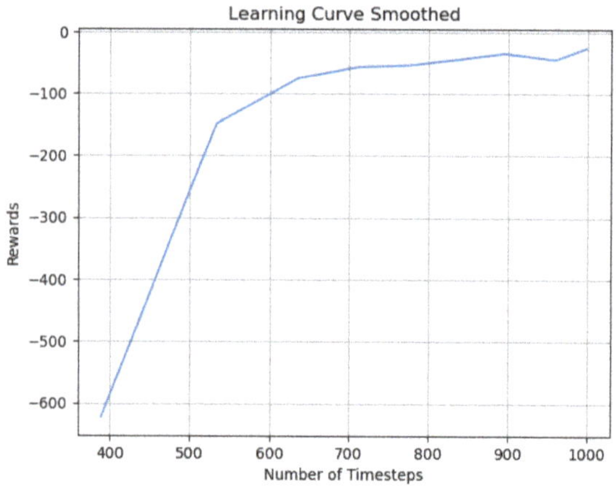

Figure 5. Agent learning curve.

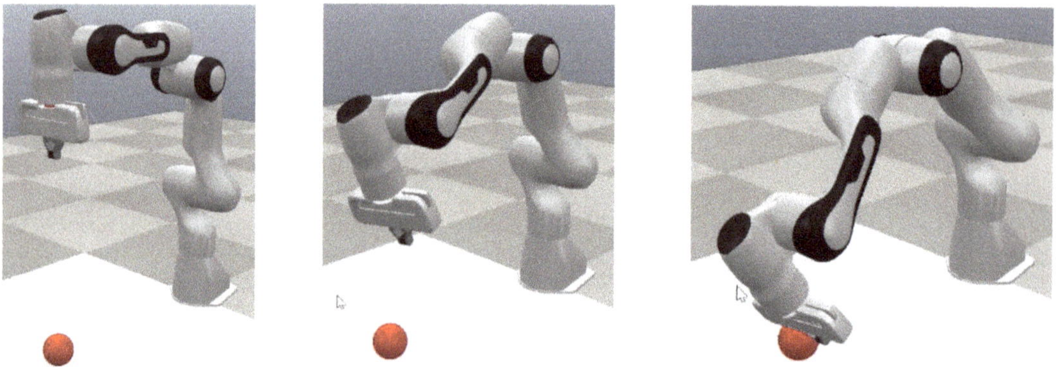

(a) (b) (c)

Figure 6. The sequence of manipulator movement to reach the target: (**a**) The initial position of the robotic manipulator; (**b**) End-effector moves about half the distance to the target; (**c**) End-effector arrives at the target goal.

The performance evaluation of the A2C method for this robotics application is based on the average reward; this parameter is a common metric for the evaluation of RL algorithms. The task to be performed in this work was the reaching of a target by the end-effector of a robotic manipulator, and according to the reward function designed in this system, the reward value should increase as the measured distance between end-effector and target decreases. This shows that the performance is satisfactory.

The evaluation of the trained DRL model was performed by running it with 100 episodes. Two important parameters are presented to measure the agent's performance based on A2C as shown in Figure 7; the curve in red is the reward value accumulated during that number of episodes, and the curve in blue represents the measured distance of the end-effector tip when approaching the target.

With the results obtained from the training process in Table 1 and the evaluation process in Figure 7, it is shown that the reward value increases when the measured distance between the tip positioned at the extreme of the end-effector and the target sphere decreases, with each action predicted by the DRL model. This indicates that the trained agent satisfies the execution of the task during the evaluation of the system in simulation.

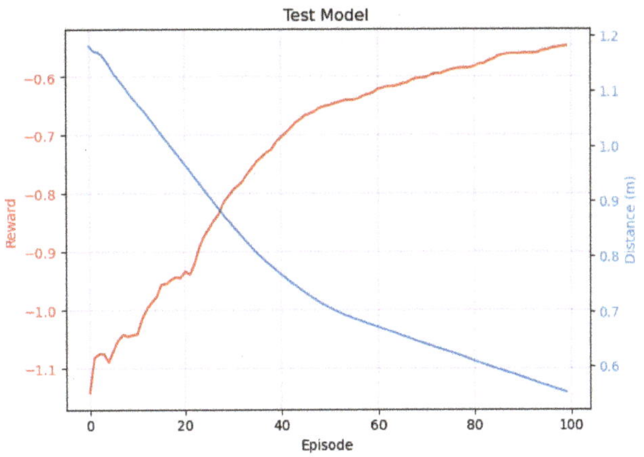

Figure 7. The evaluation parameters of the DRL Agent. The red curve is the reward accumulated when the end-effector approaches the target sphere position. The blue curve is the distance between the end-effector and the target sphere.

In addition to the metrics (reward) and measured distance (m) of the robotic system training, the use of the PC hardware resources of some software components was checked. Table 2 shows the hardware resources used by the software components of the robotic manipulator system during the training process and verifies that the highest consumption occurs by the following: the simulation platform consumes 35% of the total GPU, the Python programming language consumes 39% of the RAM memory, and finally, when using Anaconda to implement the different scripts in Firefox, it uses 9% of the RAM.

Table 2. Hardware performance.

Software	PC Component				
	CPU (%)	**GPU (%)**	**VRAM (%)**	**RAM (%)**	**FPS**
CoppeliaSim 4.5.1	2	35	37	1	59
Python 3.11.4	1	0.4	-	39	-
Firefox 118.0.2	0.2	0.1	-	9	-

6. Conclusions

In this paper, the A2C-based deep reinforcement learning method to solve the task of reaching a goal for a robotic manipulator is presented; simulation results in CoppeliaSim validate the performance of the proposed system. In addition to the training and evaluation processes of the DRL model, the impact of this DRL control method on the performance of a PC hardware is verified.

Reward calculation is mainly based on the measured distance between the end-effector position and the target to be reached, as well as a penalty to the agent when making movements that are too large. During the training and evaluation processes of the agent, each parameter of the reward function is monitored in order to determine if the task to be executed is being fulfilled; these data are evidenced in Table 1 and Figure 6. Therefore, as the end-effector-target distance decreases, the reward has a higher value.

The agent is composed of artificial neural networks that are integrated in a critical actor architecture and learned based on sensory information of the end-effector's state with respect to the target. Its policy based on direct feedback from the environment with data is evaluated with the following parameters: position, distance, and motion cost.

In Section 5, some quantitative results were presented such as the agent's learning curve, rewards obtained during a certain number of episodes, and the distance measured between the end-effector and the target. These values were obtained from the training and evaluation of the system and show that the agent was learning to perform the task of reaching a target within the work zone in the CoppeliaSim simulation scene.

The simulation experiments performed showed that the training process was performed in a short time due to GPU acceleration. Current implementation used data from various sources such as joint positions given in angles, data related to the distance from the end-effector to the target, and the distance between the end-effector mark and the target sphere. The effect of the processing performance can be varied by scaling the system via increase of the number of sensing sensors and/or complexity of the deep neural network.

The performance of the DRL agent depends on many factors, and we can emphasize these: the number of episodes for training the model, the framework selected to implement the DRL algorithm, the design of the neural network architecture, and process logic for the actions at each step to be taken by the agent.

Author Contributions: Conceptualization, methodology, C.C.-C. and R.S.; software, R.S.; validation, R.S.; writing—original draft preparation, C.C.-C. and R.S.; writing—review and editing, R.S.; supervision, C.C.-C.; project administration and funding acquisition, C.C.-C. All authors have read and agreed to the published version of the manuscript.

Funding: This research was funded by Universidad Tecnica Particular de Loja, grant number PROY_ARTIC_CE_2022_3667.

Institutional Review Board Statement: Not applicable.

Informed Consent Statement: Not applicable.

Data Availability Statement: Data are contained within the article.

Conflicts of Interest: The authors declare no conflict of interest.

References

1. del Real Torres, A.; Andreiana, D.S.; Ojeda Roldan, A.; Hernandez Bustos, A.; Acevedo Galicia, L.E. A Review of Deep Reinforcement Learning Approaches for Smart Manufacturing in Industry 4.0 and 5.0 Framework. *Appl. Sci.* **2022**, *12*, 12377. [CrossRef]
2. Bhuiyan, T.; Kästner, L.; Hu, Y.; Kutschank, B.; Lambrecht, J. Deep-Reinforcement-Learning-based Path Planning for Industrial Robots using Distance Sensors as Observation. In Proceedings of the 2023 8th International Conference on Control and Robotics Engineering (ICCRE), Niigata, Japan, 21–23 April 2023.
3. Jiang, R.; Wang, Z.; He, B.; Di, Z. Vision-Based Deep Reinforcement Learning For UR5 Robot Motion Control. In Proceedings of the 2021 IEEE International Conference on Consumer Electronics and Computer Engineering (ICCECE), Guangzhou, China, 15–17 January 2021; pp. 246–250.
4. Collins, J.; Chand, S.; Vanderkop, A.; Howard, D. A review of physics simulators for robotic applications. *IEEE Access* **2021**, *9*, 51416–51431. [CrossRef]
5. Gupta, S.; Singal, G.; Garg, D. Deep reinforcement learning techniques in diversified domains: A survey. *Arch. Comput. Methods Eng.* **2021**, *28*, 4715–4754. [CrossRef]
6. Nguyen, H.; La, H. Review of deep reinforcement learning for robot manipulation. In Proceedings of the 2019 Third IEEE International Conference on Robotic Computing (IRC), Naples, Italy, 25–27 February 2019; pp. 590–595.
7. Stooke, A.; Abbeel, P. Accelerated methods for deep reinforcement learning. *arXiv* **2018**, arXiv:1803.02811.
8. Gym, O.; Sanghi, N. *Deep Reinforcement Learning with Python*; Springer: Berlin/Heidelberg, Germany, 2021.
9. Dong, H.; Ding, Z.; Zhang, S. *Deep Reinforcement Learning—Fundamentals, Research and Applications*; Springer: Berlin/Heidelberg, Germany, 2020.
10. Liu, L.-L.; Chen, E.-L.; Gao, Z.-G.; Wang, Y. Research on motion planning of seven degree of freedom manipulator based on DDPG. In Proceedings of the Advanced Manufacturing and Automation VIII 8, Changzhou, China, 20–21 September 2018; pp. 356–367.
11. Robotics, C. Robotics Simulator CoppeliaSim. Available online: https://www.coppeliarobotics.com/ (accessed on 1 July 2023).
12. Towers, M.; Terry, J.K.; Kwiatkowski, A.; Balis, J.U.; Cola, G.d.; Deleu, T.; Goulão, M.; Kallinteris, A.; Arjun, K.G.; Krimmel, M.; et al. Gymnasium. Available online: https://github.com/Farama-Foundation/Gymnasium (accessed on 1 July 2023).
13. Raffin, A.; Hill, A.; Gleave, A.; Kanervisto, A.; Ernestus, M.; Dormann, N. Stable-baselines3: Reliable reinforcement learning implementations. *J. Mach. Learn. Res.* **2021**, *22*, 12348–12355.

14. Mnih, V.; Badia, A.P.; Mirza, M.; Graves, A.; Lillicrap, T.; Harley, T.; Silver, D.; Kavukcuoglu, K. Asynchronous methods for deep reinforcement learning. In Proceedings of the International Conference on Machine Learning, New York, NY, USA, 20–22 June 2016; pp. 1928–1937.
15. Han, D.; Mulyana, B.; Stankovic, V.; Cheng, S. A Survey on Deep Reinforcement Learning Algorithms for Robotic Manipulation. *Sensors* **2023**, *23*, 3762. [CrossRef] [PubMed]

engineering proceedings

Proceeding Paper

Computational Representation of Cellular Lines: A Text Mining Approach †

Ivan Carrera *, Henry Guanoluisa and Alexis Miranda

Departamento de Informática y Ciencias de la Computación, Escuela Politécnica Nacional, Quito 170525, Ecuador; henry.guanoluisa@epn.edu.ec (H.G.); alexis.miranda@epn.edu.ec (A.M.)
* Correspondence: ivan.carrera@epn.edu.ec
† Presented at the XXXI Conference on Electrical and Electronic Engineering, Quito, Ecuador, 29 November–1 December 2023.

Abstract: In the rapidly evolving landscape of cancer drug research, cellular lines serve as invaluable tools for understanding drug-sensitive and drug-resistant tumors. The computational representation of cellular lines is usually based on genomic profiling, even though this method cannot be applied in a large scale. This study introduces a novel approach to the computational representation of cellular lines using text mining techniques. By meticulously extracting and analyzing textual data from the scientific literature, we developed a computational representation of these cellular lines. Our methodology encompassed advanced Natural Language Processing (NLP) for text extraction and machine learning models for predictive analysis. We achieved a comprehensive description of each cellular line. To validate our findings, we generated a distance matrix for all cellular lines, leading to the construction of a dendrogram representing cellular line relationships. This dendrogram shows a resemblance with the established cell line ontology from CLO. Our results bridge the gap between cellular line representation and text mining, offering a robust computational model that can significantly impact cancer drug research.

Keywords: text mining; natural language processing; predictive modeling; machine learning; cellular line representation; drug response prediction; personalized medicine

Citation: Carrera, I.; Guanoluisa, H.; Miranda, A. Computational Representation of Cellular Lines: A Text Mining Approach. *Eng. Proc.* **2023**, *47*, 13. https://doi.org/10.3390/engproc2023047013

Academic Editor: Soraya Lucia Sinche Maita

Published: 4 December 2023

1. Introduction

Cellular lines are indispensable tools in cancer research. These in-vitro models offer a controlled environment to study the biology of cancer cells, test potential therapeutic agents, and understand drug resistance mechanisms. Cellular lines have been instrumental in elucidating the molecular pathways of carcinogenesis and have paved the way for the development of targeted therapies. Their significance is underscored by the vast body of scientific literature dedicated to their study, emphasizing their role in advancing our understanding of cancer and its treatment.

The drug discovery process can be leveraged by Artificial Intelligence [1]. Furthermore, AI fosters the emergence of Computational Precision Medicine, allowing the design of therapies tailored to individual patients' physiology, disease features, and environmental exposures [2]. The combination of AI with big data and advanced computing has the potential to revolutionize evidence-based, personalized medicine, making treatments more efficient and tailored to individual needs [3].

As the volume of data related to cellular lines continues to grow, there is an increasing need for efficient computational representation to manage and analyze this information. Text mining, a subset of data mining, offers a promising solution. By extracting valuable insights from vast amounts of textual data, text mining can help researchers identify patterns, trends, and relationships that might otherwise go unnoticed [4]. In the context of cellular lines, text mining can facilitate the identification of novel drug targets, elucidate mechanisms of drug resistance, and streamline the drug discovery process.

Despite the wealth of information available on cellular lines, there remains a gap in the systematic computational representation of these data.

This research aims to present a comprehensive computational representation of cellular lines. Our aims are to: (1) extract relevant information from textual data sources related to cellular lines, (2) develop a computational model that captures the intricacies of cellular line data, and (3) validate the relevance of the extracted information in the context of cancer drug research.

2. Motivation

The computational representation of cellular lines has been a topic of interest in the scientific community in the last 15 years. However, there is no consensus on the most effective method for achieving this representation. The most prevalent technique employed to date is genomic representation. Though this approach offers valuable insights into the genetic makeup of cellular lines, it falls short of providing a comprehensive view. For instance, a study by Li et al. found that established cell lines are generally a poor representation of primary tumor biology, indicating that genomic representation might not capture the full spectrum of cellular characteristics [5]. Genomic representation primarily focuses on the genetic sequences and variations, often overlooking other crucial aspects such as cellular behaviors, interactions, and responses to various treatments.

In contrast, other domains within bioinformatics have witnessed significant advancements through the integration of text mining techniques. Text mining has been employed to explore publication trends in various biomedical areas, facilitating the discovery of new insights and relationships [6]. Domains such as proteomics, genomics, and systems biology have successfully employed text mining to uncover patterns, trends, and relationships, leading to novel insights and a deeper understanding of complex biological systems. Kitano discussed the integration of mathematical modeling, molecular interaction networks, and cellular structure physics, suggesting that achieving computational models predicting cellular system behaviors is increasingly feasible [7]. Yun et al. combined text mining with gene expression analysis to reveal a relationship between a specific molecule and the invasiveness of a glioblastoma cell line [6].

Thus, given the successes of text mining in these domains, it presents a compelling unaddressed approach for its application in the computational representation of cellular lines. By leveraging text mining, there is potential to extract a more holistic and nuanced understanding of cellular lines, bridging the gap left by genomic representation and offering a more comprehensive view of cellular dynamics and characteristics.

3. Materials and Methods

In our research, we employed a systematic methodology to achieve a comprehensive representation of cellular lines. Our data sources were the Cellosaurus database and PubMed, from which relevant information on cellular lines was extracted. To process these data, we utilized text mining techniques, specifically Term Frequency-Inverse Document Frequency (TF-IDF) and Support Vector Domain Description (SVDD). Following the extraction of features, we computed similarities between cell lines based on their SVDD representations. This facilitated the construction of a dendrogram, providing a hierarchical representation that elucidates the relationships between various cell lines. This structured approach ensured both the depth and breadth of our analysis, leading to meaningful insights into cellular line representations. Our methodology is represented in Figure 1.

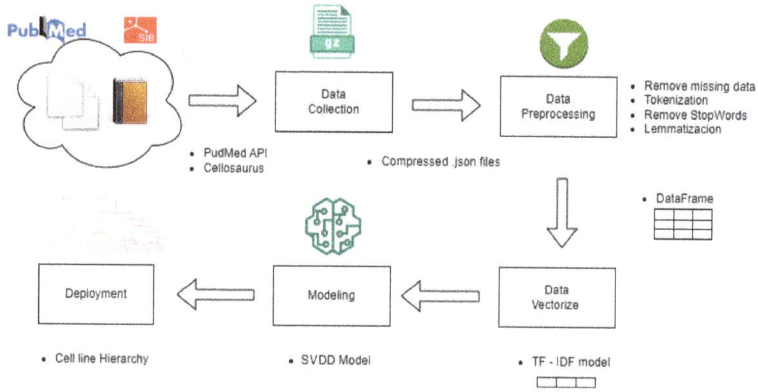

Figure 1. Methodology for data extraction and representation.

3.1. Data Sources

The data for this research were sourced from two major databases. The list of cellular lines was extracted from Cellosaurus, a comprehensive cell line database that presents a large collection of cell line names and their synonyms [8]. For the descriptions of these cellular lines, we turned to PubMed, a database of scientific papers in the biomedical domain [9]. In PubMed, we searched for all the papers that refer to all known names and synonyms of cellular lines.

In our research, the Cellosaurus database was pivotal in providing both the primary names and associated synonyms of cellular lines. Upon extraction, all synonyms for a given cellular line were consolidated into a unified list. This ensured that any mention of the cellular line, regardless of the specific name or synonym used, could be identified and processed. Recognizing the variability in naming conventions, we consolidated all synonyms for each cellular line into a unified list. During data extraction from sources like PubMed, this list ensured the comprehensive capture of relevant data, regardless of naming variations. All synonyms underwent normalization to a standard format, removing discrepancies like special characters or capitalization differences. By systematically managing and incorporating these synonyms, we ensured thorough and accurate data extraction, capturing the entirety of available knowledge on the studied cellular lines. This approach was taken aiming to capture a broad spectrum of the literature, encompassing most cellular lines and their descriptions. In Figure 2, we can see an example for a query using the PubMed API. Our corpus comprises 266,790 papers from 21,844 cell lines.

```
for index, row in df_queries.iterrows():
    try:
        string_search = Entrez.read(Entrez.esearch(db="pubmed", term=row['query'] +
                                                   ' AND ("cell line" OR "cell-line" OR "cellular line")'))
        time.sleep(1)
        pubmed_ids = [int(id) for id in string_search['IdList']]
        paper_id_list.append({'cell_line': row['cell_line'], 'pubmed_ids': pubmed_ids, 'num_ids': len(pubmed_ids)})
        print("Linea Celular " + row['cell_line'])
    except Exception as e:
        print(f"Error Linea Celular {row['cell_line']}: {str(e)}")
        error_paper_id_list.append({'cell_line': row['cell_line'], 'pubmed_ids': [], 'num_ids': 0})
```

Figure 2. An example of a query using the PubMed API.

3.2. Text Mining Techniques

After extracting abstracts related to cell lines from PubMed, we utilized the Term Frequency-Inverse Document Frequency (TF-IDF) method to transform each abstract into a numerical vector, as depicted in Figure 3. The selection of the TF-IDF method for our research was driven by its ability to efficiently evaluate the importance of words within text descriptions relative to their frequency across multiple documents. TF-IDF not only emphasizes the relevance and significance of terms, but also offers scalability and dimensionality reduction. This ensures that the most pertinent terms are highlighted, filtering out common terms and focusing on those that are truly significant, especially given the extensive data sourced from databases like PubMed.

	aa	aaa	aaaf	aaatcaatatt	aab	aabcl	aac	aacocf	aacr	aacrjourn	...	zymosan	zynaxl	zyx	zyxin	zz	zzr	zzui	zzuneui
0	0.0	0.0	0.0	0.0	0.0	0.0	0.0	0.0	0.0	0.0	...	0.0	0.0	0.0	0.0	0.0	0.0	0.0	0.0
1	0.0	0.0	0.0	0.0	0.0	0.0	0.0	0.0	0.0	0.0	...	0.0	0.0	0.0	0.0	0.0	0.0	0.0	0.0
2	0.0	0.0	0.0	0.0	0.0	0.0	0.0	0.0	0.0	0.0	...	0.0	0.0	0.0	0.0	0.0	0.0	0.0	0.0
3	0.0	0.0	0.0	0.0	0.0	0.0	0.0	0.0	0.0	0.0	...	0.0	0.0	0.0	0.0	0.0	0.0	0.0	0.0
4	0.0	0.0	0.0	0.0	0.0	0.0	0.0	0.0	0.0	0.0	...	0.0	0.0	0.0	0.0	0.0	0.0	0.0	0.0
...	...	...	...	...	...	...	...	...	...	...	...	...	...	...	...	...	...	...	...
266,787	0.0	0.0	0.0	0.0	0.0	0.0	0.0	0.0	0.0	0.0	...	0.0	0.0	0.0	0.0	0.0	0.0	0.0	0.0
266,788	0.0	0.0	0.0	0.0	0.0	0.0	0.0	0.0	0.0	0.0	...	0.0	0.0	0.0	0.0	0.0	0.0	0.0	0.0
266,789	0.0	0.0	0.0	0.0	0.0	0.0	0.0	0.0	0.0	0.0	...	0.0	0.0	0.0	0.0	0.0	0.0	0.0	0.0

Figure 3. Textual data are transformed into numerical data using TF-IDF. Every row corresponds to a single abstract.

Following this, the Support Vector Domain Description (SVDD) technique was applied to encapsulate the abstract instances into a unified computational representation for each cell line. Unlike traditional support vector machines, SVDD creates boundaries around data points in high-dimensional spaces, representing each cellular line as a distinct sphere, visualized in Figure 4. SVDD is a technique that allows for the creation of a boundary around data points in a high-dimensional space. By using SVDD, we were able to obtain a singular, comprehensive description for each cellular line. The synergy of TF-IDF and SVDD provided a robust methodology, delivering comprehensive insights into cellular lines.

	cell_line	center	radius
0	CVCL_0028	[0.002517910175187784, 0.007314898799435672, -0...	0.066173
1	CVCL_0080	[0.019450449413684838, 0.013177261231692031, -...	0.054771
2	CVCL_0081	[-0.009486665032411371, 0.027937856571742012, 0...	0.046550
3	CVCL_0082	[0.023437894045063443, 0.006965141811658809, 0...	0.038501
4	CVCL_0107	[0.005348316471685562, 0.01594712288027022, 0...	0.098690
...	...	...	...
9243	CVCL_ZZ58	[-0.007880641290901352, -0.0014381966014572116...	0.897165
9244	CVCL_ZZ59	[-0.007880641290901352, -0.0014381966014572116...	0.897165
9245	CVCL_ZZ60	[-0.007880641290901363, -0.0014381966014572116...	0.897165
9246	CVCL_ZZ61	[-0.007880641290901363, -0.0014381966014572116...	0.897165
9247	CVCL_ZZ70	[-0.0037931607355641476, -0.008175344523148489...	0.892744

Figure 4. Cell line spherical representation.

4. Results

4.1. Cell Line Hierarchy

Upon successfully acquiring the spherical representations of the cell lines, we proceeded to quantitatively assess the inter-relations between these spheres. This was achieved by calculating the distances between each spherical representation, providing a metric to gauge their relative proximities and distinctions.

The resulting distance metrics facilitated the construction of a hierarchical representation of the cell lines. Intriguingly, this derived hierarchy exhibited a striking resemblance to the established Cell Line Ontology. Such an alignment underscores the potential of our methodology to naturally capture and reflect the intrinsic classifications and relationships inherent to cellular lines within recognized biological frameworks. We employed the `scipy.cluster.hierarchy` library in Python. This library provided the necessary tools for hierarchical clustering, enabling us to create the dendrogram based on the computed distances between cellular lines, ensuring both precision and reproducibility in our representation. We obtained a dendogram for showing the relations between cell lines, shown in Figure 5. The different colors in the figure represent clusters of cellular lines.

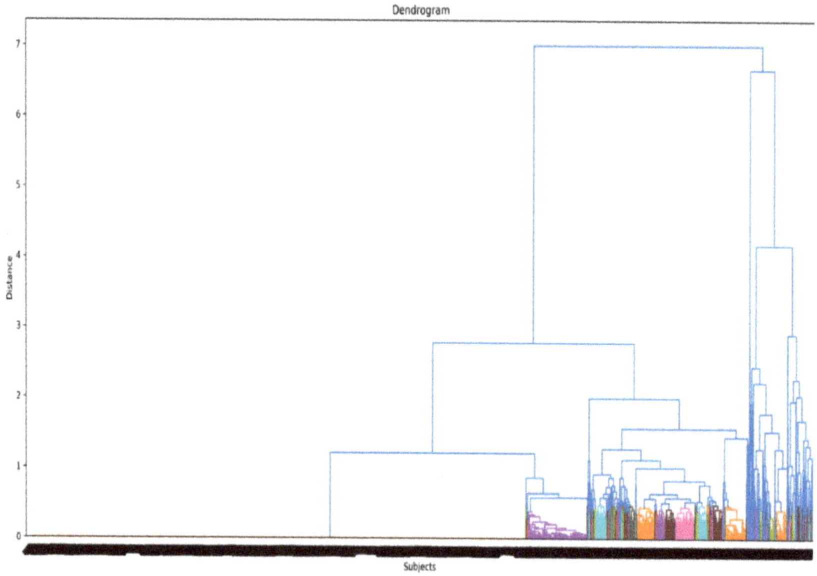

Figure 5. Cell line hierarchical representation.

4.2. Discussion

The computational representation of cellular lines, as visualized through spheres in a hyperspace, provided profound insights into the inherent relationships and distinctions among various cellular lines. By quantifying the distances between these spheres, we were able to discern patterns and clusters, indicating groups of cellular lines with similar characteristics and behaviors. This spatial representation not only facilitated a deeper understanding of the cellular lines in isolation, but also in relation to one another, highlighting potential synergies and differences that might not be evident through traditional analytical methods.

Our findings have significant ramifications for drug testing. The ability to computationally represent and compare cellular lines provides a foundation for predicting their responses to various drugs. By understanding the inherent characteristics of a cellular line, researchers can potentially forecast its reaction to a specific therapeutic agent, streamlining the drug testing process. This could lead to more efficient drug trials, as compounds can be tested on cellular lines that are computationally predicted to be responsive, thereby reducing the number of ineffective trials and accelerating the discovery of potent drugs.

Though our methodology and findings are promising, they are not without limitations. The reliance on textual abstracts from the PubMed library means that our data are only as comprehensive as the abstracts themselves. Important details and nuances present in the full text of articles might be overlooked. Additionally, the dynamic nature of scientific research means that new findings are continuously emerging, and our representation might not capture the very latest advancements in the field.

Our research represents a significant step forward in the computational representation of cellular lines and offers a foundation for future studies in this domain. By bridging the gap between textual data and computational analysis, we hope to drive advancements in drug testing and personalized medicine, ultimately benefiting patients and the broader medical community.

5. Conclusions and Future Work

Our research demonstrates that text mining is not only feasible, but also effective in deriving a computational representation of cellular lines. The techniques employed allowed for the extraction, processing, and representation of vast amounts of textual data related to cellular lines.

The similarity between the derived hierarchy from our text mining approach and the Cell Line Ontology underscores the validity of our methodology. It suggests that text-based data, when processed appropriately, can yield representations that are in alignment with established biological classifications.

As the demand for analyzing vast sources of text continues to grow, text mining applications are poised to play an increasingly pivotal role in biomedicine. Future work can improve the presented methodology, exploring additional text mining tools and techniques that can enhance the accuracy and depth of the extracted information.

Author Contributions: Conceptualization, I.C.; methodology, I.C.; software, H.G. and A.M.; writing—original draft preparation, I.C.; writing—review and editing, I.C.; project administration, I.C.; funding acquisition, I.C. All authors have read and agreed to the published version of the manuscript.

Funding: This research was funded by Escuela Politécnica Nacional grant number PII-DICC-2023-01.

Data Availability Statement: Data and codes are available at https://github.com/ivan-carrera/engproc2023 (accessed on 12 October 2023).

Conflicts of Interest: The authors declare no conflict of interest.

Abbreviations

The following abbreviations are used in this manuscript:

NLP	Natural Language Processing
CLO	Cell Line Ontology
TF-IDF	Term Frequency-Inverse Document Frequency
SVDD	Support Vector Domain Description

References

1. Ho, D. Artificial intelligence in cancer therapy. *Science* **2020**, *367*, 982–983. [CrossRef] [PubMed]
2. Moingeon, P.; Kuenemann, M.A.; Guedj, M. Artificial Intelligence-enhanced drug design and development: Toward a computational precision medicine. *Drug Discov. Today* **2021**, *27*, 215–222. [CrossRef] [PubMed]
3. Dilsizian, S.E.; Siegel, E.L. Artificial Intelligence in Medicine and Cardiac Imaging: Harnessing Big Data and Advanced Computing to Provide Personalized Medical Diagnosis and Treatment. *Curr. Cardiol. Rep.* **2013**, *16*, 441. [CrossRef] [PubMed]
4. Madhugiri, V.S.; Ambekar, S.; Strom, S.F.; Nanda, A. A technique to identify core journals for neurosurgery using citation scatter analysis and the Bradford distribution across neurosurgery journals. *J. Neurosurg.* **2013**, *119*, 1274–1287. [CrossRef] [PubMed]
5. Li, A.; Walling, J.; Kotliarov, Y.; Center, A.; Steed, M.; Ahn, S.; Rosenblum, M.; Mikkelsen, T.; Zenklusen, J.; Fine, H. Genomic Changes and Gene Expression Profiles Reveal That Established Glioma Cell Lines Are Poorly Representative of Primary Human Gliomas. *Mol. Cancer Res.* **2008**, *6*, 21–30. [CrossRef] [PubMed]
6. Yun, C.; Katchko, K.M.; Schallmo, M.S.; Jeong, S.; Yun, J.; Chen, C.H.; Weiner, J.A.; Park, C.; George, A.; Stupp, S.I.; et al. Aryl Hydrocarbon Receptor Antagonists Mitigate the Effects of Dioxin on Critical Cellular Functions in Differentiating Human Osteoblast-Like Cells. *Int. J. Mol. Sci.* **2018**, *19*, 225. [CrossRef] [PubMed]
7. Kitano, H. Computational cellular dynamics: A network–physics integral. *Nat. Rev. Mol. Cell Biol.* **2006**, *7*, 163–163. [CrossRef]
8. Bairoch, A. The Cellosaurus, a Cell-Line Knowledge Resource. *J. Biomol. Tech. JBT* **2018**, *29*, 25–38. [CrossRef] [PubMed]
9. National Center for Biotechnology Information. 2023. Available online: https://www.ncbi.nlm.nih.gov/pubmed/ (accessed on 31 August 2023).

engineering proceedings

MDPI

Proceeding Paper

Cluttered Environment and Target Simulator to Evaluate Primary Surveillance Radar Processors †

Fernando Lara [1,*], Ricardo Mena [1], Antonio Flores [1], Felipe Grijalva [2] and Roman Lara-Cueva [3,4]

1 Departamentro de Electrónica, Telecomunicaciones y Redes de Información (DETRI), Escuela Politécnica Nacional, Quito 170525, Ecuador; ricardo.menav@epn.edu.ec (R.M.); luis.flores04@epn.edu.ec (A.F.)
2 Colegio de Ciencias e Ingenierías "El Politécnico", Universidad San Francisco de Quito USFQ, Quito 170157, Ecuador; fgrijalva@usfq.edu.ec
3 Grupo de Investigación en Sistemas Inteligentes (WiCOM-Energy), Departamento de Ingeniería Eléctrica, Electrónica y Telecomunicaciones, Universidad de las Fuerzas Armadas ESPE, Sangolquí 171103, Ecuador; ralara@espe.edu.ec
4 Ad Hoc Networks Research Center (CIRAD), Departamento de Ingeniería Eléctrica, Electrónica y Telecomunicaciones, Universidad de las Fuerzas Armadas ESPE, Sangolquí 171103, Ecuador
* Correspondence: marco.lara@epn.edu.ec
† Presented at the XXXI Conference on Electrical and Electronic Engineering, Quito, Ecuador, 29 November–1 December 2023.

Abstract: This research article presents a comprehensive study focusing on advancing radar systems for unmanned aerial vehicle (UAV) surveillance in cluttered environments. The proliferation of UAV technology and its diverse applications have raised concerns about airspace security. To tackle this issue, this article introduces a novel simulator designed to evaluate the performance of primary monopulse radar processors. The simulator accurately replicates scenarios involving clutter Weibull distributions, stationary and moving targets, as well as pulse compression situations, thereby enabling precise and controlled evaluations. The study employs the simulator to assess radar processors, including a variant of moving target detection (MTD) and a constant false alarm rate (CFAR) processor. By implementing a rigorous methodology, the article underscores the significance of simulating cluttered conditions in refining the effectiveness of radar processors. The results yield valuable insights, facilitating objective interpretations. The proposed simulator and its implications contribute to enhancing UAV surveillance and airspace security, thereby pushing forward the capabilities of radar systems.

Keywords: cluttered environments; radar simulator; MTD and CFAR processors

Citation: Lara, F.; Mena, R.; Flores, A.; Grijalva, F.; Lara-Cueva, R. Cluttered Environment and Target Simulator to Evaluate Primary Surveillance Radar Processors. *Eng. Proc.* **2023**, *47*, 14. https://doi.org/10.3390/engproc2023047014

Academic Editor: Fernando Carrera Suarez

Published: 4 December 2023

1. Introduction

The new technological advancements in UAVs and the increasing number of civil or non-military applications [1] threaten the security of the airspace [2]. This underscores the imperative need to enhance the capabilities of surveillance and object detection systems. Radar systems (radio detection and ranging), known for their robust performance in complex environments by accurately determining the azimuthal position, distance, and velocity of UAVs, as discussed in [3], assume a pivotal role. Furthermore, their real-time responsiveness makes these systems a fundamental tool for governmental entities when making decisions regarding UAV intrusion into restricted aerial spaces.

UAVs are known to have a limited radar cross-section (RCS), which is a measure of how detectable an object is by radar systems [4]. Objects with a smaller RCS are harder to detect with radar because they reflect less radar energy back to the receiver. UAVs are designed to have a small RCS to minimize their radar signature and make them less visible to radar systems. Therefore, it is necessary to develop new processors capable of detecting targets in environments with a signal to clutter ratio (SCR) close to 0. Traditional processors like the moving target indicator (MTI) described in [5] and the moving target detection (MTD) discussed in [6] are presently employed in primary surveillance systems, as referenced in [7–9]. For the detection of UAVs in urban settings, novel processors based on artificial intelligence, as mentioned in [10], and multiple-input–multiple-output (MIMO) systems, discussed in [11,12], among other approaches, are being tested.

To enhance and assess the performance of the processors, controlled environments are essential. Furthermore, training machine learning and deep learning systems requires substantial volumes of data, which, due to operational costs, prove challenging to acquire.

Hence, various simulated environments have been proposed. In [13], the simulation of cloud position acquisition through radar systems is presented, aimed at training prediction systems. In [14], a 77 GHz radar simulator for adaptive cruise control is introduced, where the complete frequency modulated continuous wave (FMCW) radar platform is simulated and different algorithm enhancements are tested. In [15], a radar simulator for human activity is presented, considering micro-Doppler effects. Finally, in [16], a maritime environment simulator for primary radars is showcased. This simulator incorporates empirical sea clutter parameters into a model used to simulate a maritime setting. In [17], a radar signal generation tool based on weather time series is presented to test radar processors. In [18], the use of a radar for the recognition of human activities is presented; synthetic data are generated from initial spectrograms with image transformation.

The accurate simulation of cluttered environments offers significant advantages in terms of economic savings, time efficiency, and effectiveness. This enables the evaluation of radar processor performance and the adjustment of key parameters before the physical implementation of those processors. To the best of our knowledge, no simulators have been found for primary monopulse radars in environments with ground clutter.

This article proposes the configuration of a clutter and target simulator to assess primary monopulse surveillance radar processors. The proposed system simulates Weibull clutter, stationary targets, and moving targets with or without pulse compression (Swerling 0) [19]. The generated simulator is tested with two radar processors, a variant of moving target detection (MTD) [20] and a constant false alarm rate (CFAR) processor [21], in order to develop a tool that permits the assessment and testing of radar design before its implementation.

Section 2 outlines the implementation methodology and the tests conducted on the simulator. In Section 3, the results and discussion are presented. Finally, Section 4 offers conclusions and outlines future work.

2. Materials and Methods

The objective of this article is to generate two matrices that simulate the signals acquired by monopulse surveillance radars. These arrays contain simulated in-phase (I) and quadrature (Q) signals, considering Gaussian noise, ground clutter, and targets. The implementation of the radar signal simulator is described in Section 2.1. The tests conducted with the simulated I/Q signals are presented in Section 2.2.

2.1. Simulator

2.1.1. Clutter model

The Weibull clutter model is represented by Equation (1):

$$c[n] = w_c[n] + \alpha c[n-1], \quad (1)$$

where $c[n]$ represents the clutter signal obtained, $w_c[n]$ corresponds to additive Gaussian noise, and α is the correlation coefficient, which depends on the radar's integration pulses. For short-range surveillance radars (with high antenna rotation speeds), it is recommended to have $\alpha > 0.8$, while for long-range surveillance radars, $\alpha < 0.8$ is advised.

In Figure 1, an example of a generated clutter signal is depicted. This clutter corresponds to ground reflections of the generated signal.

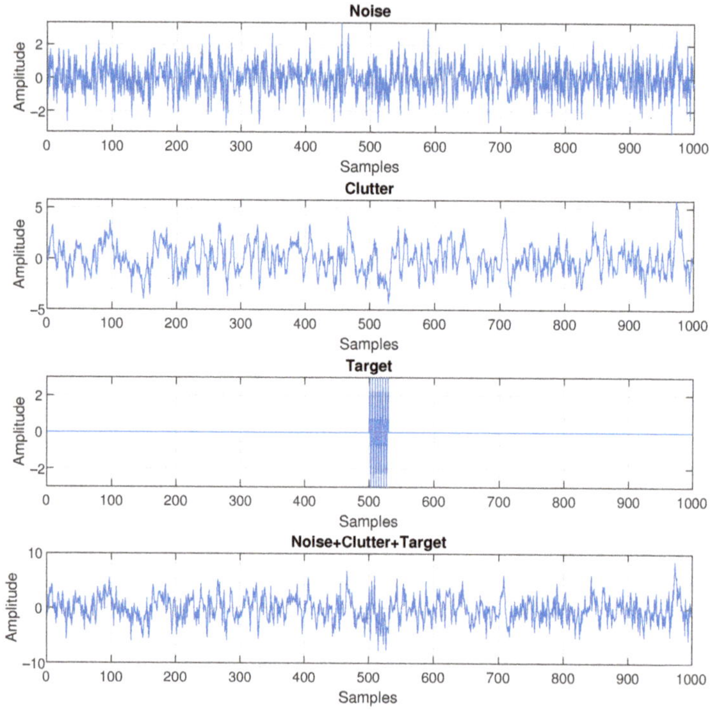

Figure 1. Simulated radar ring (Gaussian noise, clutter, and target).

2.1.2. Target Model

The number of pulses to be integrated (pulses that illuminate a target in a resolution cell) is defined by Equation (2).

$$N = \frac{\phi}{6\omega T} = \frac{\phi \times PRF}{6\omega}, \tag{2}$$

where N is the number of pulses to be integrated, ϕ is the beamwidth of the radar antenna's main lobe (measured at the half-power points of the beam), ω is the antenna rotation speed in RPM (revolutions per minute), and T is the pulse repetition period, which can be expressed as $PRF = \frac{1}{T}$ (pulse repetition frequency).

With these parameters, a number of samples can be determined for simulating the target. The target model used is presented in Equation (3), which corresponds to a target with Doppler shift and Swerling 0 characteristics [3,19].

$$s[n] = A \sin(2\pi f_d n), \tag{3}$$

where A is the amplitude of the target and f_d is the Doppler frequency. An example of a target signal is shown in Figure 1.

The simulator includes a module for simulating expanded pulses with bi-phase codes, such as Barker codes [22]. To achieve this, the code vector needs to be provided. For instance, $b[n]$ is a Barker code of length 5: $b[n] = [+1, +1, +1, -1, +1]$. To obtain the expanded pulse, Equation (4) is used.

$$S_E[n] = [s_1[n]b[2], s_2[n]b[2], \ldots, s_M[n]b[M]], \tag{4}$$

where M represents the pulse length.

2.1.3. Integration of the System

The radar target detection problem can be formulated as a binary hypothesis, as shown in Equation (5). The null hypothesis represents the absence of a target and includes the addition of noise $w[n]$ and clutter $c[n]$. The alternative hypothesis includes the target $s[n]$ in addition to noise and clutter.

$$\begin{aligned} H_0 &: r[n] = w[n] + c[n], n = 0, 1, 2, 3, ..., N \\ H_1 &: r[n] = w[n] + c[n] + s[n], n = 0, 1, 2, 3, ..., N. \end{aligned} \tag{5}$$

This concept from Equation (5) is used to generate the simulated I/Q signals, where noise, clutter, and/or target signals can be added to each cell (see Figure 1). The aggregation of cells generates a radar ring, which is integrated as shown in Figure 2.

In addition, Figure 2 demonstrates an example of ring integration, illustrating correlated clutter within cells of the same ring and no correlation between rings. This background noise is complemented by cells with target signals. The number of integration pulses depends on the physical characteristics of the radar, as discussed in the previous section.

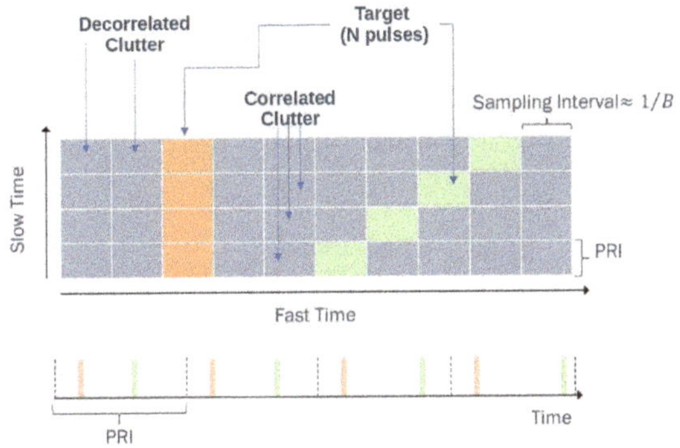

Figure 2. Data matrix integration.

The number of rings is related to the resolution cells, radar range (Equation (6)), the cone of silence, and the sampling frequency. The number of samples per ring depends on the PRF.

$$R_d = \frac{c \times \delta}{2}, \tag{6}$$

where R_d is the length of the cell, c is the speed of light, and δ is the pulse duration. In the case of coded pulses, δ represents the pulse duration relative to the code length.

Controlling the signal-to-clutter ratio (SCR) is an advantage of using a simulated environment, considering Equation (7).

$$SCR = \frac{P_{target}}{P_{noise} + P_{clutter}} = \frac{P_S}{P_W + P_C}, \tag{7}$$

where P_S is the power of the signal reflected from the target, P_W is the power of the noise signal, and P_C is the power of the clutter signal.

The generated test environments correspond to a Skyguard Oerlikon radar [20] (surveillance radar with a range of 16 km) and an AN/TPS 70 radar [23] (surveillance radar with a range of 440 km). These environments are used to evaluate the MTD and CFAR processors.

2.2. Testing

Firstly, the simulated signals undergo a high-pass filter designed to accentuate high-frequency components and focus on relevant target characteristics, thereby enhancing the quality of the observed signals. Subsequently, a decompression process is implemented using known codes, enabling the recovery of efficiently encoded target information for transmission or storage. By reconstructing the original signal, subsequent processing and analysis are facilitated. Following this, a magnitude detector is introduced, identifying significant peaks in the processed signals, a pivotal step for accurate target detection in the presence of clutter. Additionally, the constant false alarm rate (CFAR) algorithm is implemented, automatically adjusting the detection threshold based on local noise levels while maintaining a constant false alarm rate, further enhancing the detection system's reliability.

Lastly, the persistent challenge of stationary clutter in cluttered environments is tackled via a suppression algorithm rooted in target detection. This implementation significantly contributes to the reduction in false alarms and, importantly, optimizes precision in identifying moving targets.

Processors

In order to verify the environment generation, two traditional radar processors are employed, the MTD processor, for monopulse targets without pulse compression [6,20,24] (as shown in Figure 3a), and the CFAR processor, for pulse-compressed targets [19,21], using Barker code [22,25] (as depicted in Figure 3b).

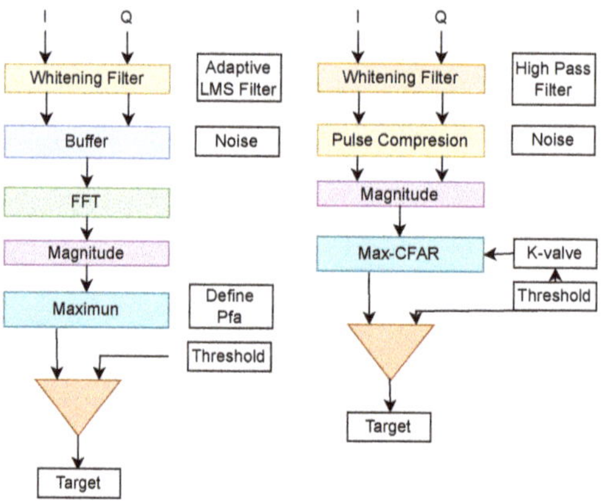

(a) (b)

Figure 3. Radar processors. (**a**) MTD. (**b**) CFAR.

These processors generate a detection vector with information on azimuth, distance, and whether the target is stationary or mobile. This information is displayed on a plan position indicator (PPI) display.

3. Results and Discussion

To evaluate the controlled SCR simulated environment, two recordings are generated: one simulating acquisition parameters for an AN/TPS 70 radar and the second simulating acquisition with an Oerlikon Skyguard radar. The respective processors are then applied. For the AN/TPS 70 case, a Barker code pulse compression magnitude detector CFAR is used, while for the Oerlikon Skyguard case, an MTD processor is employed.

3.1. AN/TPS 70

In Figure 4, the phase of the simulated 10 s record (one rotation) is presented, considering a PRF of 250 Hz and a sampling frequency in mega samples per second equal to 4. The entire environment is assumed to exhibit reflections with a correlation coefficient of 0.6, in addition to three regions of elevated ground clutter magnitude. The environment contains six targets (see Table 1), which were simulated with different integration numbers, equivalent to simulating targets of varying sizes. All simulated targets have an SCR of 3 dB and three concentrations of ground clutter (azimuth: $0.012\pi, 1.92\pi, 0.92\pi$; and range: 27.5 km, 192.5 km, 421.7 km) within a 3 km dimension.

Table 1. Simulated targets for AN/TPS 70.

Target Number	Range (km)	Azimuth (rad)	Integration Pulse
1	37.4	0.0π	8
2	11.7	1.9π	12
3	184.1	1.0π	7
4	274.3	0.6π	5
5	367.4	0.1π	10
6	402.6	1.6π	9

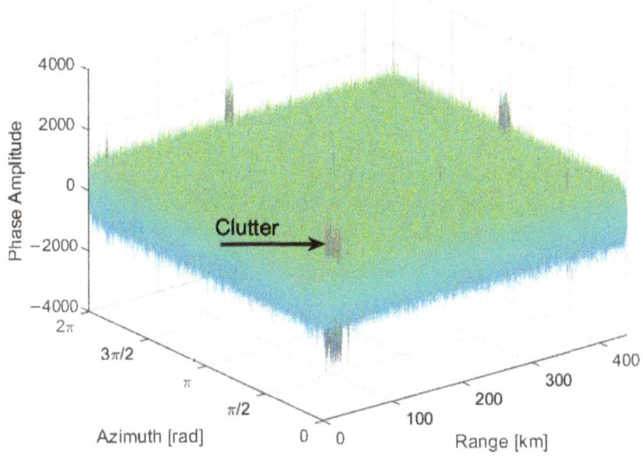

Figure 4. Simulated environment with targets and clutter (AN/TPS 70).

In Figure 5a, the output of a high-pass filter (clutter whitening filter) is presented. As observed in the figure, clutter is attenuated, though the targets are not yet discernible. The simulator's purpose is to test the capabilities of each stage of a radar processor.

The next stage involves a compression process using a Barker code. In Figure 5b, peaks with magnitudes higher than the noise level can be observed. These peaks are associated with the targets. The presented information only represents the phase. Combining the phase and quadrature information, the magnitude is obtained (see Figure 5c).

To distinguish targets from noise, a max CFAR processor [21] is applied. The output of this processor is shown in Figure 5d.

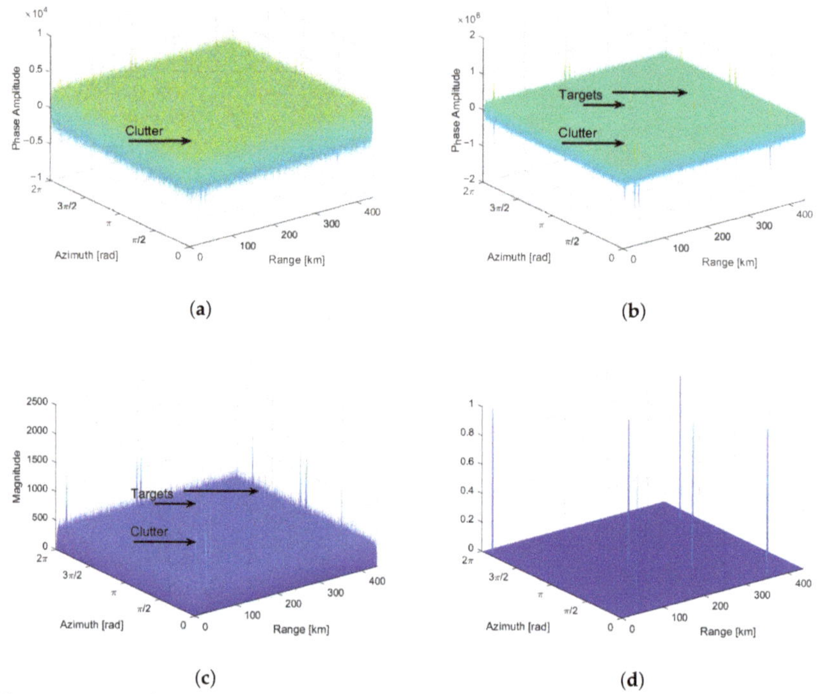

Figure 5. Target detection. (**a**) Simulated environment passed through a clutter whitening filter. (**b**) Environment passed by pulse compression. (**c**) Magnitude of simulated environment. (**d**) Target detection with threshold.

As a final stage, stationary clutter is removed by utilizing the simulator's capability to generate moving targets. This allows for varying the target's position during each rotation (10 s). In this case, information from four rotations (120,000,000 complex I/Q data points) is used. If a target remains in the same position for at least three rotations, it is considered a stationary target. Otherwise, it is considered a moving target and displayed on the PPI display, as seen in Figure 6.

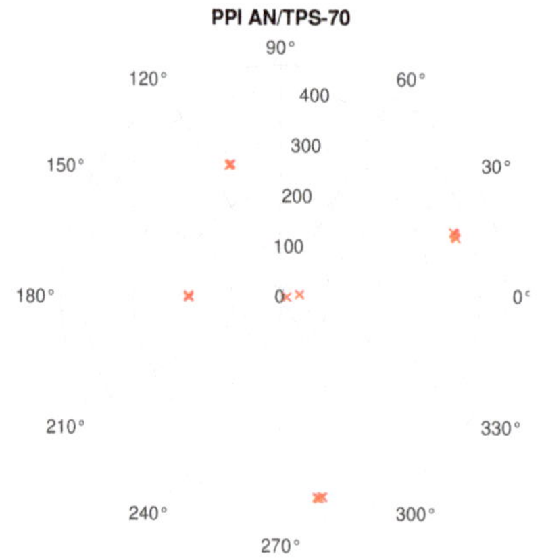

Figure 6. PPI simulation of AN/TPS 70.

3.2. Oerlikon Skyguard

A real application of the applicability of these simulators is in the Oerlikon Skyguard air defense system. The developed simulator generates various matrices with noise, clutter, and targets. Additionally, it associates target information and presents it on a PPI display. Targets can be labeled as either stationary or mobile based on their presentation on the PPI.

In the case of simulating phase and quadrature signals from Oerlikon Skyguard, a PRF of 6900 Hz is considered, a sampling frequency of 5 mega samples per second is considered, with a rotation speed of 1 RPM. With these data for the Oerlikon Skyguard radar, 5,000,000 complex I/Q data points are simulated per second. The targets on the Oerlikon Skyguard radar do not have pulse compression; considering Equation (2) with the characteristics of Oerlikon Skyguard there are 33 integration pulses and 150 m cells per simulated objective.

Figure 7a presents the radar environment for the Oerlikon Skyguard system. It displays radar returns over time or azimuth, with prominent peaks indicating detected targets, such as aircraft, and smaller, scattered returns representing clutter from the surrounding environment. The graph provides insights into the radar's ability to distinguish between targets and clutter.

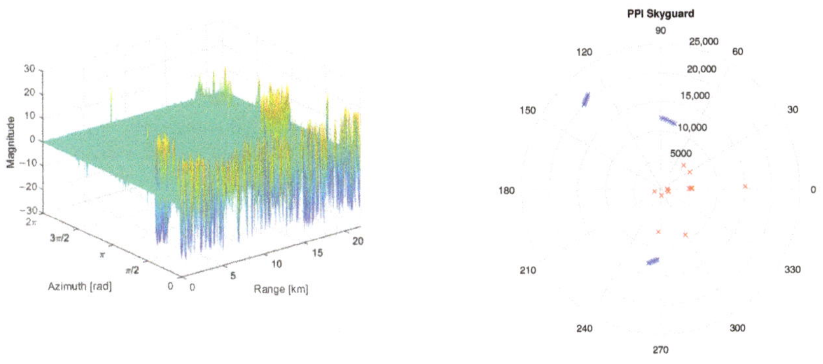

(a) (b)

Figure 7. Environment for Oerlikon Skyguard. (**a**) Environment with clutter and targets; (**b**) PPI.

In Figure 7b, the Oerlikon Skyguard radar's coverage area is presented in the PPI diagram. It centers around the radar's location, with concentric circles representing distance. Detected targets are symbolized on the display, showing their positions in relation to the radar's azimuth and range.

The original Oerlikon Skyguard system has an MTI processor, with a detection probability of 80%, with an SCR of 9 dB, and a false alarm probability of 10^{-5} [20,24]. By using simulated environments to train processors with machine learning and deep learning, the theoretical capabilities are increased, with a detection probability of 90%, an SCR from -3 dB to 0 dB, and a false alarm probability of 10^{-5} [10]. With these improvements in processing it is possible to detect targets with lower RCS, such as UAVs, with traditional radar systems.

4. Conclusions

The developed simulator presents a controlled platform that facilitates the evaluation of primary monopulse radar processors, enabling precise predictions of their real-world performance. By utilizing this simulator to assess various radar processor variants, such as MTD and CFAR, the study underscores the pivotal role of simulation in advancing radar technology for practical applications.

The research highlights the importance of incorporating fundamental physical radar parameters, including PRF, frequency, and range. These parameters are meticulously integrated to simulate the acquired phase and quadrature data corresponding to single or multiple radar rotations. This approach ensures the simulator's accurate replication of real-world conditions, providing an authentic assessment of primary monopulse radar processors for UAV surveillance. By incorporating these radar parameters into the simulation, the study enhances evaluation reliability and advances radar technology for practical applications, encompassing enhanced UAV surveillance and elevated airspace security.

A significant contribution of this study is the integration of controlled objectives within the context of SCR. These objectives encompass both mobile and fixed entities. By incorporating controlled objectives into the simulation, the study conducted a comprehensive evaluation of primary monopulse radar processors. This integration facilitates a meticulous assessment of how radar systems interact with dynamic and stationary targets, further refining the simulator's capability to emulate real-world scenarios. Consequently, the research offers a pragmatic avenue for refining radar technologies, their practical application, and the subsequent enhancement in UAV surveillance capabilities. This integration underscores the importance of an inclusive evaluation approach that encompasses diverse objectives and contributes to the ongoing innovation of radar system design and evaluation practices.

In the future, it is planned to improve the target model with different Swerlings, as well as different clutter models for sea clutter, among others. Additionally, it is intended to train processors based on machine learning to improve the detection capacity of current systems.

Author Contributions: Conceptualization, F.L.; methodology, F.L., F.G. and R.L.-C.; formal analysis, F.L.; investigation, F.L., F.G. and R.L.-C.; resources, F.L., F.G. and R.L.-C.; data curation, F.L. and R.M.; writing—original draft, F.L., R.M. and A.F.; writing— review and editing, F.L., R.M. and A.F.; visualization, R.M. and A.F.; supervision, F.G. and R.L.-C.; project administration, F.L. and R.L.-C.; funding acquisition, F.L. All authors have read and agreed to the published version of the manuscript.

Funding: This research received no external funding.

Institutional Review Board Statement: Not applicable.

Informed Consent Statement: Not applicable.

Data Availability Statement: No new data were created or analyzed in this study. Data sharing is not aplicable to this article.

Conflicts of Interest: The authors declare no conflict of interest.

References

1. Wellig, P.; Speirs, P.; Schuepbach, C.; Oechslin, R.; Renker, M.; Boeniger, U.; Pratisto, H. Radar systems and challenges for C-UAV. In Proceedings of the 2018 19th International Radar Symposium (IRS), Bonn, Germany, 20–22 June 2018; IEEE: Piscataway Township, NJ, USA, 2018; pp. 1–8.
2. Peacock, M.; Johnstone, M.N. Towards Detection and Control of Civilian Unmanned Aerial Vehicles. In Proceedings of the 14th Australian Information Warfare Conference, Perth, Australia, 2–4 December 2013; Edith Cowan University Security Research Institute: Perth, Australia; pp. 9–15.
3. Skolnik, M.I. *Radar Handbook*; McGraw-Hill: New York, NY, USA, 1970.
4. Farlik, J.; Kratky, M.; Casar, J.; Stary, V. Radar cross section and detection of small unmanned aerial vehicles. In Proceedings of the 2016 17th International Conference on Mechatronics-Mechatronika (ME), Prague, Czech Republic, 7–9 December 2016; pp. 1–7.
5. Tomita, Y.; Irabu, T.; Kiuchi, E. Moving Target Indication Radar. US Patent 4,053,885, 28 December 1977.
6. O'Donnell, R.; Muehe, C.; Labitt, M.; Drury, W.; Cartledge, L. Advanced signal processing for airport surveillance radars. In *IEEE Electronic and Aerospace Systems Convention (EASCON'74)*; IEEE: Piscataway Township, NJ, USA, 1974; p. 71.
7. Hyun, E.; Jin, Y.S.; Ju, Y.; Lee, J.H. Development of short-range ground surveillance radar for moving target detection. In Proceedings of the 2015 IEEE 5th Asia-Pacific Conference on Synthetic Aperture Radar (APSAR), Singapore, 1–4 September 2015; IEEE: Piscataway Township, NJ, USA, 2015; pp. 692–695.
8. Santi, F.; Pastina, D.; Bucciarelli, M. Maritime moving target detection technique for passive bistatic radar with GNSS transmitters. In Proceedings of the 2017 18th International Radar Symposium (IRS), Prague, Czech Republic, 28–30 June 2017; IEEE: Piscataway Township, NJ, USA, 2017; pp. 1–10.
9. Nag, S.; Barnes, M. A moving target detection filter for an ultra-wideband radar. In Proceedings of the 2003 IEEE Radar Conference (Cat. No. 03CH37474), Huntsville, AL, USA, 8 May 2003; IEEE: Piscataway Township, NJ, USA, 2003; pp. 147–153.

10. Carrera, E.V.; Lara, F.; Ortiz, M.; Tinoco, A.; León, R. Target detection using radar processors based on machine learning. In Proceedings of the 2020 IEEE ANDESCON, Quito, Ecuador, 13–16 October 2020; IEEE: Piscataway Township, NJ, USA, 2020; pp. 1–5.
11. Yuan, Y.; Yi, W.; Hoseinnezhad, R.; Varshney, P.K. Robust power allocation for resource-aware multi-target tracking with colocated MIMO radars. *IEEE Trans. Signal Process.* **2020**, *69*, 443–458. [CrossRef]
12. Shi, Z.; Wang, H.; Leung, C.S.; So, H.C.; Eurasip, M. Robust MIMO radar target localization based on lagrange programming neural network. *Signal Process.* **2020**, *174*, 107574. [CrossRef]
13. Haynes, J.M.; Marchand, R.T.; Luo, Z.; Bodas-Salcedo, A.; Stephens, G.L. A multipurpose radar simulation package: QuickBeam. *Bull. Am. Meteorol. Soc.* **2007**, *88*, 1723–1728. [CrossRef]
14. Kärnfelt, C.; Péden, A.; Bazzi, A.; Shhadé, G.E.H.; Abbas, M.; Chonavel, T. 77 GHz ACC radar simulation platform. In Proceedings of the 2009 9th International Conference on Intelligent Transport Systems Telecommunications, (ITST), Lille, France, 20–22 October 2009; IEEE: Piscataway Township, NJ, USA, 2009, pp. 209–214.
15. Vishwakarma, S.; Li, W.; Tang, C.; Woodbridge, K.; Adve, R.; Chetty, K. SimHumalator: An open-source end-to-end radar simulator for human activity recognition. *IEEE Aerosp. Electron. Syst. Mag.* **2021**, *37*, 6–22. [CrossRef]
16. Rosenberg, L.; Watts, S.; Bocquet, S. Scanning radar simulation in the maritime environment. In Proceedings of the 2020 IEEE Radar Conference (RadarConf20), Florence, Italy, 21–25 September 2020; IEEE: Piscataway Township, NJ, USA, 2020; pp. 1–6.
17. Schvartzman, D.; Curtis, C.D. Signal processing and radar characteristics (SPARC) simulator: A flexible dual-polarization weather-radar signal simulation framework based on preexisting radar-variable data. *IEEE J. Sel. Top. Appl. Earth Obs. Remote Sens.* **2018**, *12*, 135–150. [CrossRef]
18. Hernangómez, R.; Visentin, T.; Servadei, L.; Khodabakhshandeh, H.; Stańczak, S. Improving Radar Human Activity Classification Using Synthetic Data with Image Transformation. *Sensors* **2022**, *22*, 1519. [CrossRef]
19. Rahman, H. *Fundamental Principles of Radar*; CRC Press: Boca Raton, FL, USA, 2019.
20. Lara, F.; Ortiz, M.; Carrera, E.V.; Tinoco, A.F.; Moya, H.; León, R. Bayesian Processors of Radar Signals for the Skyguard Radar System of the Ecuadorian Army. *IEEE Lat. Am. Trans.* **2021**, *20*, 153–161. [CrossRef]
21. Mansouri, H.; Hamadouche, M.; Ettoumi, F.Y.; Magaz, B. Performance analysis of a Weighted Max CFAR processor. In Proceedings of the 2008 International Radar Symposium, Wroclaw, Poland, 21–23 May 2008; IEEE: Piscataway Township, NJ, USA, 2008; pp. 1–4.
22. Wang, S.; He, P. Research on Low Intercepting Radar Waveform Based on LFM and Barker Code Composite Modulation. In Proceedings of the 2018 International Conference on Sensor Networks and Signal Processing (SNSP), Xi'an, China, 28–31 October 2018; pp. 297–301. [CrossRef]
23. Pike, J. AN/TPS-70. Federation of American Scintists, Military Analysis Network. 1999. Available online: https://man.fas.org/dod-101/sys/ac/equip/an-tps-70.htm (accessed on 1 September 2023).
24. Lara, F.; Ortiz, M.; Carrera, E.V.; Tinoco, A.; León, R. Filtros adaptativos en el procesador bayesiano mti (moving target indicator) para el sistema de radar skyguard del ejército ecuatoriano. *Revista Ibérica de Sistemas e Tecnologias de Informaçao* **2020**, *E29*, 650–664. (In Spanish)
25. Turyn, R. Four-phase Barker codes. *IEEE Trans. Inf. Theory* **1974**, *20*, 366–371. [CrossRef]

engineering proceedings

Proceeding Paper

Industrial Application of Photovoltaic Systems with Storage for Peak Shaving: Ecuador Case Study †

Jesús Guamán-Molina [1,*], Patricio Pesantez [2,*], Carla Chavez-Fuentez [1] and Alberto Ríos [1]

1 Faculty of Engineering in Electronic and Industrial Systems, Technical University of Ambato, Ambato 180104, Ecuador; fchavez2602@uta.edu.ec (C.C.-F.); a.rios@uta.edu.ec (A.R.)
2 School of Electrical and Electronic Engineering, National Polytechnic School, Quito 170525, Ecuador
* Correspondence: jguaman0585@uta.edu.ec (J.G.-M.); patricio.pesantez@epn.edu.ec (P.P.); Tel.: +59-3995005649 (J.G.-M.)
† Presented at the XXXI Conference on Electrical and Electronic Engineering, Quito, Ecuador, 29 November–1 December 2023.

Abstract: Decentralized generation has gained importance in the energy industry, since self-consumption with renewable resources presents attractive costs and allows load management actions. In this sense, photovoltaic generation systems are a promising technology. This work presents a proposal for a peak shaving system using solar photovoltaic (PV) energy and a battery storage system, known as battery energy storage systems (BESS), to be installed by an industrial customer to reduce energy consumption during peak hours. For the study, a hybrid approach is presented, starting from deterministic variables, such as the demand curve of the industry under study, and the generation of stochastic variables, such as the energy production of the photovoltaic system. For the analysis of the proposed peak shaving system, the design and sizing of the photovoltaic systems are developed in a base case of self-generation and an optimized system for the system to cover the energy demand generated during peak hours. The technical–economic study carried out in the research allowed us to determine the optimal power of the photovoltaic system with the storage system. The proposed system allows the integration of a peak shaving strategy from a certain power limit, in order to cover the peak demand over this power limit, allowing the system to be profitable under the current regulations and standards in Ecuador.

Keywords: peak shaving; industry; photovoltaic system; BESS; optimization

Citation: Guamán-Molina, J.; Pesantez, P.; Chavez-Fuentez, C.; Ríos, A. Industrial Application of Photovoltaic Systems with Storage for Peak Shaving: Ecuador Case Study. *Eng. Proc.* **2023**, *47*, 15. https://doi.org/10.3390/engproc2023047015

Academic Editor: Fernando Carrera Suarez

Published: 4 December 2023

1. Introduction

In recent years, seeking a balance between demand and generation is ideal for properly functioning the power electric system (PES). The gradual increase in electricity demand worldwide makes the installation of alternative energy generation systems indispensable [1]. Technological development has allowed a large penetration of renewable generation systems to displace fossil resources [2,3]. Following the important objectives of the Paris Agreement, a massive reduction in CO_2 emissions should be achieved by reducing energy consumption by implementing frugality measures, energy efficiency, rapid diffusion, and the optimal use of renewable energies [4]. On the other hand, in the last decade, traditional electrical grids have undergone a migration to a more modern network in which a complex communications network is included that allows the monitoring, control, and management of electrical systems, which is known as a Smart Grid [4].

The increasing integration of renewable energy resources (RES) in distribution systems with high and inherent variable production, with the inclusion of new assets, such as electric vehicle charging stations, and the incorporation of new industries worldwide, is leading to increasing challenges in grid management. The loads presented today in the country's different sectors, in addition to having a volatile demand profile, can also require large amounts of power from the grid in various and short periods of the day, to which a traditional power grid is not prepared to respond [5].

Electric power is one of the main inputs for economic and social development in the different regions of the world as it drives the growth of industrial production. To ensure the best conditions for productivity and competitiveness, it is necessary to propose guidelines allowing scenarios of greater energy efficiency in which the participation of renewable energies is more actively involved [6]. Industry is considered one of the sectors with the highest energy consumption. It has been one of the main axes for the application of energy efficiency improvements, including integrating renewable energies.

The different consumption patterns of industries, and the forms of billing considered in the different countries of the region and the world, are one of the main problems for the increase in production in industries [7,8]. The competitiveness of industrial companies depends, among other important factors, on low energy costs. Industrial customers use energy-intensive equipment at short time intervals during the day, so the additional costs of keeping up with peak demand are passed onto customers as demand charges. The problem of peak demand reduction has also been studied at the local level; thus, in 2017, in reference [9], a study of the different methods of peak demand reduction and their feasibility in Ecuador was presented. A study of the different peak shaving strategies developed at the international and regional levels is presented in ref. [9]. A regulatory analysis for implementing peak shaving strategies in Ecuador is also presented. The implementation of demand response as a peak shaving strategy is evaluated on the demand curve of the National Interconnected System and a typical load profile of the industrial sector. The displacement of demand from the peak period to the base period allows a 20% load reduction, which can be achieved using load controllers.

2. Proposal Methodology

This paper proposes optimizing a battery storage system as a peak shaving strategy for a commercial medium voltage, MV, customer. Figure 1 presents the network connection schematic of the study network. As seen in Figure 1, the schematic presents a DC system coupling connection, an industrial load, and a grid connection point [10].

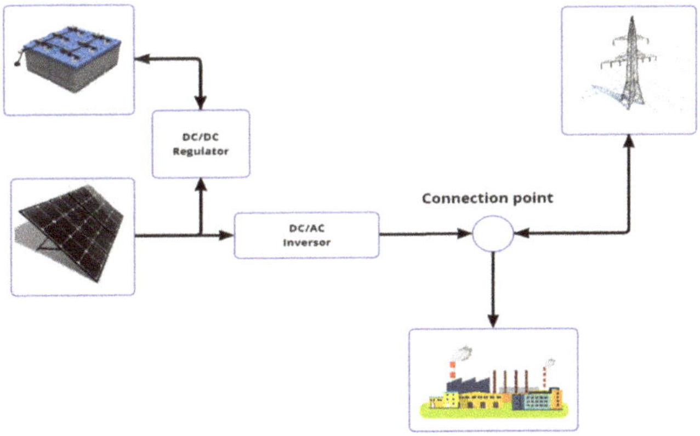

Figure 1. Network of study.

The evaluation methodology proposed for optimizing the photovoltaic generation system with battery storage as a peak shaving self-consumption system for large consumers is presented in the following paragraphs.

The methodology starts with the analysis of the demand profile of an industrial consumer in order to determine the optimal power of the photovoltaic generation system with battery storage. Once the average daily demand profile of the industry is analyzed and defined, we proceed to the sizing of the PV system with storage. The sizing of the system is carried out by checking the coupling criteria of the PV generator–inverter.

To make the system economically viable, an optimization of the photovoltaic system with storage is proposed by determining a limit power of the network, which allows for reducing the size of the installation so that the system covers the peaks of the demand in case of exceeding the limit power.

A diagram of the methodology proposed in this work is shown in Figure 2.

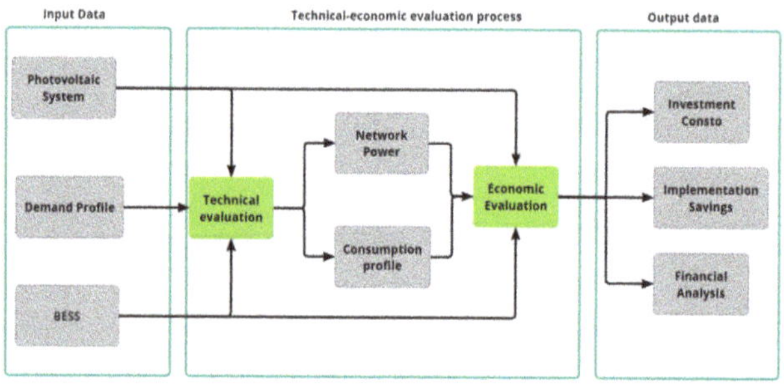

Figure 2. Proposed methodology for peak shaving system analysis.

2.1. Step 1: Determine the Demand Profile

To determine the average demand profile, a classification of working days, weekends, and holidays was made to have a total reading of the industry's behavior. This allows us to have the average demand profile as a starting point and identify the maximum demand for the sizing of the photovoltaic system with BESS.

Once the classes were determined, the maximum demand curve for the most critical day was identified, allowing the photovoltaic installation's initial sizing (Figure 3). The energy consumed in the industry on the most critical day is 2.7 MWh/day.

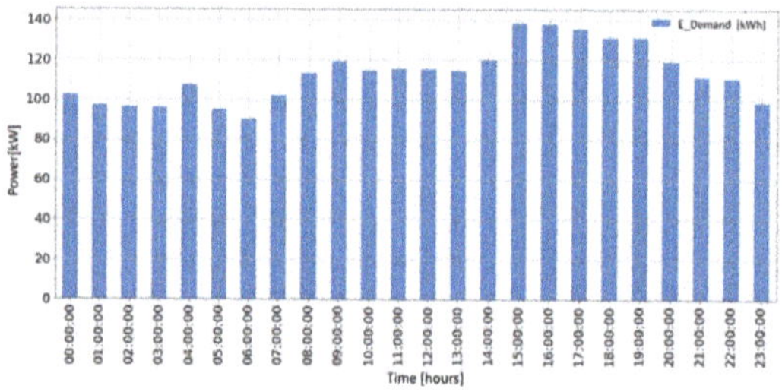

Figure 3. Maximum daily demand profile of the industry with an hourly frequency.

2.2. Step 2: Sizing the Solar Photovoltaic System

Once the maximum daily demand curve has been obtained, the optimum power of the installation is calculated using the following Equation (1) [11,12]:

$$P_{Otp} = \frac{\sum_{i=mes\ 1}^{month\ 12} Emonth_i}{Plant_{\ Factor} \times 8760} \quad (1)$$

where the Emonthly is calculated with the maximum daily demand to supply the most critical day of demand.

With the P_{Otp} data, we proceeded to simulate the photovoltaic installation. The solar system sizing was carried out using the coupling criteria of the generated PV-inverter [13,14]. To calculate the number of strings in parallel, as well as the number of panels per string for this case study, the technical characteristics of the 340 Wp Jinko polycrystalline photovoltaic panel and the 500 kW Hyundai HPC 500HL inverter are used.

3. Case Studies

Once the simulation has been performed in the PVsyst software (Version 7.2), the energy production data for all the days of the simulation year are imported with an hourly interval for which the most critical day of the year is analyzed, determined as June 21, in which the radiation and temperature data presented in Figure 4a,b, respectively, have been obtained.

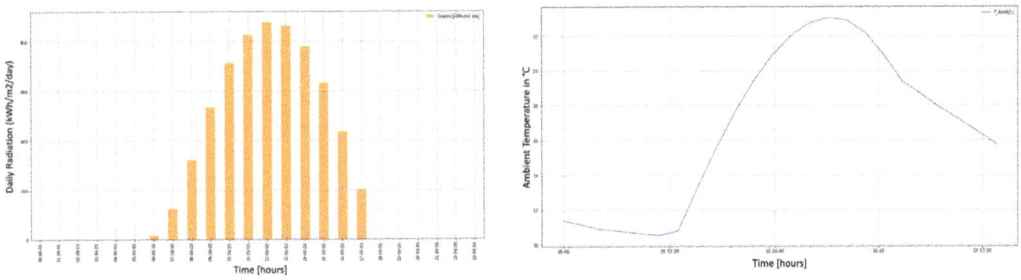

(**a**) Radiation (**b**) Temperature

Figure 4. Dailly radiation (**a**) and ambient temperature (**b**) data for June 21.

Based on the results obtained, two-scenario system analysis is proposed:

- Scenario 1 (Self-consumption): Peak shaving system with renewable self-generation and battery storage, limiting the maximum consumption of the grid and the injection of energy into the grid.
- Scenario 2 (Optimized): Peak shaving system with renewable self-generation and battery storage, limiting grid consumption to a power limit so that the photovoltaic system and storage can supply the energy needed during peak hours if the grid power limit is exceeded.

3.1. Scenario 1

For Scenario 1, self-consumption, the energy performance of the installation was evaluated based on the data presented in Table 1.

Table 1. Simulation data for Scenario 1.

Photovoltaic Modules	Jinkosolar-JKM 340PP-72
Number of Modules	1656
Modules in series	18
Modules in parallel	92
Total area of modules	3213 m^2
Inverter	Hyundai HPC-500HL-EU
Number of Inverters	1
Nominal Power of the Installation	563 kWp
Rated AC power	500 kWCA
Battery	LG Chem Rack R800
Battery Type	Li NMC
Battery in series	1
Parallel Cells	29
Depth of discharge	95%
Stored Energy	1287 kWh
Overall System Capacity	1867 Ah
Annual Industry Requirement	818 MWh/year
Industry Average Load	93.3 kW
Maximum Industry Load	143.1 kW

A photovoltaic system with battery energy storage is simulated using the self-consumption strategy. This strategy maximizes self-consumption by prioritizing the user's needs. The photovoltaic energy is used first to feed the load and then to charge the battery, which will later be used to cover the demand outside the hours of radiation. The objective of this scenario is to minimize the energy consumed by the grid; it is worth mentioning that, based on the proposed scheme, the storage system is not charged from the grid. The demand profile and the energy generated by the system are presented in Figure 5.

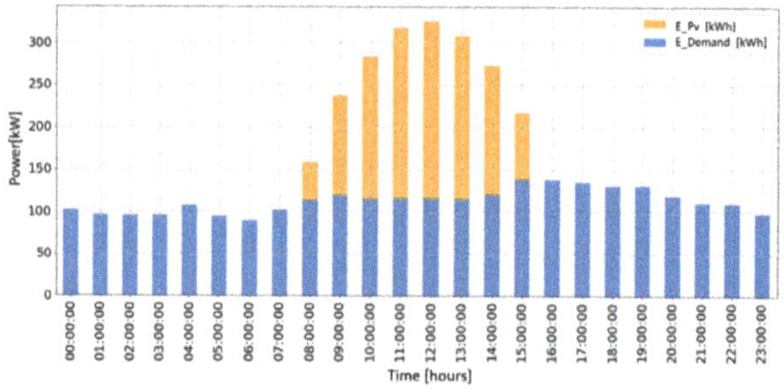

Figure 5. Energy generated by the photovoltaic system and demand, Scenario 1.

Based on the data obtained from the simulation, as well as the industry demand measurement data, the following equation can obtain the battery capacity (2) [15,16]:

$$C_{Batera} = \frac{\frac{E_{Surplus}}{Inverter\ Efficiency}}{System\ Voltage} \quad (2)$$

The batteries are sized with the help of Equation (2) and the data obtained from the equipment specification sheets. Based on the maximum and minimum DC input voltage of the inverter, which are between 450 and 850, respectively, a commercial battery voltage of 725 VDC is established.

Figure 6 shows the temporal evolution of the power balance of the system for an average week in July, the most critical month of the year. As can be seen, the most critical day of the week is when there is a participation of energy from the grid, while the other days are covered entirely by the photovoltaic system and the batteries.

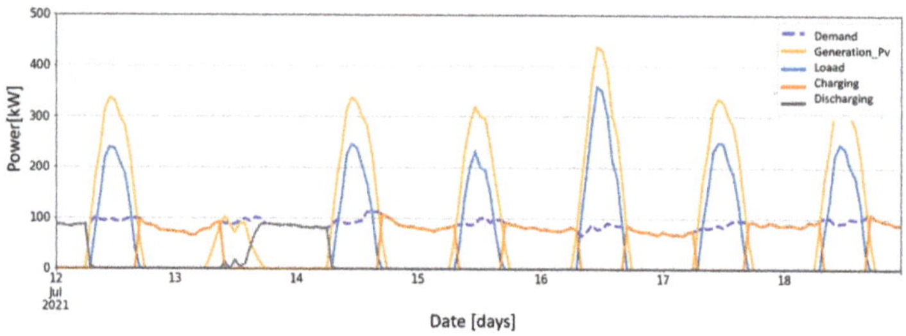

Figure 6. Temporal evolution of power balance in a week, Scenario 1.

Once the annual evaluation of the installation has been carried out, it can be observed that the system stores approximately 44% of the surplus energy, so 20.7% of the annual energy demand will have to be covered by the grid. This consumption of 142 MWh per year represents a saving of 82% of the total energy consumed annually by the industry under study.

3.2. Scenario 2

Once Scenario 1 of the solar photovoltaic system with storage has been obtained, the system optimization is carried out as a peak shaving strategy in the system under study. It is important to determine a power limit that allows the calculation of the photovoltaic system's appropriate power and storage capacity.

The PV system's optimal power should supply energy above the determined limiting power (Figure 7). This will allow the PV system to be sized in such a way that it can store only the energy needed to reduce the industry's peak demand that is within the peak hours above the determined power limit. The grid input power limit P_{lim} is defined based on billable demand, which indicates that it cannot be less than 60% of the maximum demand recorded in the previous 12 months, so the maximum demand is calculated with Equation (3) [17].

$$P_{lim} = MaximumDemand_{\text{Annual}} \times 60\% \text{ kW} \tag{3}$$

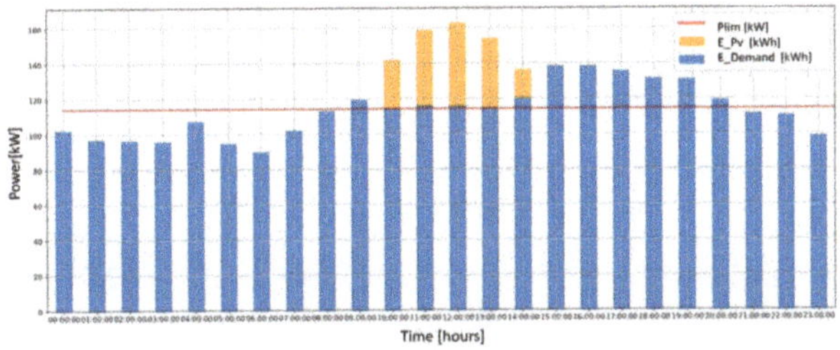

Figure 7. Energy generated by the PV system and load consumption, Scenario 2.

On the other hand, in Scenario 2 (optimized), as a peak shaving strategy, the system was optimally sized so that the system limits the power input from the grid to 60% of the maximum generated demand. Therefore, the PV system with storage overseas covers the peaks generated above the determined power limit. The simulation data of the optimized case are presented in Table 2.

Table 2. Simulation data for Scenario 2.

Photovoltaic Modules	Jinkosolar-JKM 340PP-72
Number of Modules	864
Modules in series	18
Modules in parallel	48
Total area of modules	1676 m^2
Inverter	Hyundai HPC-500HL-EU
Number of Inverters	1
Nominal Power of the Installation	297 kWp
Rated AC power	250 kWCA
Battery	LG Chem Rack R800
Battery Type	Li NMC
Battery in series	1
Parallel Cells	5
Depth of discharge	95%
Stored Energy	222 kWh
Overall System Capacity	322 Ah
Annual Industry Requirement	818 MWh/year
Industry Average Load	93.3 kW
Maximum Industry Load	143.1 kW

Based on the optimization of the solar photovoltaic system's capacity, the system's optimal capacity is 242 kW, so the evaluation of the most critical day was performed with an installation of 250 kWp based on the existing commercial equipment (Figure 7).

Figure 8 presents the time evolution of the power balance of the optimized system for an average week of July, the most critical month of the year [18]. In the figure, it is possible to appreciate the energy peak shaving from the grid solved by the storage system if the established power limit is exceeded.

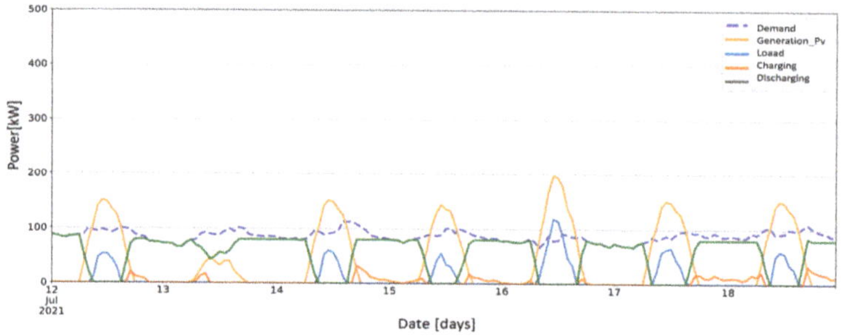

Figure 8. Temporal evolution of the power balance in one week, Scenario 2.

Once the system's annual performance has been evaluated, it can be observed that the system stores approximately 14% of the energy generated by the PV system. Therefore, the grid system is predominant.

4. Results

As can be seen in Figure 9a for Scenario 1 (self-consumption), the PV system with storage was sized to cover the entire demand of the industry. This means that the PV system will generate a total of 807 MWh/year to cover the annual demand of the industry, which is equal to 818 MWh/year. The designed system provides approximately 87% of the industry's total demand. The photovoltaic system can cover the demand during the hours when there is radiation and, in turn, generate surplus energy that is stored in the batteries to cover peak demand throughout the day [19].

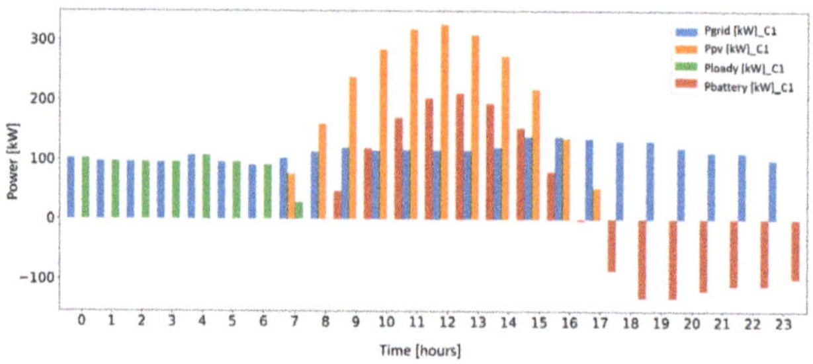

(**a**) Scenario 1

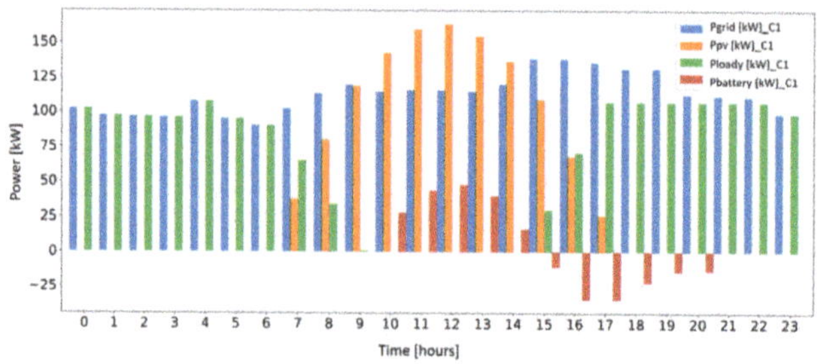

(**b**) Scenario 2

Figure 9. Temporal evolution of the power balance for the most critical day.

In case two, presented in Figure 9b, the PV system can cover the demand in the hours when there is radiation and, in turn, generate surplus energy that is stored in the batteries to cover peak demand throughout the day, when the determined power limit is exceeded.

The results obtained for the two cases evaluated are shown in Table 3. As can be seen in the table, even though the self-consumption system reduces energy consumption from the grid by 80% compared to the optimized system, which only reduces energy consumption by 43%, when evaluating the system according to the current ARCENNER 001/21 regulation, which specifies a net energy consumption, it is extremely important to evaluate the financial aspect of the installation [20].

Table 3. Energy results of the evaluated cases.

Scenarios Evaluated	Power Produced [MW/year]	Stored Energy [MW/year]	Direct Use Energy [MW/year]	Power Injected [MW/year]	Power Consumed from the Grid [MW/year]
Without Peak Shaving System	0	0		0	817
Case 1 (Self-consumption)	870	355	360	92	140
Case 2 (Optimized)	421.9	58.6	303	59	459

The most relevant results obtained in the financial analysis are presented in Table 4. The photovoltaic system with battery storage for self-consumption as a peak shaving strategy, in Scenario 1, allows for greater savings in the purchase of energy; however, it requires a higher investment and, by managing an energy netting system, it is not profitable, presenting a negative ROI. However, case two, with a lower power of the photovoltaic system as well as the storage system, presents a better solution with IER 22.7% in approximately 16 years.

Table 4. Most relevant results of the evaluated cases.

Scenarios Evaluated	Investment [USD]	Savings due to Reduced Purchase of Energy [USD/year]	Payback Period [years]	ROI [%]	LCOE [USD/kWh]
Without Peak Shaving System	0	0	0	0	0
Case 1 (Self-consumption)	1,009,328.00	55,746.06	23	−9.2	0.32
Case 2 (Optimized)	424,198.00	16,360.00	16	22.7%	0.09

5. Conclusions

One of the key points for the design of photovoltaic stations with storage is the quality of the demand profile data that can be obtained, as well as the characterization of these data to obtain a correct starting point. Likewise, the use of computer tools specialized in the design and sizing of photovoltaic installations allows the validation of mathematical models and provides support in terms of meteorological data information for the sizing of these. One of the most critical points in the development of these systems is the great variety of manufacturers of elements that exist today, so having a reference of validated elements in operation is a great solution. Even though the manufacturing costs of photovoltaic systems have been decreasing year by year, storage systems, such as lithium technology, still do not reach competitive prices in the countries of the region due to the subsidies that each country has. However, the integration of the two technologies can be economically attractive for the development of projects smaller than 2 MW for commercial and industrial consumers.

As has been evaluated in this work, a high storage capacity can double the cost of the project, due to the high prices of the elements that make up the installation. Proposing a photovoltaic system with storage as self-generation is not profitable in commercial and industrial sectors due to the energy costs in Ecuador. Despite the existence of a tariff scheme with hourly demand consumption, electricity costs are not comparable to the costs of energy produced. Therefore, the proposal of "peak shaving" with a photovoltaic system with optimal storage to reduce peak demand from limited power is a viable strategy.

The optimization of the system reduces the initial investment of the project by 50% and, since it is carried out within the regulatory framework of Ecuador, where there is a netting of energy, the proposal presents an economic feasibility from year 16 onwards. The results obtained indicate that one of the best solutions for the industry is the installation of 250 kWp and battery storage with a capacity of 222 kWh.

Author Contributions: Conceptualization, J.G.-M. and P.P.; methodology, J.G.-M. and P.P.; software, J.G.-M. and C.C.-F.; validation, J.G.-M. and P.P. and A.R.; investigation, J.G.-M.; resources, J.G.-M. and P.P.; writing—original draft preparation, J.G.-M. and C.C.-F.; writing—review and editing, J.G.-M., C.C.-F., P.P., A.R.; visualization, J.G.-M. and C.C.-F.; supervision, P.P. and A.R. All authors have read and agreed to the published version of the manuscript.

Funding: This research received no external funding.

Institutional Review Board Statement: Not applicable.

Informed Consent Statement: Not applicable.

Data Availability Statement: The authors confirm that the data supporting the findings of this study are available within the paper.

Conflicts of Interest: The authors declare that they have no conflict of interest regarding the publication of this paper.

References

1. Achara, J.P.; Le Boudec, J.-Y.; Paolone, M. *Technologies for Integration of Large-Scale Distributed Generation and Volatile Loads in Distribution Grids*; Springer: Berlin/Heidelberg, Germany, 2018.
2. Alexis, B.L.R.; Cedeño, E.A.L. La generación de energía eléctrica para el desarrollo industrial en el ecuador a partir del uso de las energías renovables. *Univ. Cienc. Tecnol.* **2020**, *24*, 36–46. [CrossRef]
3. Soto, G.J.A. *Control de Procesos Industriales Con Minimización Del Consumo Energético*; Escuela de Ingeniería Eléctrica y Mecánica: Medellín, Colombia, 2019.
4. Berr, L.H.; Zuluaga, C. Smart Grid y la energía solar fotovoltaica para la generación distribuida: Unarevisión en el contexto energético mundial. *Ing. Y Desarro.* **2014**, *32*, 369–396.
5. Bitar, S.M.; Chamas, F. *Estudio de Factibilidad Para la Implementación de Sistemas Fotovoltaicos Como Fuente de Energía en el Sector Industrial de Colombia*; CESA: Thebarton, Australia, 2017.
6. BoroumandJazi, G.; Rismanchi, B.; Saidur, R. A review on exergy analysis of industrial sector. *Renew. Sustain. Energy Rev.* **2013**, *27*, 198–203. [CrossRef]
7. Chau, T.K.; Yu, S.S.; Fernando, T.; Iu, H.H.-C. Demand-Side Regulation Provision From Industrial Loads Integrated with Solar PV Panels and Energy Storage System for Ancillary Services. *IEEE Trans. Ind. Inform.* **2017**, *14*, 5038–5049. [CrossRef]
8. Arrinda, J.; Rodriguez, M.; Leralta, J.; Lopez, A.; Barrena, J.A. Optimization of the consumption for industrial customers using battery energy storage systems. In Proceedings of the IEEE Eurocon 2015—International Conference on Computer as a Tool (EUROCON), Salamanca, Spain, 8–11 September 2015; pp. 1–6.
9. Cuasapaz, J.C.C. Estudio de los Métodos de Reducción de Demanda Eléctrica en Horas Pico 'Peak Shaving' y su Factibilidad en Ecuador. Bachelor's Thesis, Escuela Politecnica Nacional, Quito, Ecuador, 2017.
10. Ramasamy, V.; Feldman, D.; Desai, J.; Margolis, R. *US Solar Photovoltaic System and Energy Storage Cost Benchmarks: Q1 2021*; National Renewable Energy Lab. (NREL): Golden, CO, USA, 2021.
11. IRENA. *Solar Energy*; International Renewable Energy Agency: Masdar City, United Arab Emirates, 2021.

12. Troncoso, N.; Rojo-González, L.; Villalobos, M.; Vásquez, C.; Chávez, H. Economic decision-making tool for distributed solar photovoltaic panels and storage: The case of Chile. *Energy Procedia* **2019**, *159*, 388–393. [CrossRef]
13. Duffie, J.A.; Beckman, W.A.; Blair, N. *Solar Engineering of Thermal Processes, Photovoltaics and Wind*; John Wiley & Sons: Hoboken, NJ, USA, 2020.
14. Rout, K.C.; Kulkarni, P. Design and performance evaluation of proposed 2 kW solar PV rooftop on grid system in Odisha using PVsyst. In Proceedings of the 2020 IEEE International Students' Conference on Electrical, Electronics and Computer Science (SCEECS), Bhopal, India, 22–23 February 2020; pp. 1–6.
15. Mermoud, A.; Villoz, A.; Wittmer, B.; Apaydin, H.; PVsyst, S. Economic Optimization of PV Systems with Storage. In Proceedings of the 37th European Photovoltaic Solar Energy Conference, Online, 7–11 September 2020.
16. Oudalov, A.; Cherkaoui, R.; Beguin, A. Sizing and optimal operation of battery energy storage system for peak shaving application. In Proceedings of the 2007 IEEE Lausanne Power Tech, Lausanne, Switzerland, 1–5 July 2007; pp. 621–625.
17. Venu, C.; Riffonneau, Y.; Bacha, S.; Baghzouz, Y. Battery storage system sizing in distribution feeders with distributed photovoltaic systems. In Proceedings of the 2009 IEEE Bucharest PowerTech, Bucharest, Romania, 28 June–2 July 2009; pp. 1–5.
18. Reihani, E.; Sepasi, S.; Roose, L.R.; Matsuura, M. Energy management at the distribution grid using a Battery Energy Storage System (BESS). *Int. J. Electr. Power Energy Syst.* **2016**, *77*, 337–344. [CrossRef]
19. Tran, Q.T.; Davies, K.; Sepasi, S. Isolation Microgrid Design for Remote Areas with the Integration of Renewable Energy: A Case Study of Con Dao Island in Vietnam. *Clean Technol.* **2021**, *3*, 804–820. [CrossRef]
20. Guamán-Molina, J.I. Design of a "Peak Shaving" System Based on Renewable System and Battery Storage for Large Electricity Consumers in Ecuador, Application for an Industrial Client. Bachelor's Thesis, Escuela Politecnica Nacional, Quito, Ecuador, 2022.

engineering proceedings

MDPI

Proceeding Paper

Evaluating Scalability, Resiliency, and Load Balancing in Software-Defined Networking †

Pablo Barbecho Bautista [1,*], Jaume Comellas [2] and Luis Urquiza-Aguiar [3]

1 Departamento de Ingeniería Eléctrica, Electrónica y Telecomunicaciones, Universidad de Cuenca, Ave. 12 de Abril y Agustin Cueva, Cuenca 010101, Ecuador
2 Optical Communications Group (GCO), Universitat Politècnica de Catalunya (UPC), Jordi Girona 1-3, 08034 Barcelona, Spain; jaume.comellas@upc.edu
3 Departamento de Electrónica, Telecomunicaciones y Redes de Información, Escuela Politécnica Nacional (EPN), Ladrón de Guevara E11-253, Quito 170525, Ecuador; luis.urquiza@epn.edu.ec
* Correspondence: pablo.barbecho@ucuenca.edu.ec
† Presented at the XXXI Conference on Electrical and Electronic Engineering, Quito, Ecuador, 29 November–1 December 2023.

Abstract: With emerging technologies like cloud computing and big data, managing traditional networks has become more demanding. Software-defined networking (SDN) promises faster implementation, flexibility, and simplified network management. However, due to SDN's centralized nature, it encounters limitations. SDN controllers should have enough processing power to deal with a high amount of flow. In addition, a single point of failure may affect the network's resiliency. For these issues, multi-instance implementation enables distributed control. However, this solution implies an intrinsic controller-to-controller synchronization channel. In this article, we propose different failure scenarios in both the data and control planes to provide network administrators with a clear view of the constraints of network reliability, load balancing, and scalability in SDN environments. The simulation results show that, regarding resiliency, SDN networks require half the time compared to traditional networks in order to recover from a link failure. Regarding load-balancing capabilities, load balancing is not guaranteed with the reactive forwarding approach (on-demand flow installation). Lastly, the SDN multi-instance solution impacts the network performance by between 1% and 21% compared to the single-instance case.

Keywords: software-defined networking; network evaluation; ONOS; ODL

Citation: Bautista, P.B.; Comellas, J.; Urquiza-Aguiar, L. Evaluating Scalability, Resiliency, and Load Balancing in Software-Defined Networking. *Eng. Proc.* **2023**, 47, 16. https://doi.org/10.3390/engproc2023047016

Academic Editor: Jackeline Abad

Published: 4 December 2023

1. Introduction

Software-defined networking (SDN) has emerged as an architecture that centralizes a network's intelligence, breaking the control plane out of the switch and delegating this control to a central control element. In contrast to legacy networks' distributed nature, the SDN architecture centralizes network management in one device: the SDN Controller.

OpenDayLight (ODL) and the Open Network Operating System (ONOS) are the leading open-source controllers. This work starts by evaluating ODL and ONOS controllers to provide a reference point for the following testing scenarios. This work assesses how the SDN controller deals with different failure scenarios in their data and control planes.

First, we evaluate the data plane network resiliency regarding path computation and link failover recovery time. For this, we use the Intents Subsystem implemented in the ONOS. Using intent on computer networking is a new concept used in SDN. The objective is to translate high-level policies into the network configuration [1]. Then, load-balancing capabilities are evaluated regarding the network's reaction time. Here, directives are configured in the ONOS controller, considering proactive and reactive forwarding modes in charge of distributing flows between alternative paths. The Dijkstra algorithm is used to perform the routing process.

Later, at the control plane, it is critical for the controller to present enough processing power, ensuring continuity of service despite failures. Further, a single-server instance results in a single point of failure. One of the mechanisms used to manage SDN scalability issues is the multi-instance controller. The latter implies an intrinsic controller-to-controller synchronization channel. In this regard, this work evaluates the impact of this solution by increasing the number of switches and controllers (multi-instance scenario) and assessing the network's throughput (flows/s).

The main contributions of this work are as follows:

- Producing comprehensive failure scenarios in control and data planes. The proposed scenarios are intended to be easily reproduced to assess future approaches in SDN.
- Providing network administrators with a clear view of SDN reliability, load-balancing, scalability issues, and strategies to handle burdens.
- Providing ODL and ONOS open-source controller evaluation.

The remainder of this paper is structured as follows. Section 2 summarizes the related works. Section 3 introduces evaluated network features. Then, the system model is described in Section 4, including the performance evaluation of ONOS and ODL controllers. Then, the simulated scenarios and results are discussed in Section 5. Finally, conclusions are presented in Section 6.

2. Literature Review

In [2], the authors explore machine learning techniques applied in SDN. The authors optimize delay and connectivity for SD-WAN environments using a multi-agent deep reinforcement learning algorithm. Load balancing at the data plane is not considered. In this sense, SDN controllers, with a global view of network resources, may produce optimal solutions compared to traditional networks, where the load balancing is uniquely based on local information. In this regard, in this work, we present a straightforward load-balancing scenario based on the Intents Subsystem of the ONOS. Further, in [3], the authors include a load balancing evaluation of traffic among Customer Premise Edges (CPEs) in a distributed control scenario for SD-WANs.

In [4], the authors evaluate the robustness of ODL and ONOS SDN controllers, including distributed control plane failure scenarios. Nevertheless, the performance degradation due to controllers' east–west communication is not considered. Therefore, we include an evaluation of a multi-instance system, analyzing the impact of the intrinsic controller synchronization channel in distributed control solutions with an incremental number of switches at the data plane.

Finally, Table 1 summarizes the differences between other works compared to the present work.

Table 1. Overview of related works.

Author(s)	Main Objective	Scalability	Resiliency	Load Balancing
[2]	Optimization of delay and connectivity for SD-WAN scenarios.	Evaluated in terms of the number of switches.	Does not address directly.	Does not address.
[3]	Optimization of load balancing in distributed control scenarios based on deep reinforcement learning.	Evaluated considering a multi-instance scenario and CPEs.	Does not address.	Evaluated in terms of migrated flows between controllers.
[4]	Evaluate ODL and ONOS controllers robustness.	Evaluated in terms of recovery time.	Evaluated in terms of forwarded packets.	Does not address.
The present work	Evaluate SDN scalability, resiliency and load balancing.	Evaluated considering a multi-instance scenario and incremental switches.	Evaluated in terms of packet losses and recovery time.	Evaluated considering proactive and reactive forwarding in terms of the round trip time.

3. SDN Features

3.1. Network Resiliency

OpenFlow-based controllers have a centralized view of the whole network and orchestrate all the elements. Therefore, solutions with optimal path computation should be included in new network operation systems, providing minimal convergence times, network resiliency, and efficient use of the available resources. This work evaluates an online path computation solution for a resilient SDN network based on the ONOS.

3.2. Load Balancing

The multiple spanning tree protocol (MSTP) is commonly used in traditional networks to implement a trunk load-balancing solution in a loop-free topology with alternative paths. Later in this work, we implement an SDN deterministic load balancing scenario based on OpenFlow and the proactive forwarding mechanism for the ONOS controller.

3.3. Scalability

In SDN, abstracting the lower-level functionalities allows the control plane to manage the overall network through the controller. However, the paradigm of centralized control results in scalability issues. The latter means that the SDN controllers must have enough processing power to deal with the high amount of flow and be aware of the number of devices a single controller can handle.

4. System Model Description

In Figure 1, the system model is presented. First, an Ubuntu Server, version 20.04 LTS, 64 bits, 2 GB of RAM, and two CPUs is used for each controller (ONOS, ODL). Furthermore, two machines with Ubuntu 18.04, 64 bits, 1 GB RAM, and two CPUs are used for the Carleton Benchmark Suite (Cbench) and Mininet. The former refers to a benchmarking framework used for evaluating OpenFlow Controllers. The latter allows the instantiation of Open vSwitches and virtual hosts.

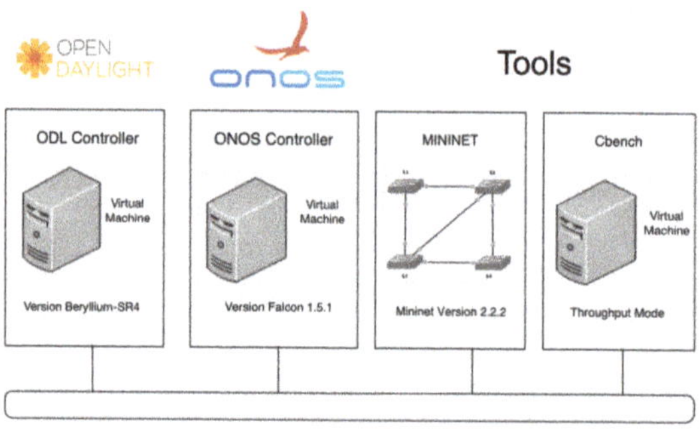

Figure 1. System model configuration includes (i) two OpenFlow SDN controllers, (ii) one Mininet instance used to implement the network topology (Open vSwitches and hosts), and (iii) Cbench used to evaluate the SDN controllers.

The pre-eminent open-source controller deployments are ODL and the ONOS. It is crucial to compare ODL with the emerging open-source ONOS controller to provide a point of reference between controllers and select one controller for the failure scenarios.

Figure 2a presents the process to evaluate SDN controllers (ONOS and ODL). First, the topology is installed using the Cbench tool, including an Open Vswitch and 1000 virtual hosts. Then, the switch creates as many packets as the switch's buffer allows, without waiting for a reply; while the buffer is not full, each packet is queued in the controller and generated outgoing packets (*Flow_Mod*) are counted. We want to measure how many packets can be handled/processed by each controller so that the controller's throughput (flows/s) is calculated.

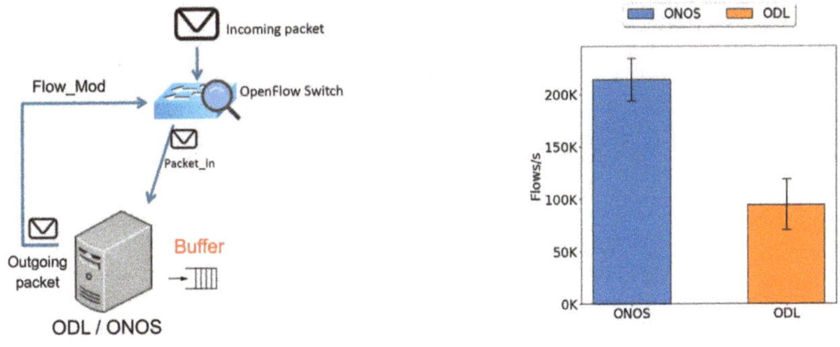

(a) (b)

Figure 2. (**a**) Cbench tool process used to evaluate the performance of SDN controllers. The topology includes an Open Vswitch and 1000 virtual hosts. (**b**) SDN controller's throughput (flows/s). Results are presented with a 95% confidence interval (CI) obtained from ten simulation rounds with independent seeds.

Simulation results are presented in Figure 2b. Here, the ONOS controller outperforms ODL in 119,402 flows/s, which is a considerable difference. ODL shows roughly half of the capacity when compared to the ONOS. To increase confidence in the results, we conducted ten test rounds for both ONOS and ODL controllers, discarding the first and last values by the warm-up and cool-down effect. The measurement was performed under steady-state conditions without network changes during the test.

ONOS software results in a more reliable network operating system with better performance when compared to ODL. Therefore, it is suitable for evaluating the following experiments. The following section describes configurations of the different test scenarios using the ONOS as the central SDN controller.

5. Results

5.1. SDN Resiliency Scenario

Figure 3a shows the topology to evaluate network resiliency. Note that several routes are enabled to communicate H1 and H2. The scenario uses the reactive forwarding (RF) mechanism. The RF installs flow entries on-demand after a sender starts to transmit the initial packets.

The testing scenario simulates random link failure to evaluate the SDN network resiliency. In addition, to compare results, the same network topology shown in Figure 3a is implemented using legacy switches. The Spanning Tree Protocol (STP) is also configured as a typical implementation for reliable networks at L2, with alternative loop-free paths.

Figure 3b shows the results of SDN and legacy network simulations regarding the network's recovery time and packet losses. We notice that an SDN network with an OpenFlow RF mechanism responds faster in the case of a networking event like a link failure. It requires 2 s for recovery, with only one packet lost. Conversely, in traditional networks, STP requires 34 s for link recovery with 33 packet losses. The latter is due to the transition from blocking to forwarding mode to prevent temporal loop conditions. Furthermore, if Rapid-STP is used, it requires 4 s to recover from the link failure, which doubles the time necessary compared to SDN.

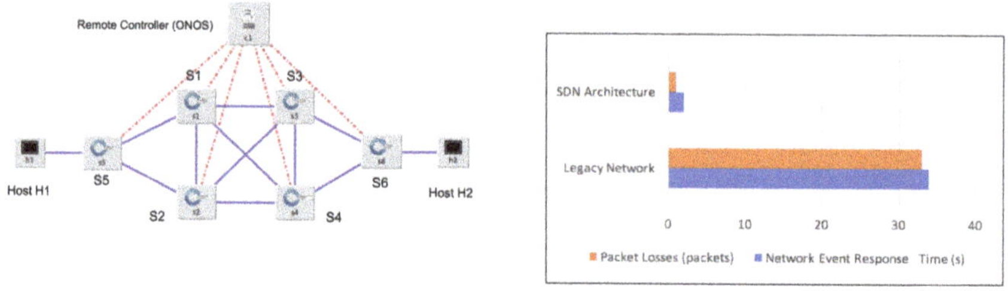

(**a**) (**b**)

Figure 3. (**a**) Network configuration for resiliency evaluation. The topology comprises six OpenFlow vSwitches (based on Mininet), a single remote controller (ONOS Controller), and two virtual hosts (H1, H2). (**b**) Network resiliency measured in terms of recovery time and packet losses for legacy and SDN networks.

5.2. SDN Load-Balancing Scenario

The network topology to evaluate the load balancing scenario is presented in Figure 4a. The idea consists of distributing traffic flows between H2 and H1 and H3 and H1 according to directives configured in the ONOS controller. Two scenarios are evaluated: (i) proactive forwarding (PF), which guarantees load balancing, and (ii) reactive forwarding (RF), where load balancing is not guaranteed.

First, the PF approach provides a deterministic load-balancing scenario. It sets end-to-end connectivity between H3 and H1, which goes through PATH 1, S6–S4–S2–S5, in Figure 4a. Traffic H2–H1 goes through PATH 2, S3–S1–S5, in Figure 4a. Second, with the RF approach, flow entries for H2–H1 and H3–H1 are automatically installed after the first packet in the network (on-demand flow installation). The optimal path computation is based on the shortest path algorithm (Dijkstra), where the number of hops is the weight metric. Note that resulting paths may use the same links, which means a congested route and non-efficient use of resources (i.e., load balancing is not guaranteed).

In Figure 4b, we use the round trip time (RTT) to assess the network's reaction time. The RF requires, on average, 18s more to establish communication between H2 and H1 and 42 s for H3 and H1. Although the PF approach outperforms RF in both cases, by considering scalability, a non-scalable solution occurs due to the limited memory at the switch tables where flows should be installed permanently.

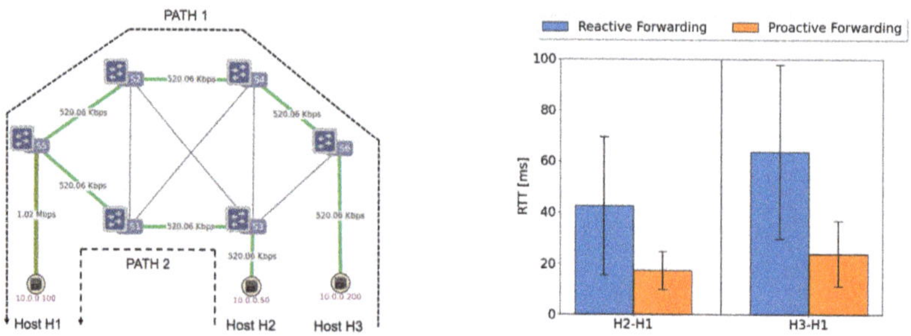

(**a**) (**b**)

Figure 4. (**a**) The load-balancing scenario includes three virtual hosts (H1, H2, H3), six OpenFlow vSwitches (based on Mininet), and a remote SDN controller (ONOS) implementing proactive and reactive forwarding. (**b**) RTT results for proactive forwarding with load balancing and reactive forwarding without guaranteed load balancing.

5.3. SDN Scalability Scenario

One of the mechanisms used to manage SDN scalability issues comprises implementing multi-server instances (i.e., distributed control). This implies an intrinsic controller of the controller channel, in charge of controller synchronization. Note that a distributed control design also increases the system's resilience at the control plane level.

Figure 5b shows the scenario to evaluate the scalability of the multi-instance approach in SDN. Here, service management is distributed between controllers in the ONOS Cluster; see the blue and brown boxes in Figure 5b. As a point of reference, the performance of a single instance (one single ONOS controller) is evaluated and compared to the multi-instance (two ONOS controllers) implementation.

Figure 5b shows the results for single and multi-instance scenarios to evaluate this solution for SDN scalability issues. We can see the multi-instance case has a significant impact on network performance. When considering a different number of switches [1–64], the SDN performance degradation begins to be noticeable when eight switches are installed in the network. Degradation in the performance ranges from 1% to 21% compared to the single-instance case. This is due to the synchronization process between controllers, resulting in a degradation in the whole system performance.

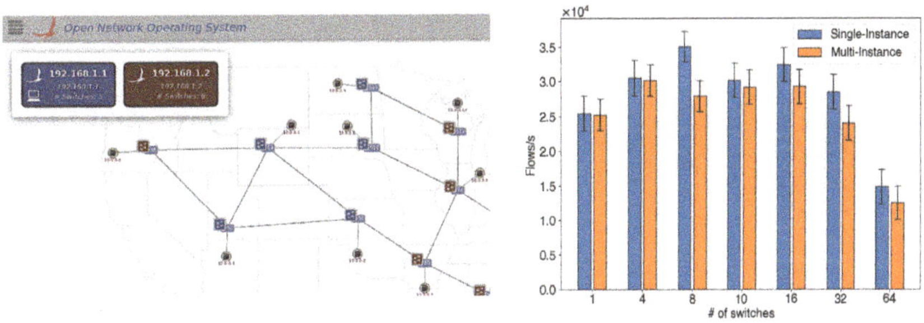

(**a**) (**b**)

Figure 5. (**a**) Distributed control design to evaluate scalability. Topology is composed of 1 to 64 switches (based on Mininet) and two controllers (ONOS) configured in a cluster (multi-instance). (**b**) Performance of single and multiple instance implementation.

6. Conclusions

This work evaluates different failure scenarios in both data and control planes. The simulation results show that the ONOS controller handles double the throughput compared to the ODL controller. Regarding the network's resiliency, at the data plane, SDN requires half the time compared to traditional networks. The SDN load-balancing feature can be implemented by setting deterministic directives in the controller. However, this leads to scalability issues at the data plane due to the limited memory in switches. Lastly, the SDN multi-instance solution impacts the network performance by between 1% and 21% compared to the single-instance case. In this regard, the multi-instance scenario improves resiliency and scalability, although the controllers' synchronization channel limits the maximum number of cluster instances.

Author Contributions: Conceptualization, P.B.B. and J.C.; methodology, P.B.B. and L.U.-A.; software, P.B.B.; validation, P.B.B. and L.U.-A.; formal analysis, P.B.B. and L.U.-A.; investigation, P.B.B. and J.C.; data curation, P.B.B.; writing—original draft preparation, P.B.B. and J.C.; writing—review and editing, P.B.B., J.C. and L.U.-A. All authors have read and agreed to the published version of the manuscript.

Funding: This research received no external funding.

Institutional Review Board Statement: Not applicable.

Informed Consent Statement: Not applicable.

Data Availability Statement: Data are contained within the article.

Conflicts of Interest: The authors declare no conflict of interest.

References

1. Park, K.; Sung, S.; Kim, H.; Il Jung, J. Technology trends and challenges in SDN and service assurance for end-to-end network slicing. *Comput. Netw.* **2023**, *234*, 109908. [CrossRef]
2. Ouamri, M.A.; Azni, M.; Singh, D.; Almughalles, W.; Muthanna, M.S.A. Request delay and survivability optimization for software defined-wide area networking (SD-WAN) using multi-agent deep reinforcement learning. *Trans. Emerg. Telecommun. Technol.* **2023**, *34*, e4776. [CrossRef]
3. Ouamri, M.A.; Barb, G.; Singh, D.; Alexa, F. Load Balancing Optimization in Software-Defined Wide Area Networking (SD-WAN) using Deep Reinforcement Learning. In Proceedings of the 2022 International Symposium on Electronics and Telecommunications (ISETC), Timisoara, Romania, 10–11 November 2022; pp. 1–6. [CrossRef]
4. Ruchel, L.V.; Turchetti, R.C.; de Camargo, E.T. Evaluation of the robustness of SDN controllers ONOS and ODL. *Comput. Netw.* **2022**, *219*, 109403. [CrossRef]

engineering proceedings

Proceeding Paper

Utility of Field Weakening and Field-Oriented Control in Permanent-Magnet Synchronous Motors: A Case Study †

Jorge Medina [1,*], Christian Gómez [1], Marcelo Pozo [1], William Chamorro [1] and Victor Tibanlombo [2]

1 Departamento de Automatización y Control Industrial, Facultad de Ingeniería Eléctrica y Electrónica, Escuela Politécnica Nacional, Quito 170525, Ecuador; christian.gomez@epn.edu.ec (C.G.); marcelo.pozo@epn.edu.ec (M.P.); william.chamorro@epn.edu.ec (W.C.)
2 Departamento de Energía Eléctrica, Facultad de Ingeniería Eléctrica y Electrónica, Escuela Politécnica Nacional, Quito 170525, Ecuador; victor.tibanlombo@epn.edu.ec
* Correspondence: jorge.medinaj@epn.edu.ec
† Presented at the XXXI Conference on Electrical and Electronic Engineering, Quito, Ecuador, 29 November–1 December 2023.

Abstract: The significance of electric vehicles is progressively escalating, underscoring the criticality of technologies underpinning the functionality of their propulsion systems. This particular case study delves into the simulation of field-oriented control, coupled with field weakening, aimed at regulating a salient-pole permanent-magnet synchronous motor (PMSM). This approach involves the utilization of two cascaded loops for current and voltage, each employing PI controllers. The fine-tuning of these controllers' parameters hinges on the motor characteristics, as well as the desired response bandwidth.

Keywords: PMSM; field weakening; BLAC; FOC; SVPWM

Citation: Medina, J.; Gómez, C.; Pozo, M.; Chamorro, W.; Tibanlombo, V. Utility of Field Weakening and Field-Oriented Control in Permanent-Magnet Synchronous Motors: A Case Study. *Eng. Proc.* **2023**, 47, 17. https://doi.org/10.3390/engproc2023047017

Academic Editor: Jackeline Abad

Published: 5 December 2023

1. Introduction

Permanent-magnet synchronous motors (PMSMs) have gained prominence in electric mobility due to their impressive power density and capacity to surpass nominal speeds, necessitating the implementation of flux-weakening techniques, [1,2]. Research highlighted in [3] demonstrates that integrating a PMSM into an electric vehicle elevates its autonomy by 15% compared to an induction motor with equivalent power ratings. Articles such as [4–7] indicate that with the right control strategy, PMSM can provide a high torque-to-current ratio, a high power-to-weight ratio, and high efficiency and robustness.

This work introduces a comprehensive case study focused on a nested control architecture based on a field-oriented controller (FOC) tailored for a salient-pole permanent-magnet synchronous motor with a brushless alternating current (PMSM-BLAC), and a speed control loop. This case study provides details of a simulation that it is open to the community, using real specifications from a commercial motor. To evaluate the FOC controller design, a detailed model of the electrical and mechanical motor components expressed in a dq0 reference frame is proposed; in addition, the controller is tuned using a dedicated approach that is used in real synchronous-based motor systems of BLAC type so that we can produce as realistic a simulation as possible for further real implementation. The controller along the power electronics and the motor are mounted on a MATLAB/SIMULINK, where a space-vector pulse width modulation signal (SVPWM) drives the inverter, and field weakening is emulated to achieve speeds higher than the nominal one. The controller plus the motor system are evaluated in different scenarios, like locked rotor tests, speed and current control with different variable velocity profiles, experiments with different load–velocity profiles, and system evaluations with field weakening in constant load and torque conditions. The following sections present details about the controller and motor modelling methodology, experiments, results, and conclusions.

2. Materials and Methods

The proposed simulation model includes two control loops. The inner loop assumes the responsibility of managing the oriented field, while the outer loop governs the motor velocity. This controller intricately interfaces with a three-phase PMSM-BLAC motor, with its operations orchestrated by an SVPWM signal as shown in Figure 1. The comprehensive simulated setup encapsulates the three-phase inverter's provision of voltage, thereby engendering outcomes that mirror real-world dynamics with remarkable fidelity. To further enhance the realism, dynamic modulation of the Setpoint is employed, facilitating a diverse array of load profiles.

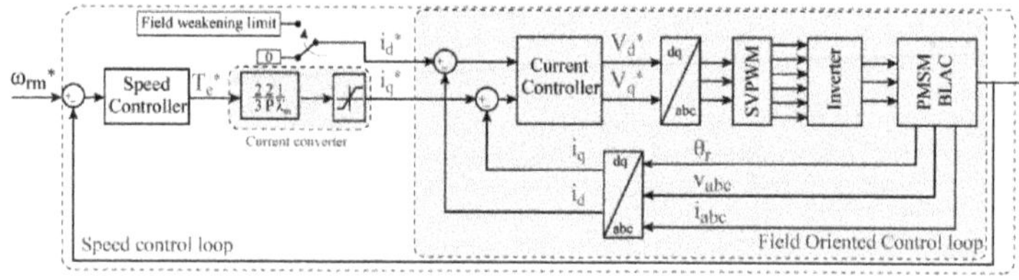

Figure 1. Overall control scheme implemented for the PMSM-BLAC motor (In control context "*" means desired value).

The motor parameters used in the simulated model were obtained from a commercial one, and are listed in Table 1. The following sections will present specific details of the motor and controller modelling.

Table 1. PMSM-BLAC motor specifications (N/A = Not Applicable).

Characteristic	Value	Unit	Characteristic	Value	Unit
Topology	Surface-mounted permanent-magnet synchronous machine	N/A	Voltage	120	[V]
Pole pairs	7	N/A	Synchronous inductance	0.344	[mH]
Connection type	Star	N/A	Magnetic flux constant	39.6	[mWb]
Resistance/phase	22.2	[mΩ]	Nominal velocity	1590	[rpm]
Inertia	0.008	[Kgm^2]	Max. Current	121	[A]

2.1. PMSM-BLAC Motor Modelling

The voltage and current expressions on the $dq0$ axes are derived from the simplified circuits shown in Figure 2a,b, where the voltage, magnetic flux, and current components of the direct axis and quadrature axis are V_x, ψ_x, and i_x, with $x \in \{d,q\}$, *respectively*. Ohm's law, Kirchhoff's law, and Maxwell equations yield the following:

$$\begin{bmatrix} V_d \\ V_q \end{bmatrix} = R\begin{bmatrix} i_d \\ i_q \end{bmatrix} + \frac{d}{dt}\begin{bmatrix} \psi_d \\ \psi_q \end{bmatrix} + w_e\begin{bmatrix} -\psi_q \\ \psi_d \end{bmatrix} \quad (1)$$

$$\begin{bmatrix} \psi_d \\ \psi_q \end{bmatrix} = \begin{bmatrix} L_d & 0 \\ 0 & L_q \end{bmatrix}\begin{bmatrix} i_d \\ i_q \end{bmatrix} + \begin{bmatrix} \psi_m \\ 0 \end{bmatrix} \quad (2)$$

$$T_e = \frac{3p}{2}(\psi_d i_q - \psi_q i_d) \quad (3)$$

$$\omega_m = \int \left(\frac{T_e - T_m - Bw_m}{J}\right)dt, \quad (4)$$

where p is the pole pair number, and R is the resistance of the stator windings. Recall that the inductances in the $dq0$ frame, L_q and L_d, are the same and equal to L for the PMSM-BLAC motor [8–15]. The remaining terms belongs to the magnetic torque T_e, mechanical and electrical velocity ω_m, ω_e, respectively, load torque T_m, coefficient of viscous friction B, and motor inertia J.

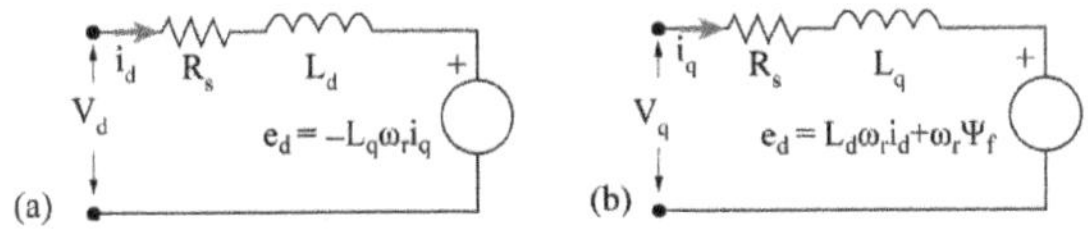

Figure 2. Equivalents of a PMSM motor on: (**a**) d axis and (**b**) q axis.

In order to analyse the three-phase stator voltage behaviour in the dq frame, we must depart from $\alpha\beta$ coordinates using a transform matrix that holds peak convention and a-phase-to-d-axis alignment [9]. The notation ${}^{\mathcal{B}}_{\mathcal{A}}(\cdot)$ stands for a transform matrix that maps a value from the $\mathcal{A}$ to $\mathcal{B}$ reference; in this case, it maps from abc to dq, yielding

$$ {}^{dq}_{abc}C = {}^{dq}_{\alpha\beta}C^{\alpha\beta}_{abc}C = \frac{2}{3}\begin{bmatrix} \cos(\theta) & \cos(\theta - 120^\circ) & \cos(\theta + 120^\circ) \\ -\sin(\theta) & -\sin(\theta - 120^\circ) & -\sin(\theta + 120^\circ) \end{bmatrix} \tag{5} $$

Recall that the implemented MATLAB model makes use of Equations (1)–(4), and the currents were transformed from a static reference frame abc to rotational axes $dq0$ using the transform matrix stated in Equation (5).

2.2. Vectorial Field-Oriented Control

In this work, we implemented a vectorial field-oriented control (FOC) scheme due to its excellent dynamics to address, with great accuracy, transient and steady-state responses [8,16,17]. The main objective of vectorial control relies on controlling the space vectors concerning the stator current [18,19]; furthermore, it allows for the de-coupling of speed and torque to achieve independent control of both variables.

Recall that in an AC motor, the space angle between the rotating stator field and the rotor flux changes due to the load, which causes an oscillatory response. In contrast, in a DC motor, the armature current directly controls the torque and the rotor field current, and they can be accessed independently. The angle between these variables is held orthogonally through a mechanical switching system, such as brushes and commutators.

The vectorial FOC emulates the performance of a DC machine because it observes the position of the rotor field and directs the stator field to achieve a constant 90-degree angle between both of them. The last condition is used to reach the maximum torque while we control the rotor speed independently. FOC control requires a position sensor to know the rotor angular pose at all time, which is also related to the rotor flux. It is worth mentioning that the stator field is directed through the three-phase current's phases [20–23]. In order to figure the open-loop current model out, and represent it on the $dq0$ axes, we depart from Equation (1), which allows us to determine the motor inductances as follows:

$$ L_d\frac{di_d}{dt} = v_d - R_s i_d + w_e L_q i_q \;\; \& \;\; L_q\frac{di_q}{dt} = v_q - R_s i_q - w_e(L_d i_d + \Psi_m), \tag{6} $$

from which we can obtain the stator currents in $dq0$ and translate them into the Laplace domain as follows:

$$ i_d(s) = \frac{V_d + \omega_e L_q i_q}{L_d s + R_s} \;\; \& \;\; i_q(s) = \frac{V_q - \omega_e L_d i_d - \omega_e \Psi_m}{L_q s + R_s} = \frac{V_q - \omega_e L_d i_d - \omega_e k_e}{L_q s + R_s}. \tag{7} $$

The torque can be derived from Equation (2), and because $L_d = L_q$, the PMSM-BLAC motor yields $T_e = \frac{3p}{2}\psi_m i_q = k_t \times i_q$. The motor's angular speed is computed from Equation (4), which can be expressed in the frequency domain as

$$\omega_m(s) = \frac{T_e - T_m}{Js + B}. \tag{8}$$

The above-mentioned equations are placed together in the simulation diagram showed in Figure 3, where we can notice the correlation between currents on the *d* and *q* axes, which forces us to implement a decoupling technique for implementing independent current controllers.

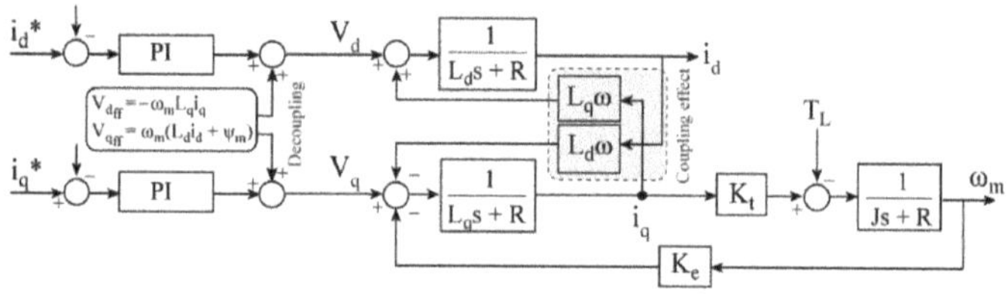

Figure 3. Simulation diagram of the PMSM-BLAC motor on *dq*0 axis.

The currents i_d and i_q are decoupled using a feedforward approach with EMF compensation [9,20–23], as described in the blocks of Figure 3, which will result in independent controlled current variations, where i_d is set to 0, and i_q varies according to the torque [24]. The current control selected in this work was a PI, due to its fast response to changes in the setpoint, its capability of reducing the error in steady state to zero [25–27], and the fact that it is the most common controller used at an industrial level. The PI control is modelled with no relation to any current *i* and its desired value noted as *i**, which can be expressed as a first-order system in the Laplace domain to reduce overshooting and oscillations, leading to

$$\frac{i(s)}{i^*(s)} = \frac{K_p s + K_i}{Ls^2 + (K_p + R)s + K_i} \approx \frac{W_{cc}}{s + W_{cc}}, \tag{9}$$

where W_{cc} denotes the cutoff frequency in the closed loop of the current controller. The tuning approach selected to compute the constants k_p and k_i was the principle of the optimal magnitude criterion, introduced by Sartorius and Oldenbourg. This principle states that a controller should have a closed-loop magnitude in the frequency domain as close to the unit as possible, in the widest possible frequency range [25–27]. In this sense, the controller constant can be computed as follows:

$$K_{i,c} = RW_{cc}, \ \& \ K_{p,c} = LW_{cc}. \tag{10}$$

2.3. Speed Control

The speed control for the PMSM-BLAC motor was achieved with an external loop, over the previously mentioned current loop. The implemented controller was a PI as in the previous section to avoid issues arising due to the derivative component. In this sense, the noise introduced by the power electronic devices during the switching phase may have negative effects on the derivative component in any controller architecture.

PI control was carried out by simplifying the mechanical motor model, without considering the viscosity coefficient [9,28–30], yielding the open loop transfer function, where the PI constant for the speed control loop has an *s* subindex, as shown in Figure 4.

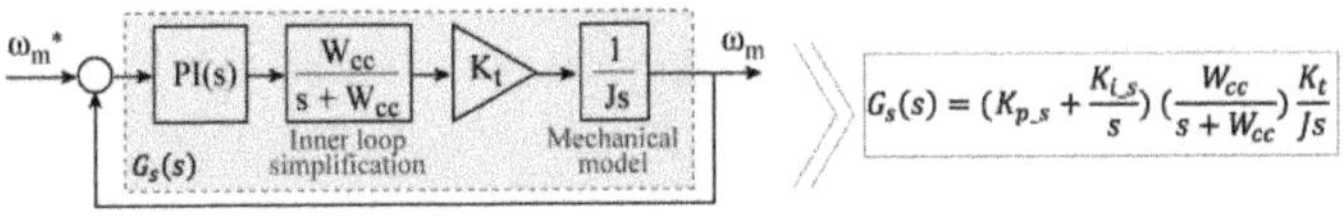

Figure 4. Simplified model of the external speed loop.

It is relevant to mention that in cascaded systems, the inner control variable has a larger bandwidth than the outer one, which results in a difference in response speed, where the inner variable is the faster one [26,30]. In this sense, the tuning bandwidth for the speed control is smaller than the one of the current control. Under the assumption that the cutoff frequency of the speed controller given by $W_{PI} = \frac{K_{i,s}}{K_{p,s}}$ is significantly smaller than the current control loop bandwidth W_{CC}, then the open loop transfer function can be approximated using the Bode diagram, as shown in Figure 5.

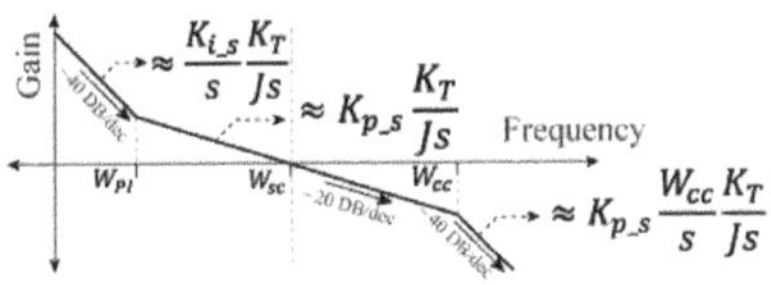

Figure 5. Bode diagram of speed control transfer function in open loop.

$G_s(s)$ can be approximated based on the dominant term, i.e., if $W < W_{PI}$, the integral term becomes dominant, yielding $G_s(s) \approx \frac{K_{i,s}}{s}\frac{K_T}{Js}$. The approximation for each region is listed in Figure 5. Since W_{PI} should be lower than W_{sc}, we can select a relationship of $W_{PI} = W_{sc}/4$, and with the aid of the Bode diagram in Figure 5, the PI constants can be computed as follows.

$$K_{p\_s} = \frac{JW_{sc}}{K_T} \ \& \ K_{i\_s} = K_{p\_s}\frac{W_{sc}}{4}. \quad (11)$$

It is worth mentioning that the speed control will include an anti-wind-up configuration to avoid saturations in the integral component.

As mentioned before, field weakening allows for higher velocities than the nominal one to be reached. In order to prevent damage to the motor during the field weakening, we must apply two constrains: ${i_d}^2 + {i_q}^2 \leq {I_{max}}^2$ and ${V_d}^2 + {V_q}^2 \leq {V_{max}}^2$. This represents a circle in the $dq0$ frame [9,31,32]. Combining the decoupling expressions in Figure 3 with these constrains, it is possible to determine the current limits to prevent damage when applying field weakening, which leads to

$$i_d \leq \frac{\left(\frac{V_{max}}{w_e}\right)^2 - {\Psi_m}^2 - (LI_{max})^2}{2L\Psi_m} \quad (12)$$

2.4. Vector Space Pulse Width Modulation

Vector space PWM (SVPWM) allows for the minimization of harmonic distortion, delivering a higher voltage compared to other techniques and also reducing the losses produced by switching. In this modulation technique, any voltage vector V_{ref} can be computed from V_α and V_β, which are voltage components in the $dq0$ frame estimated using Park inverse transform. The angle α allows us to identify the sector number of any voltage reference, where it can be decomposed using adjacent commutation configurations, i.e., V_1 and V_2 in sector 1 and the null configuration V_0, as shown in Figure 6 [8,9,33,34]. Figure 6 indicates how to compute the duration of each configuration, where the total commutation time is $T_s = T_1 + T_2 + T_0$, n stands for the sector number that yields Equation (13), and the commutation configurations for other sectors can be found in [33].

$$\int_0^{T_s} \bar{V}_{ref} dt = \int_0^{T_1} \bar{V}_1 dt + \int_{T_1}^{T_1+T_2} \bar{V}_2 dt + \int_{T_1+T_2}^{T_s} \bar{V}_0 dt \quad (13)$$

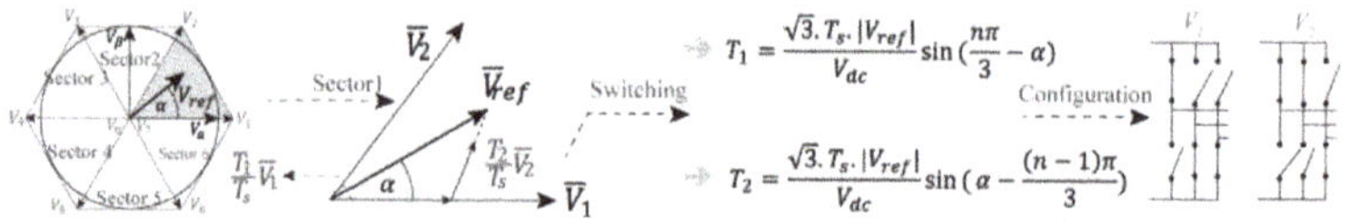

Figure 6. Voltage vector through different SVPWM sectors. Sector 1 example.

3. Results and Discussion

The proposed simulation model was implemented in MATLAB/SIMULINK using the motor parameters from Table 1, and it was evaluated in different scenarios, like the response of the current controller with different load profiles, including speeds over the nominal one, and locked rotor tests to assess dynamic changes in the setpoint of the current. It is worth mentioning that the switching frequency used in all tests was 10 kHz, and the current controller was assessed with different bandwidths, whose values were related to their constants according to Equations (10) and (11).

3.1. Internal Current Control Loop Tests

A locked rotor test ($J \to \infty$) allows us to assess the current loop, where we test different current controller bandwidths of f_{cc} ($w_{cc} = 2\pi f_{cc}$) of 500, 800, and 1000 Hz, and estimate the currents I_d and I_q settling time and error, as shown in Figure 7b. The computed PI constants allow for a reduced settling time of about 25 ms with mean current of zero in steady state. The controller performs better at higher current controller bandwidths, as shown in Table 2.

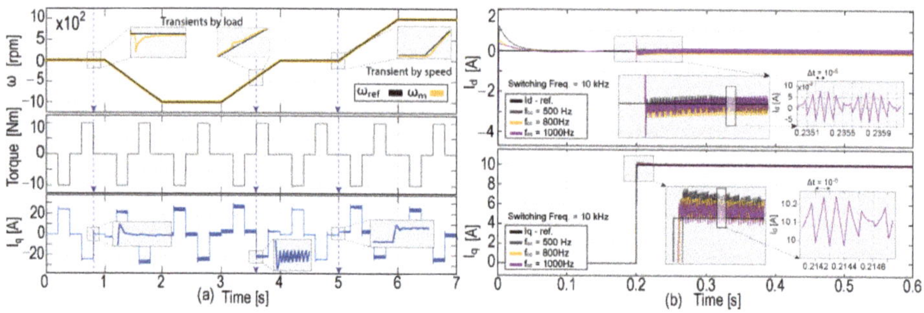

Figure 7. Control loop tests: (**a**) speed control, (**b**) current control in locked rotor.

Table 2. Performance benchmarking results for locked rotor and speed control experiments.

Locked Rotor Test					Speed Control Test						
Fcc (Hz)	Id		Iq		Fcc (Hz)	Speed		Id		Iq	
	ISE	Settling Time	ISE	ISE		ISE	Settling Time	ISE	Settling Time	ISE	Settling Time
500	2.68	35 ms	0.018	15 ms	800	0.14	120 ms	2.52	25 ms	0.012	11 ms
800	2.52	25 ms	0.012	11 ms	1000	0.125	103 ms	2.5	20 ms	0.009	8 ms
1000	2.5	20 ms	0.009	8 ms							

3.2. Velocity Control Loop Tests

The speed control loop makes use of a speed profile that remains unaffected by torque profile disturbance, as shown in Figure 7a. For visualizing purposes, we highlighted the transients during the load or velocity changes (see Figure 7a). This test was carried out with two current controller bandwidths of 800 and 1000 Hz. The error results are detailed in Table 2.

3.3. System Tests including Field Weakening

In this experiment we used the best-achieving current controller bandwidth of 1000 Hz and set a reference speed over the nominal one (1590 rpm). With no field weakening, the speed becomes saturated at a value closer to the nominal one (see yellow line in Figure 8a). Applying the field weakening constraint in Equation (13), the motor can exceed the nominal speed without surpassing the maximum current, retaining the nominal power, as shown in the orange plot in Figure 8a and the current plot on Figure 8b.

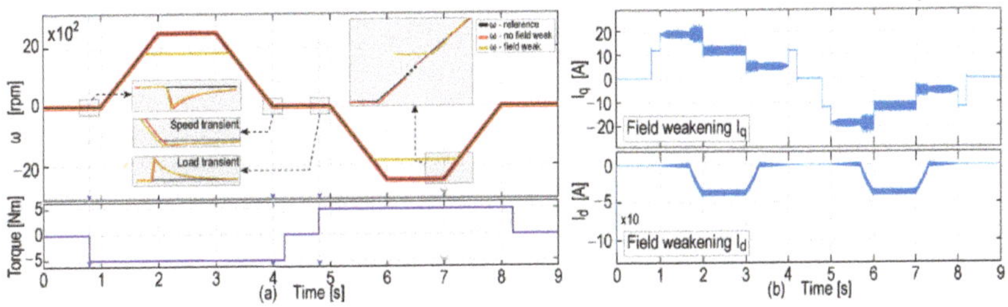

Figure 8. System response with field weakening: (**a**) speed response; (**b**) current response.

4. Conclusions

This study emphasizes the importance of the implementation of FOC in conjunction with field weakening in permanent-magnet synchronous machines. This approach grants these motors the capability to operate effectively across a wide range of speeds, offering a commendable torque response at low speeds while maintaining robust performance at high velocities.

Through the correct implementation of FOC with current and speed control, it can be ensured that a salient-pole PMSM has a stable response to speed and load profiles, with low response times and reduced over-peaks, and by using SVPWM in the three-phase inverter bridge, the available DC bus can be used in a more efficient way.

During the speed test, it was found that the integrity of control remained intact amidst dynamic shifts in speed setpoints and the introduction of load perturbations. This robustness was maintained as long as the specified speed references adhered to the desired values, irrespective of whether they fell below or soared above the nominal threshold. Notably, the utilization of field weakening played a critical role in ensuring control resilience across velocities surpassing the nominal threshold.

Author Contributions: Conceptualization, J.M. and C.G.; methodology, W.C., M.P.; software, J.M., C.G.; validation, J.M., W.C. and M. P.; formal analysis, W.C.; investigation, J.M, C.G.; data curation, J.M. and W.C.; writing—original draft preparation, J.M, W.C., V.T., C.G., M.P.; writing—review and editing, J.M. W.C. V.T All authors have read and agreed to the published version of the manuscript.

Funding: This research received no external funding.

Institutional Review Board Statement: Not applicable.

Informed Consent Statement: Not applicable.

Data Availability Statement: Not applicable.

Conflicts of Interest: The authors declare no conflict of interest.

References

1. Gómez Misnaza, C.D. Diseño y Simulación del Control por Orientación de Campo FOC para un Motor Síncrono de Imanes Permanentes tipo BLAC. Bachelor's Thesis, EPN, Quito, Ecuador, 2020. Available online: http://bibdigital.epn.edu.ec/handle/15000/21220 (accessed on 1 August 2023).
2. Sepulchre, L.; Fadel, M.; Pietrzak-David, M.; Porte, G. Flux-weakening strategy for high speed PMSM for vehicle application. In Proceedings of the 2016 International Conference on Electrical Systems for Aircraft, Railway, Ship Propulsion and Road Vehicles & International Transportation Electrification Conference (ESARS-ITEC), Toulouse, France, 2–4 November 2016; pp. 1–7. [CrossRef]
3. Eriksson, S. Drive systems with permanent magnet synchronous motors. *Automot. Eng.* **1995**, *103*. Available online: https://www.osti.gov/biblio/37082 (accessed on 1 August 2023).
4. Loganayaki, A.; Kumar, R.B. Permanent Magnet Synchronous Motor for Electric Vehicle Applications. In Proceedings of the 2019 5th International Conference on Advanced Computing & Communication Systems (ICACCS), Coimbatore, India, 15–16 March 2019; pp. 1064–1069. [CrossRef]
5. DC, Induction, Reluctance and PM Motors for Electric Vehicles | IET Journals & Magazine | IEEE Xplore. Available online: https://ieeexplore.ieee.org/document/281755 (accessed on 1 August 2023).
6. Chalmers, B.J.; Musaba, L.; Gosden, D.F. Variable-frequency synchronous motor drives for electric vehicles. *IEEE Trans. Ind. Appl.* **1996**, *32*, 896–903. [CrossRef]
7. Liu, X.; Chen, H.; Zhao, J.; Belahcen, A. Research on the Performances and Parameters of Interior PMSM Used for Electric Vehicles. *IEEE Trans. Ind. Electron.* **2016**, *63*, 3533–3545. [CrossRef]
8. Abu-Rub, H.; Iqbal, A.; Guzinski, J. *High Performance Control of AC Drives with Matlab/Simulink Models*; John Wiley & Sons: Hoboken, NJ, USA, 2012.
9. Kim, S.-H. *Electric Motor Control: DC, AC, and BLDC Motors*; Elsevier: Amsterdam, The Netherlands, 2017; p. 426.
10. Krause, P.C.; Wasynczuk, O.; Sudhoff, S.D.; Pekarek, S.D. *Analysis of Electric Machinery and Drive Systems*; John Wiley & Sons: Hoboken, NJ, USA, 2013.
11. Odo, K.; Vincent, E.S.; Ogbuka, C. A Model-based PI Controller Tuning and Design for Field Oriented Current Control of Permanent Magnet Synchronous Motor. *IOSR J. Electr. Electr. Eng. (IOSR-JEEE)* **2019**, *14*, 35–41. [CrossRef]
12. Kim, S.A.; Song, J.H.; Han, S.W.; Cho, Y.H. An Improved Dynamic Modeling of Permanent Magnet Synchronous Machine with Torque Ripple Characteristics. *J. Clean Energy Technol.* **2018**, *6*, 117–120. [CrossRef]
13. Zhao, Y. Application of Sliding Mode Controller and Linear Active Disturbance Rejection Controller to a PMSM Speed System. Mater Thesis, Cleveland State University, Cleveland, OH, USA, 2009.
14. Mukti, E.W.; Wijanarko, S.; Muqorobin, A.; Rozaqi, L. Field oriented control design of inset rotor PMSM drive. *AIP Conf. Proc.* **2017**, *1855*, 020010. [CrossRef]
15. Taha, M.; Osama, A.M.; Zaid, S. Simulation Study of a New Approach for Field Weakening Control of PMSM. *J. Power Electron.* **2012**, *12*, 136–144. [CrossRef]
16. Chau, K.T. *Electric Vehicle Machines and Drives: Design, Analysis and Application*; John Wiley & Sons: Hoboken, NJ, USA, 2015; p. 400. [CrossRef]
17. Bida, V.; Samokhvalov, D.; Al-Mahturi, F. PMSM vector control techniques—A survey. In Proceedings of the 2018 IEEE Conference of Russian Young Researchers in Electrical and Electronic Engineering (EIConRus), Moscow and St. Petersburg, Russia, 29 January–1 February 2018; pp. 577–581. [CrossRef]
18. Amin, F.; Sulaiman, E.; Mulyo Utomo, W.; Soomro, H.; Jenal, M.; Kumar, R. Modelling and Simulation of Field Oriented Control Based Permanent Magnet Synchronous Motor Drive System. *Indones. J. Electr. Eng. Comput. Sci.* **2017**, *6*, 387–395. [CrossRef]
19. Mulyo Utomo, W.; Muhammad Zin, N.; Haron, Z.; Sy Yi, S.; Bohari, A.; Mat Ariff, R.; Hanafi, D. Speed Tracking of Field Oriented Control Permanent Magnet Synchronous Motor Using Neural Network. *Int. J. Power Electron. Drive Syst. IJPEDS* **2014**, *4*, 290. [CrossRef]
20. Boudjemaa, M.; Chenni, R. Field Oriented Control of PMSM Supplied by Photovoltaic Source. *Int. J. Electr. Comput. Eng. IJECE* **2016**, *6*, 1233. [CrossRef]
21. Ahmadian, A.; Ghanbari, M.; Rastegar, H. Field-oriented Control of Permanent Magnet Synchronous Motor with Optimum Proportional-Integrator-Derivative Controller Using Dynamic Particle Swarm Optimization Algorithm. *Sci. Int.* **2014**, *27*, 155–158.
22. Morales-Caporal, R.; Ordoñez-Flores, R.; Huerta, E.; Hernandez, J.; Sandre Hernandez, O. Simulación del Control por Campo Orientado y del Control Directo del par de un Servomotor Síncrono de Imanes Permanentes con Control Inteligente de Velocidad. In Proceedings of the IX Congreso Internacional sobre Innovación y Desarrollo Tecnológico, Cuernavaca Morelos, México, 23–25 November 2011.
23. Liu, T.; Tan, Y.; Wu, G.; Wang, S. Simulation of PMSM Vector Control System Based on Matlab/Simulink. In Proceedings of the 2009 International Conference on Measuring Technology and Mechatronics Automation, Zhangjiajie, China, 11–12 April 2009; pp. 343–346. [CrossRef]
24. Qu, L.; Qiao, W.; Qu, L. An Enhanced Linear Active Disturbance Rejection Rotor Position Sensorless Control for Permanent Magnet Synchronous Motors. *IEEE Trans. Power Electron.* **2019**, *35*, 6175–6184. [CrossRef]
25. Zigmund, B.; Terlizzi, A.A.; del T, X.; Pavlanin, R.; Salvatore, L. Experimental Evaluation of PI Tuning Techniques for Field Oriented Control of Permanent Magnet Synchronous Motors. *Adv. Electr. Electron. Eng.* **2011**, *5*, 114–119.

26. Lukichev, D.; Demidova, G. Features of tuning strategy for field oriented control of PMSM position drive system with two-mass load. *Int. J. Circuits Syst. Signal Proc.* **2016**, *10*, 88–94.
27. Papadopoulos, K.G. *PID Controller Tuning Using the Magnitude Optimum Criterion*; Springer: Cham, Switzerland, 2015; ISBN 978-3-319-07262-3.
28. Yong-Bin, W.; Xiao-Dong, H.; Yi, C.; Xin-Qi, X. A MPC Method Based on the Oval Invariant Set for PMSM Speed Control System. In Proceedings of the 2016 International Symposium on Computer, Consumer and Control (IS3C), Xi'an, China, 4–6 July 2016; pp. 575–578.
29. Sidda, S.; Kiranmayi, R.; Mandadi, P. Investigation of PI and Fuzzy Controllers for Speed Control of PMSM Motor Drive. In Proceedings of the 2018 International Conference on Recent Trends in Electrical, Control and Communication (RTECC), Malaysia, Malaysia, 20–22 March 2018; pp. 133–136.
30. Cui, B.; Han, Z.; Li, J.; Mu, J. Research on PMSM Speed Control System Based on Improved Reaching Law. In Proceedings of the 2019 3rd International Conference on Circuits, System and Simulation (ICCSS), Nanjing, China, 13–15 June 2019; pp. 19–24. [CrossRef]
31. Krishnan, R. *Permanent Magnet Synchronous and Brushless DC Motor Drives*, 1st ed.; CRC Press: Boca Raton, FL, USA, 2017; ISBN 978-1-315-22148-9. [CrossRef]
32. Phung Quang, N.; Dittrich, J.-A. *Vector Control of Three-Phase AC Machines-System Development in the Practice*; Springer: Berlin/Heidelberg, Germany, 2015; ISBN 978-3-662-46914-9.
33. Muhammad, H.R. Power electronics: Circuits, devices, and applications. *Choice Rev. Online* **1989**, *26*, 26.
34. Uçak, O.; Yilmaz, S. Implementation of Space Vector PWM In FOC of a Servo Motor Drive. Available online: www.eleco.org.tr/openconf_2015/modules/request.php?module=oc_proceedings&action=view.php&id=111&a=Accept+as+Poster (accessed on 5 August 2023).

engineering proceedings

MDPI

Proceeding Paper

Design and Construction of a High-Current Capacitor Bank for Flash Graphene Synthesis †

Juan Ramírez [1,*], Esteban Yépez [2], Fernando Pantoja-Suárez [2,3], Eliana Acurio [4], Fabian Pérez [1] and Leonardo Basile [4,*]

1 Department of Electric Energy, Faculty of Electrical and Electronic Engineering, Escuela Politécnica Nacional, Ladrón de Guevara, Quito 170143, Ecuador; fabian.perez@epn.edu.ec
2 Department of Materials, Faculty of Mechanical Engineering, Escuela Politécnica Nacional, Ladrón de Guevara, Quito 170143, Ecuador; esteban.yepez@epn.edu.ec (E.Y.); fernando.pantoja@epn.edu.ec (F.P.-S.)
3 ENPHOCAMAT (FEMAN) Group, Department of Applied Physics, Universitat de Barcelona, Martí i Franqués 1, 08028 Barcelona, Spain
4 Department of Physics, Faculty of Sciences, Escuela Politécnica Nacional, Ladrón de Guevara, Quito 170143, Ecuador; eliana.acurio@epn.edu.ec
* Correspondence: juan.ramirezd@epn.edu.ec (J.R.); leonardo.basile@epn.edu.ec (L.B.)
† Presented at the XXXI Conference on Electrical and Electronic Engineering, Quito, Ecuador, 29 November–1 December 2023.

Abstract: This paper introduces the design and fabrication of a versatile capacitor bank developed specifically for its integration within the innovative flash joule heating (FJH) technique, aimed at synthesizing graphene. The capacitor bank offers two adaptable configurations, providing options for capacitance at 180,000 μF and 68,000 μF, combined with a maximum charging capability of 400 V. This equipment facilitates the swift conversion of carbon-containing materials into graphene by harnessing transient current discharges. The proposed capacitor bank facilitates the execution of comprehensive research pertaining to graphene production from diverse carbon sources, opening an opportunity for advancing exploration within this rapidly evolving field of study in Ecuador.

Keywords: flash graphene synthesis; high-current capacitor bank; flash joule heating; material transformation; carbon-containing materials

Citation: Ramírez, J.; Yépez, E.; Pantoja-Suárez, F.; Acurio, E.; Pérez, F.; Basile, L. Design and Construction of a High-Current Capacitor Bank for Flash Graphene Synthesis. *Eng. Proc.* **2023**, *47*, 18. https://doi.org/10.3390/engproc2023047018

Academic Editor: Jaime Cepeda

Published: 6 December 2023

1. Introduction

Since the discovery of graphene in 2004 [1], research into its production methods has advanced consistently. Graphene is a material with immense potential across various technological domains, encompassing energy, electronics, manufacturing, and beyond. The global market demand for graphene is estimated to reach USD 170 million by 2022 [2]; consequently, it is imperative to develop high-scale graphene production methods. However, a challenge lies in the fact that methods developed in recent years yield low quantities of graphene, lacking an industrially viable production-to-cost ratio [2]. As a result, global graphene production remains severely constrained.

Recently, a research group introduced a method termed Flash Joule Heating (FJH) for synthesizing graphene. This method involves applying a short-duration high-current pulse to a carbon-rich material, elevating its temperature to nearly 3000 K. This process results in the conversion of the material into high-quality turbostratic graphene [3]. Notably, this approach holds promise for processing diverse carbonaceous sources, potentially recycling materials typically destined for waste, such as plastics, rubber, vehicular tires, vegetal residues, and more. Moreover, its strong advantage lies in its potential for industrial scalability at reasonable costs. It is important to acknowledge that graphene produced through contemporary methods can range in cost from USD 60,000 to USD 200,000 per ton [4].

Internationally, research initiatives have showcased the effectiveness of the FJH method in converting diverse carbonaceous materials into graphene [2]. In this context,

Ecuador possesses the capacity to employ the FJH method for the conversion of distinct waste categories into industrial-grade graphene. This could consequently contribute to circular economy initiatives spanning various industrial domains, while harnessing the substantial future potential of graphene in the realm of sustainable solid waste management [5–9].

Existing works [3,10–12] use capacitor banks of thousands of microfarads and hundreds of volts to discharge energy onto carbonaceous materials through elevated currents. Although the operational principle of capacitor banks is theoretically straightforward, designing equipment for users from diverse scientific and technical backgrounds demands meticulous attention due to the potential for generating electric discharges that could be fatal to the user.

This material-driven opportunity fostered the inception of a multidisciplinary research project at the Escuela Politécnica Nacional in Ecuador. The project's objective revolves around synthesizing graphene using the FJH technique. This article presents one of the initial steps of the project: the construction of the requisite capacitor bank, a foundational component for advancing research into graphene production.

2. Design Requirements

Based on a literature review, it has been ascertained that the construction of two distinct types of capacitor banks is indeed feasible. One type involves a low capacitance configuration designed for 0.5 g material samples, while the other type entails higher capacitance capabilities suitable for samples of up to 1 g. The search for suitable materials necessitated adaptations to align with the constraints of the Ecuadorian market. Consequently, several components had to be imported, leading to cost considerations emerging as a noteworthy constraint in the construction process. Consequently, the smaller capacitor bank is envisaged to comprise ten units of 6800 μF, 400 V capacitors (model B43310-J9688-A2), whereas the larger capacitor bank would be composed of nine units of 20,000 μF, 400 V capacitors each (model B43310-A9209-M). It is pertinent to note that these capacitors belong to the electrolytic aluminum category. Consequently, the smaller capacitor bank would yield a cumulative capacitance of 68,000 μF, while its larger counterpart would offer a capacitance of 180,000 μF.

For capacitor charging, a source for LED luminaires has been used. This source can provide a maximum direct voltage output of 435 V. The source is powered by alternating current mains spanning the voltage range of 100 to 240 V, with a maximum current rating of 700 mA. The selection of this source is particularly apt for capacitor charging, as it obviates the need for high currents during the charging phase. Furthermore, the source features a potentiometer input that facilitates voltage regulation during the charging process. The specific model attributed to this source is HLG-320H-C700B.

The power source supplies the requisite 400 V for capacitor charging through the utilization of resistor R_2, which could be comprised of two 120 Ω, 5W resistors. It is worth considering the potential to reduce the resistance value of these components, given that their influence is primarily on the charging time; a reduction in resistor value would result in expedited charging times. A 500 V DC, 6 A circuit breaker CB_1 is used to establish the connection between the power source and the capacitor bank. Moreover, the power source is outfitted with a 220 V AC, 10 A circuit breaker to facilitate on/off control. Once the capacitors are charged, CB_1 is turned off, leaving the bank charged and ready for discharge.

In parallel with the capacitors, a bleeder resistor R_1 with a resistance of 220 kΩ is interlinked to enable a gradual discharge over time. This serves to ensure that the bank is discharged in cases where it remains charged. Another 500 V DC, 6 A circuit breaker CB_3 is deployed to connect all capacitors to a set of power resistors, collectively labeled as R_3. The combined resistance of these resistors amounts to approximately 220 Ω, with a power dissipation of approximately 10 W. The inclusion of these resistors allows for swift discharge of the capacitors in instances where a reduction in output voltage or complete discharge of the bank is needed.

The voltage across the capacitor bank is measured utilizing a voltmeter. Once the desired voltage level is attained across the capacitors, the discharge of the bank can be initiated. To effectuate this, flexible output cables are affixed to a test probe that incorporates two copper electrodes. This arrangement encases a quartz tube filled with the carbon material slated for conversion into graphene. The material is compacted using a clamp, and its resistance is measured utilizing an ohmmeter.

Serving as the control unit is an Arduino board, where the discharge duration is input in milliseconds through a keyboard interface. Subsequently, a MOSFET module is triggered by a push button, thereby activating relay SW_1 through a 24 V DC signal. The relay, which boasts the capability to withstand up to 500 A, corresponds to model LEV200A5ANA. Upon startup of the discharge process, the material experiences rapid heating, culminating in the discharge of the capacitor bank.

A simulation was conducted using EMTP-ATP for the capacitive bank as show in Figure 1, encompassing both charging and discharging phases to ascertain the prototype's threshold values. The test material, referred to as a coal batch and composed of tire-derived carbon black resulting from recycling, was modeled with an exponentially decaying resistance $R(t)$ in Ω throughout the graphene conversion process; this is described by Equation (1), where R_o represents the initial value of the material resistance in Ω, and t represents the time in seconds.

$$R(t) = R_o e^{-10t} \tag{1}$$

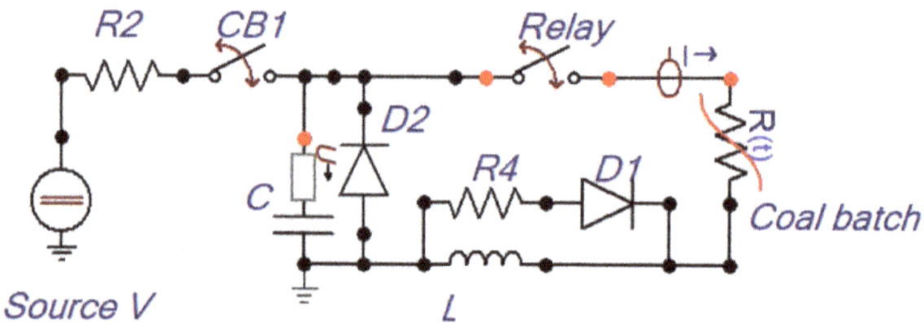

Figure 1. Simulation model for ATP.

The discharge time involving the relay was set at 500 ms. The initial resistance of the material was measured via an ohmmeter, revealing that its values can be tuned within the range of 1 to 10 Ω by adjusting the pressure on the tube clamp. Consequently, the scenario in which the highest current discharges from the capacitor bank corresponds to an initial resistance of 1 Ω. From the simulation results illustrated in Figure 2, it is discernible that the peak current for the 180,000 μF capacitive bank reached 503 A (Figure 2a), while for the 68,000 μF bank, it attained 340 A (Figure 2b).

Furthermore, as shown in Figure 2, the inclusion of an inductor resulted in a slightly reduced rise in current. This rapid current surge could potentially lead to capacitor damage or even explosions, with peak currents of 463 A (Figure 2a) and 400 A (Figure 2b) for the larger and smaller capacitor banks, respectively. On the other hand, the inductor has a beneficial effect; acting as an energy buffer, it prevents excessively rapid current surges. It also slightly affects the duration of the current pulse, which positively influences the response of the carbon material. This leads to better utilization of the energy stored in the capacitive bank. In this way, the discharge circuit encompassing the negative line incorporated an inductor L_1 of 24 mH. Additionally, a 500 A diode D_1 and a 2.5 Ω resistor R_4 were integrated into this circuit. These components collectively contribute to mitigating the discharge rate and curtailing potential current oscillations that might potentially return to the capacitors.

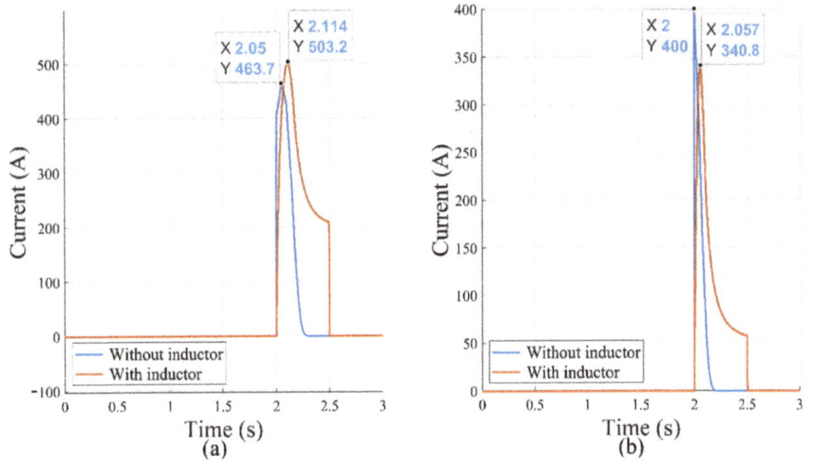

Figure 2. Simulation of the output current from capacitor banks with and without inductor: (**a**) 180,000 µF capacitive bank; (**b**) 68,000 µF capacitive bank.

As an additional safeguard, a second 500 A diode D_2 was placed in reverse bias configuration with respect to the capacitive bank, to prevent voltage spikes generated by self-induction and potential reverse voltages from causing damage to the capacitors. Removing diode D_2 revealed a negative oscillation in the currents (Figure 3), causing a momentary negative voltage across the capacitors, which could cause harm. Diode D_2 plays a crucial role in preventing negative voltage spikes from reaching the capacitors, safeguarding them from damage. Also, the simulations confirm that the THHN AWG 10 cable is well-suited for conducting the current discharge to the coal sample. It can handle over 500 A for 1 s without issues, as specified in its technical specifications [13] and the capacitor bank will provide a maximum of 500 A. In summary, the simulation demonstrates that the capacitive bank behaves satisfactorily, confirming its viability for practical utilization and construction. The proposed diagram outlining the construction of the capacitor banks is depicted in Figure 4.

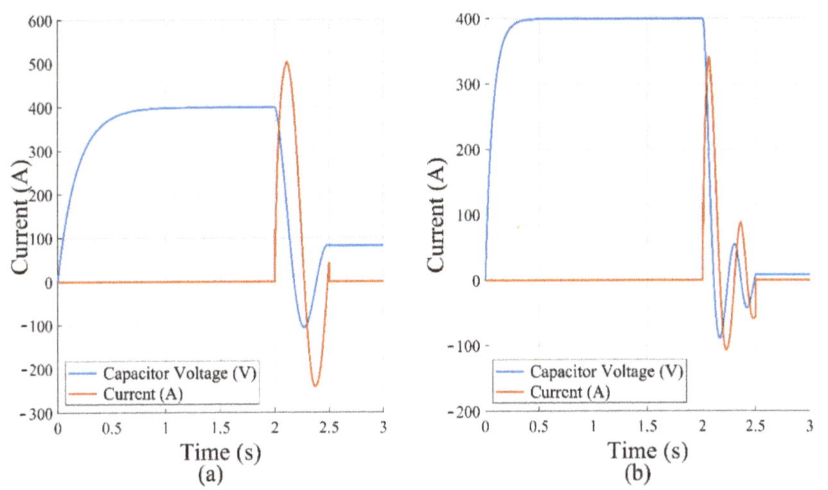

Figure 3. Simulation of the output current from capacitor banks with and without diode D_2: (**a**) 180,000 µF capacitive bank; (**b**) 68,000 µF capacitive bank.

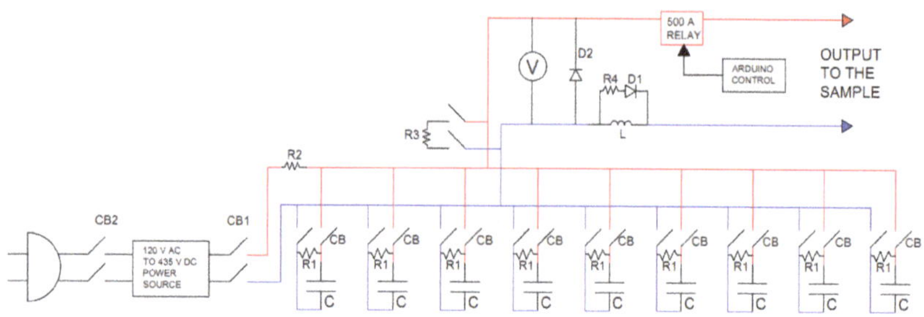

Figure 4. Schematic diagram of the capacitor bank.

3. Construction and Operational Application

In Figure 5, the capacitor bank implemented on a metallic electrical board is presented. This configuration aims to prevent any potential capacitor explosion from affecting the user. The equipment has a single-phase plug designed to connect to a standard 120 V/60 Hz power outlet. The electrical board is grounded to ensure user safety. The apparatus features a keypad, a push-button, and a voltmeter on its door panel. On the left-hand side, the circuit breaker CB_2 is situated to supply alternating voltage and energize the bank. Additionally, circuit breaker CB_1 is positioned for initiating capacitor charging, while circuit breaker CB_3, located on the right-hand side, facilitates rapid discharge of the capacitors. Also situated on the right are the conductors for connection to the test capsule of the coal batch.

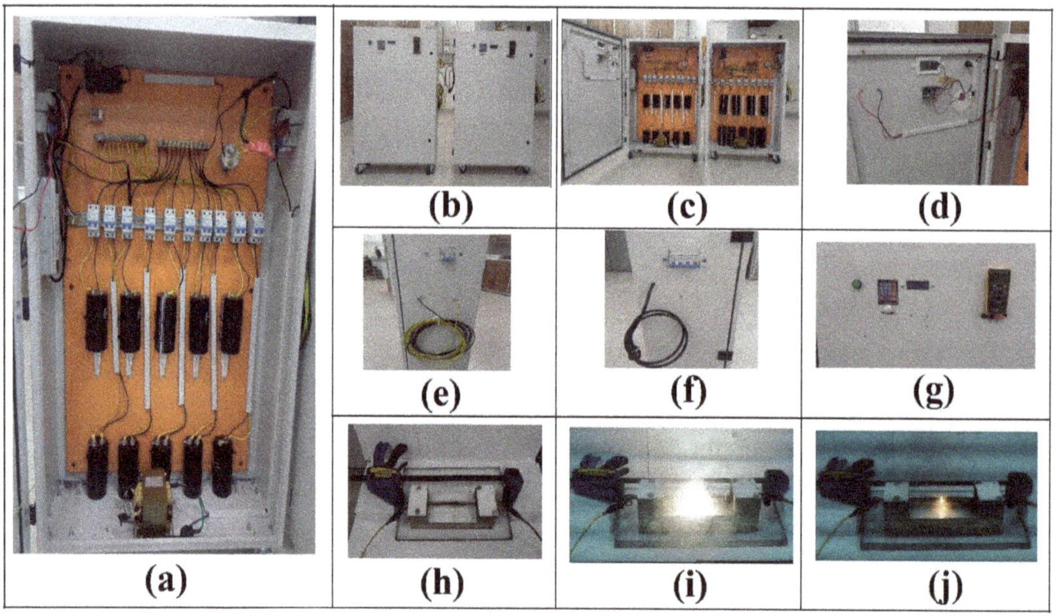

Figure 5. Final prototype of the capacitor bank. (**a**) Internal arrangement of construction elements. (**b**) Front view of capacitor banks; left: 68,000 μF, right: 180,000 μF. (**c**) Inside view of capacitor banks; left: 68,000 μF, right: 180,000 μF. (**d**) Control module with ARDUINO. (**e**) Right side of the capacitor banks: output cables and fast discharge circuit breaker CB_3. (**f**) Left side of the capacitor banks: power cord and power supply circuit breaker CB_2 (right side), and capacitor charging circuit breaker CB_1 (left side). (**g**) Front door of the bank. From left to right: discharge trigger button, time input keypad, LCD display, and voltmeter. (**h**) Clamp and quartz tube for placing the carbon material sample. (**i**) Start of the *flash* discharge. (**j**) Material at the end of the discharge.

To utilize the equipment correctly, the following procedure should be observed:

- Connect the equipment to a 120 V AC voltage source using the power cable on one side of the board.
- Utilize the external circuit breaker labeled CB_2 to power on the equipment. At this point, the LCD screen on the door will illuminate, displaying a default discharge time of 20 ms.
- Specify the discharge time on the screen using the keypad. To achieve this, press the "A" command, input the time in milliseconds, and press the "#" key. The desired time will then appear on the screen.
- Activate the external multimeter and turn the dial to the position indicated as $\overline{\overline{V}}$ for measuring continuous voltage.
- Using the external circuit breaker CB_1, charge the capacitor bank to the desired voltage, as indicated on the voltmeter. Employ the same switch to halt the charging process.

In the event of exceeding the targeted voltage, voltage reduction can be accomplished through two methods:

- Allow the voltage to regulate gradually without intervention.
- For a larger voltage excess, use the external circuit breaker CB_3, which results in faster voltage reduction, as indicated on the voltmeter. Re-engage the same switch to halt the discharge at the desired voltage.

Regarding the energy connection to the test capsule:

- Attach cables with alligator clips to the electrodes where the *flash* is intended.
- Push the green button on the door to activate the equipment.
- After pressing the button, the current will heat the material producing a flash, then inspect the voltmeter to monitor the voltage in the capacitor bank. Activate circuit breaker CB_3 to discharge if necessary.
- Disconnect the alligator clips and secure them to the cable hooks.

To ensure prolonged equipment lifespan and exercise enhanced caution, certain considerations are essential when operating the equipment:

- Always wear protective eyewear and gloves of an insulating material and maintain a safe distance from the *flash* when using the apparatus.
- Abstain from charging the equipment when the voltmeter registers a voltage exceeding 5 V.
- Fully discharge the voltage in the capacitor bank before conducting a new *flash*.

4. Energy Consumption

During the research project's development, it has been noted that the required voltage for material conversion into graphene is around 230 V across the capacitors. Within the capacitor bank, the available energy in joules E can be ascertained using Equation (2), which involves the awareness of capacitance C in farads and the capacitor voltage V in volts.

$$E = \frac{1}{2}CV^2 \tag{2}$$

For the large capacitor bank with a capacitance of 180,000 µF and a voltage of 230 V, an energy of 4761 J is made available. Subsequently, it was observed that the material undergoes various transformations depending on the initial electrical resistance value of the material, which can be adjusted by manipulating the clamping pressure of the test capsule. Therefore, through EMTP-ATP simulations utilizing the model illustrated in Figure 2, the energy consumption of the coal batch from the capacitors was tested and determined. The obtained outcomes are presented in Table 1. The findings suggest that optimal energy consumption occurs when the sample possesses an initial resistance value ranging from 4 to 7 Ω.

Table 1. Energy consumption of the coal sample based on its initial electrical resistance value.

Initial Resistance Ω	**0.25**	**0.5**	**1**	**2**	**3**	**4**	**5**	**6**	**7**	**8**	**10**	**12**	**14**	**16**
Energy consumption J	3254	4132	4536	4647	4665	4670	4671	4671	4669	4667	4661	4654	4645	4635

5. Conclusions

Through the charging and discharging of a capacitor, substantial current values can be generated, serving the purpose of converting carbon-containing materials into graphene using the FJH method. Despite the market limitations within Ecuador concerning the required materials, it remains feasible to construct a functional capacitor bank. The outcome of this endeavor has yielded a capacitor bank of 180,000 µF/400 V with a maximum capacity of approximately 500 A, and a capacitor bank of 68,000 µF/400 V with a maximum capacity of around 340 A. These banks are now ready for utilization in graphene production research.

The insights gathered from the construction of this prototype provide a clear perspective on the scalability of the equipment for the prospective development of an industrial-grade FJH system. Such a system could possess the capability to process significant material quantities.

Author Contributions: Conceptualization, J.R. and L.B.; methodology, J.R.; software, J.R., E.Y. and F.P.-S.; validation, E.Y., F.P. and E.A.; investigation, J.R.; writing—original draft preparation, J.R.; writing—review and editing, L.B.; supervision, F.P.-S.; construction, J.R. and E.Y.; project administration, L.B.; funding acquisition, L.B. All authors have read and agreed to the published version of the manuscript.

Funding: This research was funded by Escuela Politécnica Nacional, research project PIS-21-07.

Institutional Review Board Statement: Not applicable.

Informed Consent Statement: Not applicable.

Data Availability Statement: All data is included in the article.

Acknowledgments: The authors would like to express their gratitude for the collaboration of the technicians from the High Voltage Laboratory of the Escuela Politécnica Nacional, Darwin Pozo and Henry Vivas, for their assistance during the prototype construction phases and for the facilities provided to the research team.

Conflicts of Interest: The authors declare no conflict of interest.

References

1. Novoselov, K.S.; Geim, A.K.; Morozov, S.V.; Jiang, D.; Zhang, Y.; Dubonos, S.V.; Grigorieva, I.V.; Firsov, A.A. Electric field effect in atomically thin carbon films. *Science* **2004**, *306*, 666–669. [CrossRef] [PubMed]
2. Edward, K.; Mamun, K.; Narayan, S.; Assaf, M.; Rohindra, D.; Rathnayake, U. State-of-the-Art Graphene Synthesis Methods and Environmental Concerns. *Appl. Environ. Soil Sci.* **2023**, *2023*, 8475504. [CrossRef]
3. Luong, D.X.; Bets, K.V.; Algozeeb, W.A.; Stanford, M.G.; Kittrell, C.; Chen, W.; Salvatierra, R.V.; Ren, M.; McHugh, E.A.; Advincula, P.A.; et al. Gram-scale bottom-up flash graphene synthesis. *Nature* **2020**, *577*, 647–651. [CrossRef] [PubMed]
4. Wyss, K.M.; De Kleine, R.D.; Couvreur, R.L.; Kiziltas, A.; Mielewski, D.F.; Tour, J.M. Upcycling End-of-Life Vehicle Waste Plastic into Flash Graphene. *Commun. Eng.* **2022**, *1*, 3. [CrossRef]
5. Barbhuiya, N.H.; Kumar, A.; Singh, A.; Chandel, M.K.; Arnusch, C.J.; Tour, J.M.; Singh, S.P. The Future of Flash Graphene for the Sustainable Management of Solid Waste. *ACS Nano* **2021**, *15*, 15461–15470. [CrossRef] [PubMed]
6. Advincula, P.A.; Luong, D.X.; Chen, W.; Raghuraman, S.; Shahsavari, R.; Tour, J.M. Flash graphene from rubber waste. *Carbon* **2021**, *178*, 649–656. [CrossRef]
7. Wyss, K.M.; Beckham, J.L.; Chen, W.; Luong, D.X.; Hundi, P.; Raghuraman, S.; Shahsavari, R.; Tour, J.M. Converting plastic waste pyrolysis ash into flash graphene. *Carbon* **2021**, *174*, 430–438. [CrossRef]
8. Algozeeb, W.A.; Savas, P.E.; Luong, D.X.; Chen, W.; Kittrell, C.; Bhat, M.; Shahsavari, R.; Tour, J.M. Flash Graphene from Plastic Waste. *ACS Nano* **2020**, *14*, 15595–15604. [CrossRef]

9. Advincula, P.A.; Meng, W.; Eddy, L.J.; Beckham, J.L.; Siqueira, I.R.; Luong, D.X.; Chen, W.; Pasquali, M.; Nagarajaiah, S.; Tour, J.M. Ultra-High Loading of Coal-Derived Flash Graphene Additives in Epoxy Composites. *Macromol. Mater. Eng.* **2022**, *308*, 2200640. [CrossRef]
10. Wyss, K.M.; Luong, D.X.; Tour, J.M. Large-scale syntheses of 2D materials: Flash joule heating and other methods. *Adv. Mater.* **2022**, *34*, 2106970. [CrossRef] [PubMed]
11. Eddy, L.; Luong, D.X.; Beckham, J.; Wyss, K.; Cooksey, T.; Scotland, P.; Choi, C.H.; Chen, W.; Advincula, P.; Zhang, Z.; et al. Laboratory Kilogram-Scale Graphene Production from Coal. *Chemrxiv* **2023**. [CrossRef]
12. Beckham, J.L.; Wyss, K.M.; Xie, Y.; McHugh, E.A.; Li, J.T.; Advincula, P.A.; Chen, W.; Lin, J.; Tour, J.M. Machine Learning Guided Synthesis of Flash Graphene. *Adv. Mater.* **2022**, *34*, 2106506. [CrossRef] [PubMed]
13. *ICEA P-32-382*; Short Circuit Characteristics of Insulated Cable. Insulated Cable Engineers Association (ICEA): Alpharetta, GA, USA, 2018. Available online: https://standards.globalspec.com/std/13066955/ICEA%2520P-32-382 (accessed on 18 August 2023).

engineering proceedings

MDPI

Proceeding Paper

Statistical Analysis of Handover Process Performance in a Cellular Mobile Network in the City of Quito, Ecuador †

Ramiro Espinosa, Pablo Lupera-Morillo *, Valdemar Farre, Roberto Maldonado and Ricardo Llugsi Cañar

Electronics, Telecommunications, and Information Network Department, Escuela Politécnica Nacional, Quito P.O. Box 17-01-2759, Ecuador; ramiro.espinosa@epn.edu.ec (R.E.); valdemar.farre@epn.edu.ec (V.F.); roberto.maldonado01@epn.edu.ec (R.M.); ricardo.llugsi@epn.edu.ec (R.L.C.)

* Correspondence: pablo.lupera@epn.edu.ec

† Presented at the XXXI Conference on Electrical and Electronic Engineering, Quito, Ecuador, 29 November–1 December 2023.

Abstract: This paper presents an overview of the findings in the handover (HO) process performance within three routes in Quito, Ecuador. We used the Net-Monitor Software to gather information from one of the three national mobile operators. Then, we used the R tool to analyze the HO performance. We analyze several performance metrics, such as HO types, HO conditions, and the ping-pong process. Analysis of the results of the outdoor drive tests demonstrate that the radio frequency (RF) parameters, such as Received Signal Strength Indicator (RSSI), Reference Signal Received Quality (RSRQ), power margin, times radio frequency measurements repeats, and HO percentage to nearest BS, are extremely important during different HO types and ping-pong processes because there are statistical differences in these measured RF parameters. The main measurement results demonstrate that RSSI difference between inter HO and intra HO is 20 dB, whereas HOs are performed when the mobile device (MS) gets farther from the base station (BS), approximately 50% of total HOs. Operator achieves a high ping-pong rate of approximately 10% of total HOs.

Keywords: handover; LTE; cellular mobile communications; statistical analysis

Citation: Espinosa, R.; Lupera-Morillo, P.; Farre, V.; Maldonado, R.; Llugsi Cañar, R. Statistical Analysis of Handover Process Performance in a Cellular Mobile Network in the City of Quito, Ecuador. *Eng. Proc.* **2023**, *47*, 19. https://doi.org/10.3390/engproc2023047019

Academic Editor: Soraya Lucia Sinche Maita

Published: 6 December 2023

1. Introduction

Mobile network evolution aims to enhance mobile broadband capabilities according to the higher service demands. Mobile Network Operators (MNOs) continuously monitor the quality of service (QoS) to guarantee high-network performance. HO is essential in mobile network performance since it refers to transferring connections from one serving BS to another while allowing the MS to maintain continuous communication (call or data session) while moving between different cell areas, Base Stations (BSs), and different technologies within the network. It reflects the QoS, network performance, user experience, and other crucial indicators. In any technology (2G to 5G), typically, the HO process follows these steps [1]:

Measurement: The MS acquires measures of the signal strength quality of nearby BSs and from its current serving BS.

Triggering: The HO process starts when the signal strength and (or) other parameters of the current serving cell drop below certain thresholds or when a neighboring cell's parameters become significantly better.

HO Decision: Based on the MS shared measurements, the network operator performs an HO decision-making process to determine whether an HO is necessary and which neighboring cell the MS should connect to. The decision considers factors such as signal strength, quality, available resources, hysteresis, and HO policies. It is important to note that the network operator defines these. Due to this, the HO process does not have to be to the nearest BS.

Preparation: Once the HO decision is made, the target BS is informed of the incoming HO. Resources are allocated and prepared to accommodate the incoming MS call or data session.

Execution: The MS is instructed to switch its connection from the current serving cell to the target cell.

HO Completion: The HO process is complete after the mobile device successfully connects to the target base station. The new serving cell now serves the MS, and the communication continues.

The HO can be classified as "hard" (if it disconnects the MS from the current serving cell before connecting to the target cell) or "soft" (if there is a smooth transition. This happens when there is an overlap between the two cells' coverage) [2]. There are other HO classifications, such as intra-cell HO (between different sectors of the same BS), inter-cell HO (between neighboring BSs within the same network), and inter-system HO (between different cellular technologies, e.g., 4G to 5G). The specific HO procedures and parameters may vary depending on the used technologies (i.e., GSM, CDMA, WCDMA, LTE, 5G) and the network architecture. The aim is to provide the best possible QoS and maintain an uninterrupted user experience as the MS moves throughout the network [3]. One key objective of the 4G and 5G networks is to improve mobile broadband capabilities with higher demands among service consumers, such as high-speed Internet connections for urban, suburban, and rural areas. Many mobile network operators (MNOs) monitor the quality of service in terms of multiple services to guarantee high-network performance [4]. In [5], the authors provide an extensive performance evaluation of five national MNOs in Malaysia using metrics such as Reference Signal Received Power (RSRP), RSRQ, signal-to-noise ratio (SNR), throughput in downlink (DL)/uplink (UL), ping, and HO.

The 5G network spectral efficiency may improve by reducing the coverage of BSs, which reduces the number of users by each BS and enhances frequency reuse; however, this increases the HO rate, i.e., the successive change in handling BS for a mobile user. Accordingly, the MNO may implement a capacity gain but at the cost of increased HO rates and higher signaling overheads caused by the HO procedure. In [6], the study found that the basic HO scenario in 5G new radio (NR) is very similar to the long-term evolution (LTE) except for the involved entities and a slight change in HO steps. For instance, the basic HO procedure in NR is completed in 12 steps, while in LTE, it goes through 18 steps.

HO between several networks or cells of the same network is a problem for mobile nodes. In [7], the authors proposed a mathematical model for Wrong Decision Probability (WDP) and Handover Probability (HP) to better understand the behaviors of QoS parameter-based HO algorithms under different network conditions and end-user requirements; they also suggest a WDP based HO algorithm to improve the HO performance. Therefore, making HO decisions based on WDP can provide better performance. Similar work has been carried out in [8], using the same data collection methodology that we describe in this paper, plus Decision Tree Algorithms, to obtain a predictive HO approach in LTE networks.

Recent research has focused on optimizing HO control parameters appropriately for efficiently address HO issues during user mobility [9–11]. In [9], the authors propose a fuzzy-coordinated, self-optimizing HO scheme to achieve a seamless HO while users move in multi-radio access networks. The load balancing optimization function adaptively adjusts the settings of HO control parameter to achieve balance uneven loads between adjacent cells. In [10], the authors have focused on analyzing the performance of load balancing self-optimization within 5G cellular networks. The conflict resolution technique is introduced in the self-optimization network, which is responsible for addressing contraction between mobility robustness optimization and load balancing optimization [11]. This technique performs optimization by obtaining the weight function for input parameter and then monitoring the HO types occurring during HO performance.

The rest of this work is organized as follows: Section 2 provides information about the data recollection methodology and geographic description of the 3 routes. Section 3 describes the data preprocessing for further statistics, statistical analysis of the different

RF parameters at the HO process, and a brief discussion about the findings of the results. Finally, it presents conclusions.

2. Data Measurements

There are several tools for mobile cellular communications network data collection. After a careful comparison and investigation, Net Monitor was selected for the data measurements since it allows monitoring and logging of mobile network parameters without using specialized equipment (1 sample per second of 33 variables: report, sys_time, sim_state, net_op_name, net_op_code, roaming, net_type, call_state, data_state, data_act, data_rx, data_tx, gsm_neighbors, umts_neighbors, lte_neighbors, rssi_strongest, tech, mcc, mnc, lac_tac, node_id, cid, psc_pci, rssi, rsrq, rssnr, slev, gps, accuracy, lat, long, band, arfcn). It allows log exportation in .kml and .csv formats.

The data measurements were performed in Quito/Ecuador from 28 November 2022 to 18 December 2022 and were taken on three routes. The collected data were saved on several Microsoft Excel software program spreadsheets. These routes were Quitumbe (37,382 samples taken at 5.872 km/h on average), San Bartolo (15,260 samples at 10.6 km/h), and La Floresta (20,000 samples at 10.8 km/h).

It must be noted that Quito has an elevation of 2850 m asl. Its shape is very long (40 km) and very narrow (4 km), and its population is approximately 3 million people (Density = 7500/km^2). Its buildings have ten floors on average, so it is suitable for the urban environment of a small or medium-sized city in the Okumura–Hata propagation model [12]. Figure 1 displays where the data measurements were taken (yellow routes) and Figure 2 shows the points where the HO events occur (heightened in red color). These three routes were defined according to their high usage of the HO process. The Quitumbe and San Bartolo routes were selected since many mobile subscribers take these routes to travel from home to work and vice versa. The route, named La Floresta, was chosen due to its high traffic of pedestrian users. In this work, we select only one MNO. In Ecuador, 5G networks have not been put into commercial operation yet, only trial sites have been deployed. No 2G data were collected since this network is poorly used and the HO process mainly occurs between 3G and 4G technologies. Thus, the tested area mostly coverage with the 4G network. We carried out a methodology through walk tests for measuring the cellular network parameters using a test terminal.

Figure 1. Data collection routes in the measurement area.

Similarly, we conducted drive-tests along the main streets and avenues around the cell sites. The goal was to gather samples of the network's behavior, MS performance, and signal quality vs. mobility, mainly by observing different HO. At each GPS location, the Net Monitor tool simultaneously measures the parameters around the cell sites of the MNO in the three routes described earlier.

Figure 2. HO events in: (**a**) Quitumbe Route, (**b**) Floresta Route, (**c**) San Bartolo Route.

3. Sample Processing

Some additional information needs to be included in the gathered data and must be included to have a complete data set for HO analysis. This additional information was obtained with the data set applying algorithms and getting out parameters like a power margin, HO type, ping-pong effect, etc. The BS coordinates were gathered from the Cell Mapper website "https://www.cellmapper.net (accessed on 15 June 2023)" and verified during the drive tests. We pre-processed the data collected from field tests using the output files exported from the Net Monitor tool in the .csv extension. We merged all the samples from individual files into a single file. After that, we transformed raw data into valid numerical data by modifying cell formats for numerical values, dates, times, and signs. Then, we carried out specific algorithms for each column and related the measured parameters at each location point for each route. We separated the "sys_time" variable into several columns for precise date and time pointers. We added columns for general HO types (HO yes/no, Intra-cell/Inter-cell, Inter-Radio Access Technologies (RAT)) and another column for the "power margin" value when there is a handover between cells, represented as the difference between the RSSI levels (rssi–rssi_strongest). Also, we added a column to calculate the power difference by subtracting each interval's new RSSI level from the previous one. Using the Haversine formula, we added columns to calculate the distance between each BS and the MS within the mobile network. The following flowchart describe how to find new features obtained from the samples of the observations for the HO process.

Figure 3 shows the flowchart of the feature "HO_yes_no" to find the HO existence. It compares 2 variables: contiguous "node_id" (BS node id) and "cid" (BS cell id); when these are the same, then "no_HO", and when these contiguous values are different, then "yes_HO".

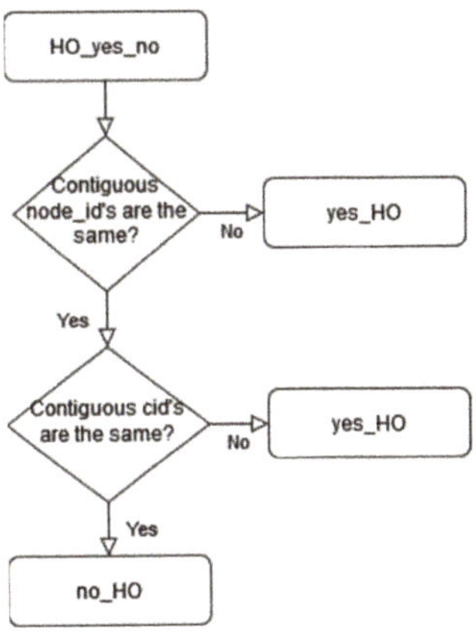

Figure 3. Flowchart of the HO in feature "HO_yes_no".

Figure 4 shows the flowchart of the feature "HO_intra_inter_BS". This feature differentiates the HO types, like intra-cell or inter-cell. It compares the neighbor's node ID during the HO process. If these are the same, there is "intra_HO"; if these are different, there is "inter_HO".

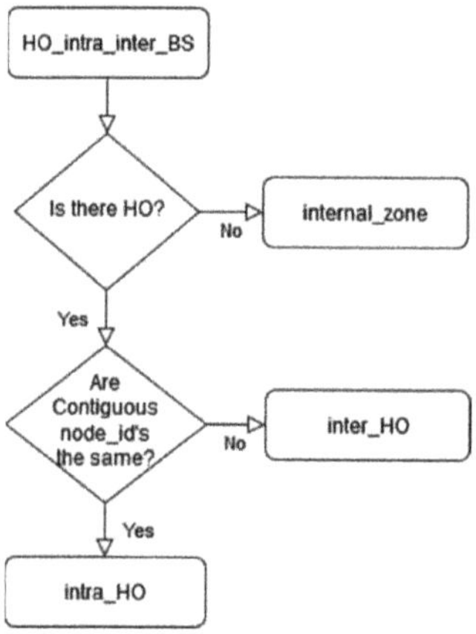

Figure 4. Flowchart of the HO in feature ""HO_intra_inter_BS".

Figure 5 shows the flowchart of the feature "HO_tech". This feature differentiates the HO between RAT. It compares two neighboring fields in the "tech" variable when the HO process is present.

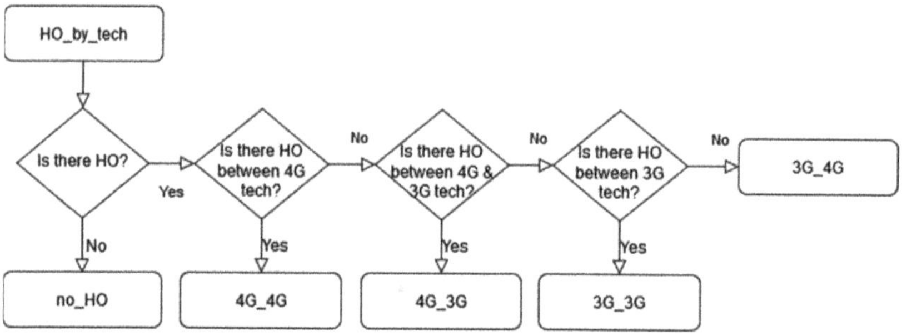

Figure 5. Flowchart of the HO by RAT in feature "HO_by_tech".

Figure 6 shows the flowchart of the feature "HO_to_nearest_BS". This feature shows if the HO process connects the MS to the nearest BS. First, we calculate the distance between BS and each MS in the route and compare if the connection establishment during the HO process is to the nearest BS.

Figure 7 shows the flowchart of the feature "times_rssi_repeats_before_HO" and "times_rsrq_repeats_before_HO". When a HO happens, these features show if the RSSI and RSRQ values repeat themselves during the previous 20 measurements and count the occurrences.

Figure 8 shows the flowchart of the feature "ping_pong". This feature depicts the "ping-pong" phenomenon in the HO process. For this, we count HO in 10 s previous, determine if MS is farther from the HO BS, and count HO in the same BS.

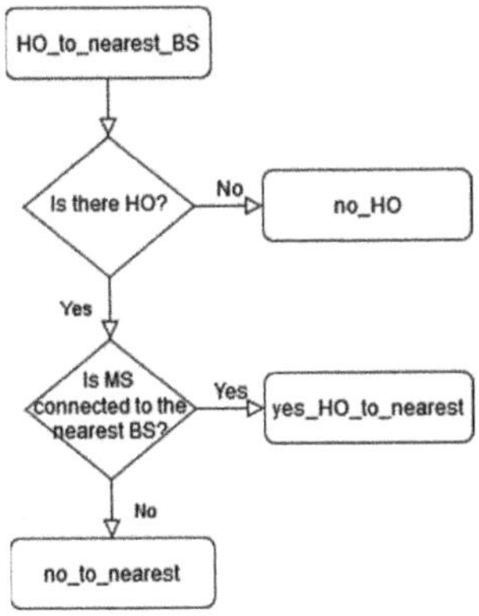

Figure 6. Flowchart of the HO establishment to nearest BS.

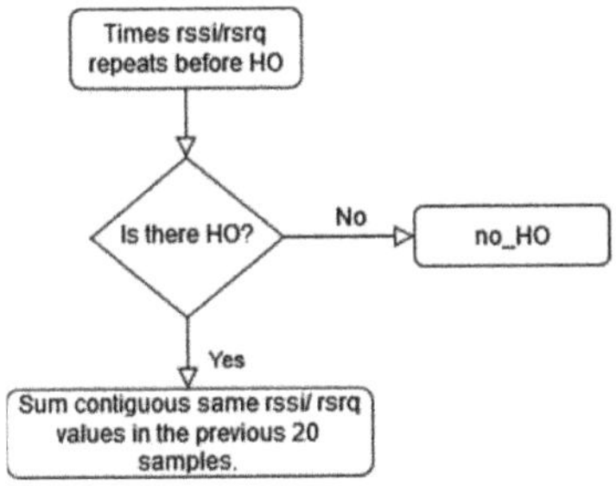

Figure 7. Flowchart of the times RSSI and RSRQ repeat previous HO.

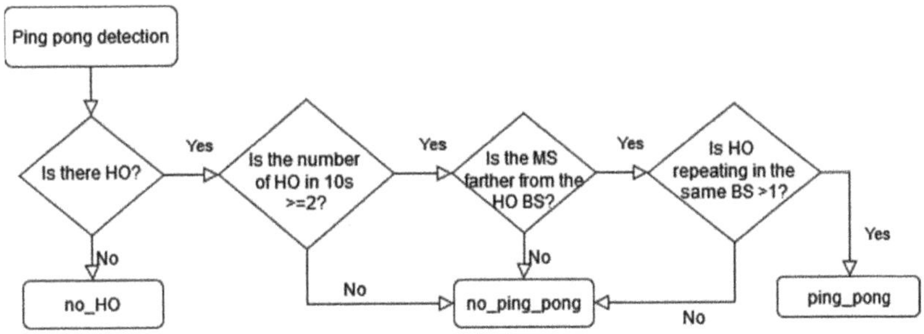

Figure 8. Flowchart of the ping-pong detection during the HO process.

4. Results and Discussions

We post-processed the data collected using the RStudio 2023.06.0 Build 421 version software. We implemented box plots and statistical analyses to obtain the main radio frequency parameters included in the HO scenarios and mobility in the target area. We analyzed and deduced the performance and behaviors for each radio parameter in the measurements and samples for each route. Figure 9 shows the HO type (intra and inter) vs. RSSI for the 3 routes.

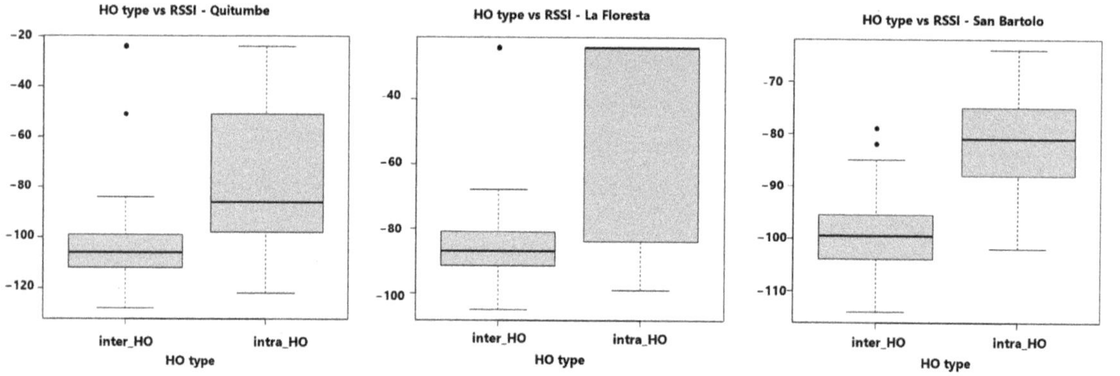

(a) (b) (c)

Figure 9. HO type vs. RSSI: (**a**) Quitumbe, (**b**) Floresta, (**c**) San Bartolo.

If the RSSI decreases to −81 and −86 dBm in the Quitumbe and San Bartolo routes, an intra-cell HO occurs in mobile networks. If the RSSI reduces further between −99 and −106 dBm, an inter-cell HO is performed. While the La Floresta route presents different behavior, intra-cell HO occurs in MNO when RSSI level is approximately −24 dBm, and inter-cell HO occurs when RSSI level is approximately −86 dBm. The results indicate that RSSI level is a key metric for HO type. For example, the RSSI level achieved by Quitumbe is −107 and −86 at inter HO and intra HO, respectively.

Concerning the RSRQ parameter, the HO process is distributed almost 50/50 between inter and intra-cell HO (−13/−17). Inter or intra HO are triggered with similar RSRQ levels in the Floresta route and between 1 dB difference at intra and inter in the other routes.

Figure 10 displays the HO type (intra and inter) vs. times RSSI measurements repeat before HO for the 3 routes. If a RSSI value in the La Floresta route repeats itself more than 20 times, then MNO performs an intra HO and 4 times for inter HO. The Quitumbe and the San Bartolo routes present similar behavior with six and three RSSI repeats before intra and inter HO, respectively. As in the previous case, the results indicate that RSSI repeats is a key metric for HO type.

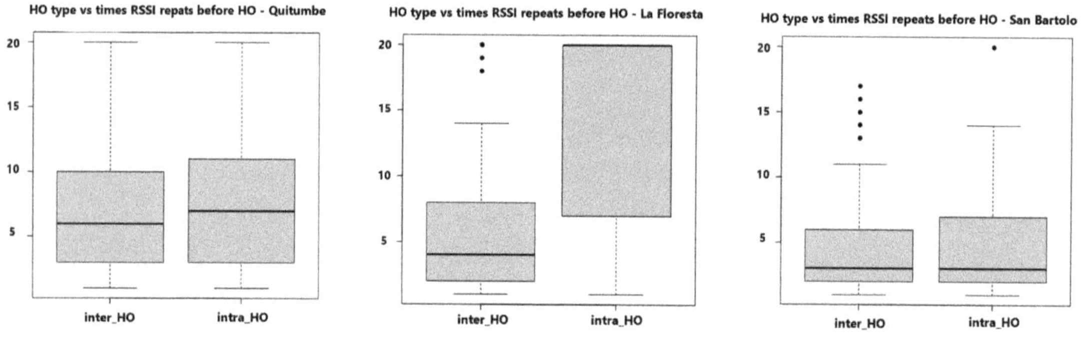

(a) (b) (c)

Figure 10. HO type vs. times RSSI repeats before HO: (**a**) Quitumbe, (**b**) Floresta, (**c**) San Bartolo.

Concerning the RSRQ repeats parameter, the Floresta route presents a singular behavior, RSRQ repeats itself 20 times for intra HO and 5 times for inter HO. The Quitumbe route presents nine repetitions for intra HO and seven for inter HO, while four repetitions for the San Bartolo route perform an inter HO and intra HO.

Figure 11 shows the HO type vs. power margin for all routes. The difference between the current and previous RSSI values is the power margin. Intra or inter HO is activated when power margins are primarily between 1 and 2 dB.

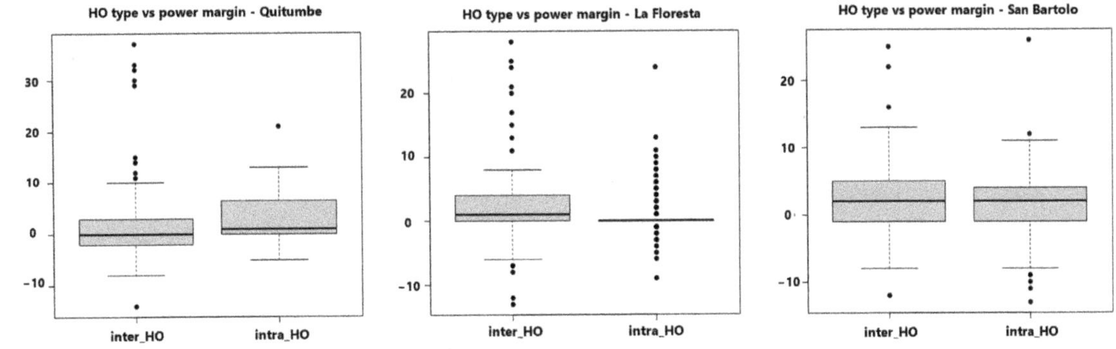

(a) (b) (c)

Figure 11. HO type vs. power margin: (**a**) Quitumbe, (**b**) Floresta, (**c**) San Bartolo.

The results reveal the HO frequency to the nearer or farther BS. In practice, the HOs are performed when the MS gets farther from the HO BS between 42% and 48%. In all routes, MNO achieve that HO occurs when the MS goes nearer the HO BS between 46% and 51%, and the remaining HOs are performed when the MS is stationary. HO to farther BS leads to possible HO failure due to physical distance from the server and the reduction in the quality of radio channel.

The results demonstrate the HO rate with ping pong and without ping pong. The majority of the HOs do not show the ping-pong behavior. Less than 10% of the HOs have this problem.

Figures 12 and 13 illustrate the RSSI and RSRQ for HO process with ping pong and without ping pong (no_ping_pong), respectively, for all routes. The Quitumbe and San Bartolo routes present similar behavior with RSSI down to −102/−107 dBm with ping-pong effects and no ping pong when RSSI is between −82 dBm and −103 dBm. The La Floresta route presents ping pong when RSSI is −24 dBm and no ping pong when RSSI is −82 dBm. An average RSRQ difference of approximately 1 to 3 dB achieves between ping pong and no ping pong. The RSSI and RSRQ may be sufficient key metrics for mobile systems to determine or evaluate the occurrence of ping pong.

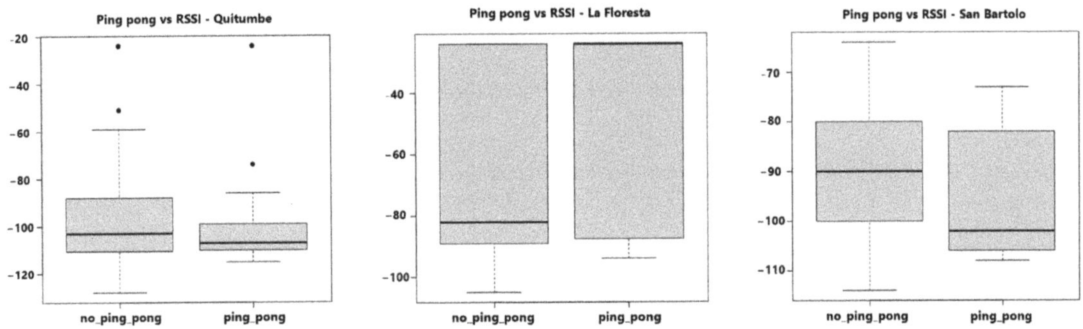

(a) (b) (c)

Figure 12. Ping pong vs. RSSI: (**a**) Quitumbe, (**b**) Floresta, (**c**) San Bartolo.

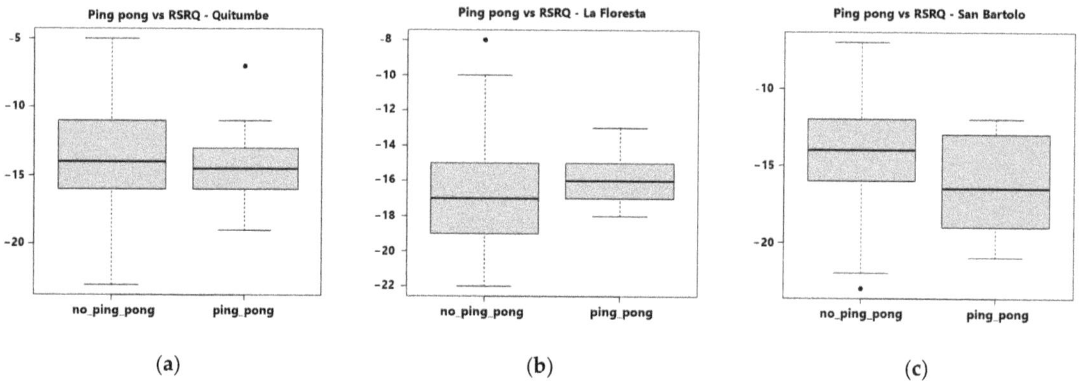

Figure 13. Ping pong vs. RSRQ: (**a**) Quitumbe, (**b**) Floresta, (**c**) San Bartolo.

5. Conclusions

This study provided an HO performance evaluation of MNO in Quito, Ecuador. The data measurements were conducted in a city with 4G and 3G technology infrastructure. The existing network was tested and evaluated with various performance metrics in two scenarios: with intra and inter HO and with the ping-pong process. Most results shown for the three routes had similar results. The difference of average performance metrics (RSSI, RSRQ, power margin, and measurement repeats) is extremely important during different HO types and the ping-pong processes. Therefore, this finding indicates that the MNO may implement algorithms to predict the HO types and the ping-pong process in a specific tested area with the use of radio frequency measurements. In addition, approximately 50% of total HO processes does not occur to nearest BS. This problem needs to be resolved to reduce the HO failure probability. Limitations not considered in this study and can be a point to future works.

Author Contributions: Conceptualization, P.L.-M. and R.L.C.; methodology, R.E., V.F. and R.M.; validation, P.L.-M. and R.L.C.; formal analysis, R.E., V.F. and R.M.; investigation, R.E., V.F. and R.M.; resources, R.E., V.F. and R.M.; writing—original draft preparation, R.E., V.F. and R.M.; writing—review and editing, P.L.-M. and R.L.C.; supervision, P.L.-M. All authors have read and agreed to the published version of the manuscript.

Funding: This research received no external funding.

Institutional Review Board Statement: Not applicable.

Informed Consent Statement: Not applicable.

Data Availability Statement: No new data were created or analyzed in this study. Data sharing is not applicable to this article.

Conflicts of Interest: The authors declare no conflict of interest.

References

1. Huawei Technologies Inc. *LTE eRAN3.0 Handover Fault Diagnosis*, 1st ed.; Huawei Technologies: Shenzhen, China, 2012; pp. 5–10.
2. Kammoun, E.; Zarai, F.; Obaidat, M.S. *Smart Cities and Homes*; Morgan Kaufmann: San Francisco, CA, USA, 2016; Chapter 6; pp. 111–129.
3. Gupta, A.K.; Goel, V.; Garg, R.R.; Thirupurasundari, D.R.; Verma, A.; Sain, M. A Fuzzy Based Handover Decision Scheme for Mobile Devices Using Predictive Model. *Electronics* **2021**, *10*, 2016. [CrossRef]
4. El-Saleh, A.A.; Al Jahdhami, M.A.; Alhammadi, A.; Shamsan, Z.A.; Shayea, I.; Hassan, W.H. Measurements and Analyses of 4G/5G Mobile Broadband Networks: An Overview and a Case Study. *Wirel. Commun. Mob. Comput.* **2023**, *2023*, 6205689. [CrossRef]
5. El-Saleh, A.A.; Alhammadi, A.; Shayea, I.; Hassan, W.H.; Honnurvali, M.S.; Daradkeh, Y.I. Measurement analysis and performance evaluation of mobile broadband cellular networks in a populated city. *Alex. Eng. J.* **2023**, *66*, 927–946. [CrossRef]

6. Tayyab, M.; Gelabert, X.; Jäntti, R. A survey on handover management: From LTE to NR. *IEEE Access* **2019**, *7*, 118907–118930. [CrossRef]
7. Chi, C.; Cai, X.; Hao, R.; Liu, F. Modeling and analysis of handover algorithms. In Proceedings of the IEEE GLOBECOM 2007—IEEE Global Telecommunications Conference, Washington, DC, USA, 26–30 November 2007; pp. 4473–4477.
8. Párraga Villamar, V.; Rocha, C.; Navarrete, H.; Lupera-Morillo, P. Modelos Predictivos de Zonas de Handover en Redes LTE con Base a Mediciones de Campo y Árboles de Decisión (Caso de Estudio Ciudad de Quito). *Rev. Politécnica* **2023**, *52*, 15–24. [CrossRef]
9. Alhammadi, A.; Hassan, W.H.; El-Saleh, A.A.; Shayea, I.; Mohamad, H.; Saad, W.K. Intelligent coordinated self-optimizing handover scheme for 4G/5G heterogeneous networks. *ICT Express* **2023**, *9*, 276–281. [CrossRef]
10. Saad, W.K.; Shayea, I.; Alhammadi, A.; Sheikh, M.M.; El-Saleh, A.A. Handover and load balancing self-optimization models in 5G mobile networks, Engineering Science and Technology. *Int. J.* **2023**, *42*, 101418. [CrossRef]
11. Alhammadi, A.; Hassan, W.H.; El-Saleh, A.A.; Shayea, I.; Mohamad, H.; Daradkeh, Y.I. Conflict resolution strategy in handover management for 4g and 5g networks, Computers. *Mater. Contin.* **2022**, *72*, 5215–5232.
12. Lupera Morillo, P. Modelo Matemático Adaptado para el Cálculo de Pérdidas de Propagación en la Banda de 900 MHz para Microceldas en la Ciudad de Quito. *Rev. Politécnica* **2018**, *41*, 29–36. Available online: https://revistapolitecnica.epn.edu.ec/ojs2/index.php/revista_politecnica2/article/view/883 (accessed on 15 June 2023).

engineering proceedings

MDPI

Proceeding Paper

Implementation of A Data-Acquisition System and Its Cloud-Based Registration Using the Unified Architecture of Open Platform Communications †

Anthony Molina *, Diego Vargas and Ana Rodas

Departamento de Automatización y Control Industrial, Escuela Politécnica Nacional (EPN), Quito 170525, Ecuador; diego.vargas@epn.edu.ec (D.V.); ana.rodas@epn.edu.ec (A.R.)
* Correspondence: anthony.molina@epn.edu.ec
† Presented at the XXXI Conference on Electrical and Electronic Engineering, Quito, Ecuador, 29 November–1 December 2023.

Abstract: Standardization and collaborative integration are fundamental for the implementation of Industry 4.0. The Open Communication Platform Unified Architecture (OPC UA) standard plays a crucial role in communication by enabling the development of heterogeneous systems and facilitating the seamless exchange of data between devices. To take full advantage of OPC UA's capabilities, it is necessary to unlock other application services, such as cloud computing. By having more tools at their disposal, industries can increase their efficiency and optimize data-driven decision making. In this study, it was proposed to use OPC UA standard in a Client/Server model to leverage OPC UA paradigms along with Software Development Kits (SDKs). The development was carried out using Python as the programming language, both to host the server on a Raspberry Pi 4B and to set up the client on a Personal Computer (PC). This client will centralize the data for sending to the Clever Cloud logging platforms and for web visualization on Tago.IO. The interoperability problem of communicating between two platforms with different Operating Systems (OS) was addressed by integrating the OPC UA standard. The configured interval of 10 s recorded an average of 11.11 s, with a maximum timeout of 59 s for data-logging temperature, pressure, humidity, and altitude. The client proved to be Hypertext Transfer Protocol Secure (HTTPS) compliant, allowing connection to Web platforms.

Keywords: OPC UA; industry 4.0; monitoring; cloud; Python

Citation: Molina, A.; Vargas, D.; Rodas, A. Implementation of A Data-Acquisition System and Its Cloud-Based Registration Using the Unified Architecture of Open Platform Communications. *Eng. Proc.* **2023**, *47*, 20. https://doi.org/10.3390/engproc2023047020

Academic Editor: Soraya Lucia Sinche Maita

Published: 7 December 2023

1. Introduction

In Industry 4.0, industrial control and monitoring systems must be at the forefront, demanding constant research [1]. Tools such as data mining, the internet of things, automation, additive manufacturing, cloud computing, and renewable energies are used to create value and improve services, work organization, and business models [2]. However, the variety of industrial communication technologies makes it difficult to integrate applications [3]. Currently, there are several industrial communication protocols based on the Industrial Ethernet (EI), such as EtherNet/IP, PROFINET, or EtherCAT, widely used but not compatible with each other [3]. This generates problems for sharing information with information technologies (IT). To address this, the OPC UA standard is adopted, which harmonizes protocols by offering equivalent functions, communication security, and flexible design [4]. From an economic perspective, it would be beneficial to apply OPC UA in open hardware devices like Arduino since, according to [5], they can drive projects due to their low cost, making them ideal for proposal development. This would enable the use of cloud-computing tools, promoting standardized communication in the field and interoperability with IT and Industry 4.0 tools.

2. General Definitions

The OPC Foundation developed two industrial standards. The first one, called OPC Classic, aims for interoperability in automation, but it relies on Windows and DCOM, which limits online communication [6]. To address this, the OPC UA standard was created, which utilizes high-performance protocols for communication [7].

OPC UA Client/Server model: OPC UA uses the Client/Server model. Clients initiate communication and connect to servers [7]. Servers provide data and respond to client requests. This is useful for collecting data in distributed control systems, where a Master Terminal Unit (MTU) acts as a client to obtain information from Remote Terminal Units (RTUs) and can also serve as a server to share data with other OPC UA clients.

OPC UA addresses: OPC UA addresses communication and information modeling. The server consists of address spaces with nodes and attributes [7]. Nodes have attributes that describe the information. The unique identification of nodes allows clients to access attribute values in service requests.

OPC UA security: According to [7], the OPC Foundation establishes different levels of security for communication between the server and client in OPC UA, ranging from no security to high security: None, Basic128Rsa15 (medium-level security), Basic256 (medium-high-level security), and Basic256Sha256 (high security). Additionally, the OPC Foundation offers various methods of authentication, including Anonymous (no user data), User (requires username and password), X509v3 (uses digital certificates), and Certificate (employs security tokens for authentication).

OPC UA data transport: The protocols for clients and servers to share data in OPC UA include TCP, which provides a reliable full-duplex communication channel that is connection oriented, and SOAP/HTTP, which operates with structured SOAP messages transmitted over HTTP, enabling a secure data exchange between devices [7].

3. State of the Art Study

Here are some similar studies to this project. A study conducted by [8] focuses on establishing a sensor network using Arduino devices and MQTT communication to manage transmission between sensors. This is integrated with cross-platform applications through the OPC UA standard, which serves as a solution to centralize communication between devices and the data they generate. Grafana is used for data analysis afterward, with Java programming language highlighted throughout the process.

In the works [9,10], the focus is on the OPC UA standard, although they address different applications and contexts. In the first work, a server is implemented on a Raspberry Pi 2, enabling communication between devices using the Modbus and OPC UA protocols. On the other hand, the second work employs a server on an Arduino Yun for communication between devices using Serial communication and the OPC UA protocol, although bottlenecks were identified on this board. In both cases, commercial OPC UA clients were used for communication with the server.

4. Development of the Proposal

Figure 1 depicts the connection diagram of the utilized devices. The Raspberry Pi 4B board (server) employs I2C communication to interface with the transducer, and utilizes OPC UA to establish a connection with a personal computer (client). This computer is responsible for reading and centralizing the data, subsequently employing cloud-computing tools such as database storage and graphical interface visualization.

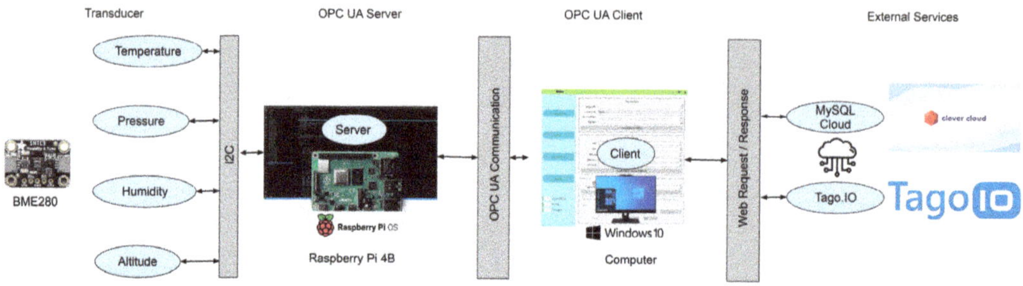

Figure 1. Proposed system architecture.

4.1. OPC UA Server Configuration

The I2C protocol is used to connect the Raspberry Pi 4B with the BME280 transducer. The transducer's parameters are configured, and values for temperature, pressure, humidity, and altitude are captured. Subsequently, an OPC UA server is created using the asyncua Server SDK. Address spaces are established for the variables. For security purposes, the Basic256Sha256 security suite is configured, utilizing AES 256-bit encryption and the SHA-256 algorithm. Secure communication is established by incorporating the certificate and key.

4.2. OPC UA Client Configuration

Creating an OPC UA client is essential to accessing data from the server and connecting the OPC UA standard with cloud services. The programming steps are outlined below.

Connecting to the OPC UA Client: OPC UA connectivity is implemented using the asyncua Client SDK. A connection class is created with a constructor that receives the server's address, security indicator, certificate path, and private key path (if necessary). The constructor initializes the server's address and creates an instance of the OPC UA client. Another method verifies the connection and returns true.

Connection for logging: PyMySQL SDK is used to connect the application to the Clever Cloud database. In the connection class, the necessary credentials for communication are configured. Then, a cursor is created to execute queries, and a connection verification method is implemented, returning a positive value if the connection is established.

Connection for visualization: A class is included to link the application with the Tago.IO platform through its SDK. Similar to other connections, an initialization method is used for the connection parameter, which is a key specifying access to Tago.IO. Connection verification is done using a method to determine the success of the connection.

Data acquisition from the server: A proposal to implement code for reading and storing OPC UA server addresses is presented. The code consists of two methods: one for initial configuration and another for reading variables. In the first method, dictionaries are used to manage variables, recording them for cloud database and visualization purposes. Attributes are also set to indicate that the attributes have not yet been recorded. The second method reads specific values on the OPC UA server, using arguments for the address and node name, along with optional configurations for recording or visualization. The presence of values in the dictionary is checked before recording, and SQL queries and lists are created for sending to Tago.IO.

Graphical user interface (GUI): A graphical interface designed in QT Designer is utilized to create a GUI that enables the configuration of the OPC UA client with cloud-computing tools. The interface comprises three sections: control buttons, an options bar, and a workspace. The options bar is divided into Configuration (for connection data and address space), Status (displaying nodes and names), Visualization (displaying data), and Help (documentation). Figure 2 displays the configuration interface with fields for input and project execution, along with OPC UA client code parameters.

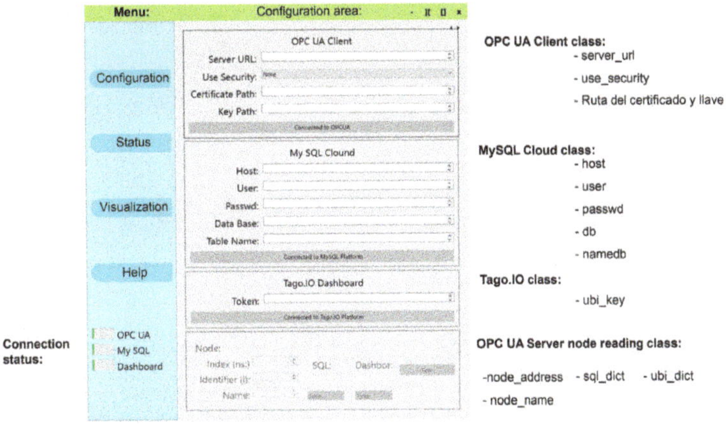

Figure 2. GUI for OPC UA Client.

4.3. Platform Configuration

Database configuration: To store data from an OPC UA client, Clever Cloud is utilized to create an SQL database integrated into the project. Management is conducted through PHPMyAdmin. A table is configured for the variables such as date, temperature, pressure, humidity, and altitude, before establishing the OPC UA client. The date field will be entered each time a value is recorded, and the variables can be nullable since the input of variables from the GUI is singular in nature.

Visualization-platform configuration: The project aims to display data from the OPC UA client in an internet-accessible interface. A HTTPS device was created in Tago.IO, linked with the OPC UA client using a token from the created device, and variables were matched with widgets for visualization. Additionally, this platform provides alert tools based on registered values.

5. Results and Discussion

The tests were conducted in a closed test environment equipped with a temperature and humidity controller, along with an RH35 data logger, to study how the proposal operates under real operating conditions. The testing process began with assessing the connection between platforms and the OPC UA standard. Subsequently, tests were conducted to capture data using cloud-computing tools, involving 8 h trials for intervals of 10, 20, and 30 s. These tests verified data and service availability. Additionally, a comparison was made between the data stored by the RH35 logger and the data recorded in Clever Cloud.

5.1. Testing Connection with OPC UA and Proposed Services

The communication involves an OPC UA server on a Raspberry Pi 4B, and a client on a PC. Communication and connection tests were carried out for recording and visualization purposes. In the first test, the server had no security level. The configuration included URLs and credentials for Clever Cloud and Tago.IO. The connection was tested with a security level, using self-signed certificates from OpenSSL on both the server and the client. The connections were successful, as indicated by the green status bars. Once the respective connections were established, platforms were usable, and nodes could be configured to avoid conflicts.

5.2. Results of OPC UA Implementation and Proposed Services

The tests verified the average recording time for different intervals and the functionality of the platforms. Variables like temperature, pressure, and humidity were monitored to assess the data accuracy. The results were presented in Figure 3, with the red line representing Clever Cloud and the blue line representing the RH35 logger.

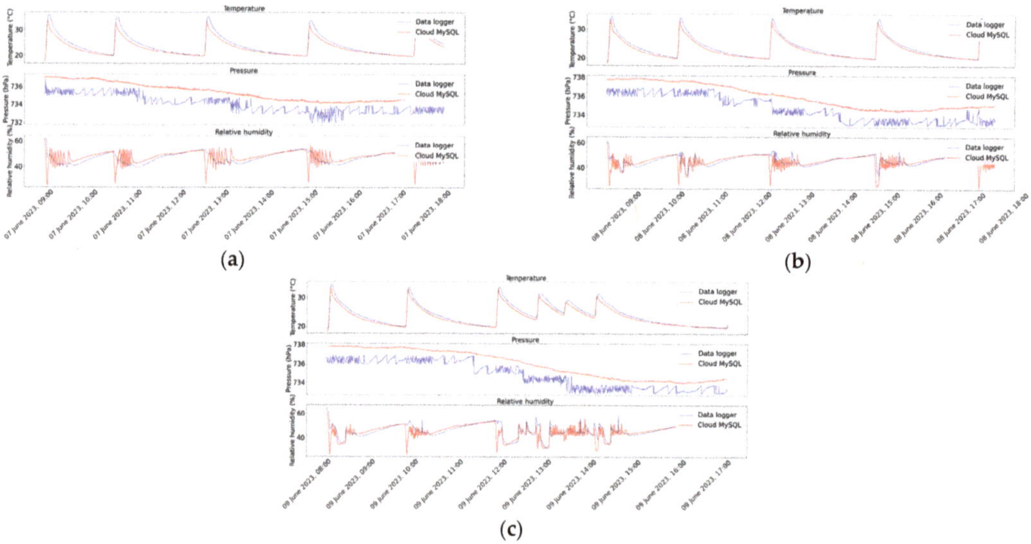

Figure 3. Implementation tests: (**a**) Timeline of the test with a 10 s interval; (**b**) timeline of the test with a 20 s interval; (**c**) timeline of the test with a 30 s interval.

Table 1 displays records and averages of temperature, pressure, and humidity, along with the Root Mean Square Error (RMSE) obtained.

Table 1. Presentation of the results achieved during the tests.

Interval	Variable	Devices	Records	Mean	RMSE
10 s	Temperature (°C)	RHT35	1128	23.85	1.44
		Cloud	2951	23.26	
	Pressure (hPa)	RHT35	1128	734.15	1.36
		Cloud	2951	735.41	
	Humidity (%)	RHT35	1128	47.86	4.52
		Cloud	2951	48.25	
20 s	Temperature (°C)	RHT35	1091	23.67	1.02
		Cloud	1543	23.13	
	Pressure (hPa)	RHT35	1091	734.77	1.33
		Cloud	1543	736.04	
	Humidity (%)	RHT35	1091	47.49	4.21
		Cloud	1543	46.64	
30 s	Temperature (°C)	RHT35	1085	24.29	0.94
		Cloud	1043	23.68	
	Pressure (hPa)	RHT35	1085	734.91	1.21
		Cloud	1043	736.06	
	Humidity (%)	RHT35	1085	46.95	4.02
		Cloud	1043	46.16	

First test (7 June 2023, 09:00): A duration of 8 h with 10 s intervals. No errors were detected. The maximum interval between records was 59 s. The average between records was 11.11 s. Second test (8 June 2023, 09:00): A duration of 8 h with 20 s intervals. No errors were recorded. The maximum interval between records was 50 s. The average between records was 21.10 s. Third test (9 June 2023, 09:00): A duration of 8 h with 30 s intervals.

There were no errors during the test. The maximum interval between records was 51 s. The average between records was 31.14 s.

The differences between the data can be attributed to the relatively lower precision but cost-effective nature of the Raspberry Pi and the BME280 transducer. The project's focus lies in monitoring and recording data in the cloud for storage and visualization through cloud computing. RMSE was employed to gauge the discrepancy between estimated (Clever Cloud) and actual (RH35) data [11]. A Python script calculated RMSE as the squared difference between these datasets, with lower values indicating higher precision.

Linear interpolation was used to estimate missing data in the logging device, which had a minimum interval of 30 s. In terms of temperature, it was observed that a greater interval between measurements led to higher precision (lower RMSE). Pressure maintained a consistent range across different intervals. Humidity also improved its precision with longer intervals. Shorter records offer a detailed view of changes, while longer records provide a general and smoothed perspective.

Tago.IO was useful for visualizing measurements and alarms, updating in sync with the database, and rapidly translating information into graphs and alerts. The platform is accessible on devices with an internet connection.

6. Conclusions

A technological solution was implemented using OPC UA to monitor variables like temperature, pressure, and humidity during 8 h tests. Despite some delays in data visualization, the Python-based solution offers flexibility for monitoring systems in industrial environments. The implementation demonstrated communication between devices with different operating systems (Raspberry Pi OS and Windows 10) and the integration of cloud-computing platforms such as Clever Cloud and Tago.IO for data recording and visualization. Although limited by using free plans, it did not affect data availability.

Prototype tests confirmed the stability and reliability of the monitoring system in a real environment for four variables. There were wait times due to network conditions and platform services, which could be improved with paid services or local servers.

OPC UA standard is compatible with HTTPS, making its implementation on Industry 4.0 platforms easier. The configurable interface connects with OPC UA clients, recording data in the cloud and visualizing it on the web.

The current project is primarily focused on monitoring, but in the future, it could incorporate the Network Time Protocol for precise synchronization in control applications alongside OPC UA. Additionally, a potential option for future updates could involve having a local visualization interface that includes iconography to enhance data comprehension within the OPC UA client application.

Author Contributions: Conceptualization, A.M. and D.V.; methodology, A.M.; software, A.M.; validation, A.M., D.V. and A.R.; formal analysis, A.M.; investigation, A.M.; resources, A.M.; data curation, A.M.; original draft writing, A.M.; writing, revision, and editing, A.M., D.V. and A.R.; visualization, A.M. and A.R.; supervision, D.V. and A.R.; project administration, A.M. All authors have read and agreed to the published version of the manuscript.

Funding: This research received no external funding.

Institutional Review Board Statement: Not applicable.

Informed Consent Statement: Not applicable.

Data Availability Statement: The following link is shared where you can find the Python script to obtain the displayed graphs, as well as the values from the table. Additionally, the link contains the data frames used in the analysis. (https://mega.nz/folder/jNpAHTyT#JySXOJ8C8rR3SX5O3ttUNQ (accessed on 25 October 2023)).

Conflicts of Interest: The authors declare no conflict of interest.

References

1. Sodhi, H. When industry 4.0 meets lean six sigma: A review. *Ind. Eng. J.* **2020**, *13*, 1–12. [CrossRef]
2. Gallo, T.; Cagnetti, C.; Silvestri, C.; Ruggieri, A. Industry 4.0 tools in lean production: A systematic literature review. *Procedia Comput. Sci.* **2021**, *180*, 394–403. [CrossRef]
3. Drahos, P.; Kucera, E.; Haffner, O.; Klimo, I. Trends in industrial communication and OPC UA. In *2018 Cybernetics & Informatics (K&I)*; IEEE: Piscataway, NJ, USA, 2018; pp. 1–5. [CrossRef]
4. Schwarz, M.H.; Borcsok, J. A survey on OPC and OPC-UA: About the standard, developments and investigations. In Proceedings of the 2013 XXIV International Conference on Information, Communication and Automation Technologies (ICAT), Sarajevo, Bosnia and Herzegovina, 30 October–1 November 2013; IEEE: Piscataway, NJ, USA; pp. 1–6. [CrossRef]
5. Kim, S.-M.; Choi, Y.; Suh, J. Applications of the Open-Source Hardware Arduino Platform in the Mining Industry: A Review. *Appl. Sci.* **2020**, *10*, 5018. [CrossRef]
6. McIlvride, B.; Thomas, A.; Real, C. *OPC Tunnelling–Know Your Options*; Cogent Real-Time Systems Inc.: Mississauga, ON, Canada, 2008.
7. Mahnke, W.; Leitner, S.-H.; Damm, M. *OPC Unified Architecture*; Springer Science & Business Media: Berlin, Germany, 2009.
8. Vimos, V.; Sacoto Cabrera, E.J. Results of the implementation of a sensor network based on Arduino devices and multiplatform applications using the standard OPC UA. *IEEE Lat. Am. Trans.* **2018**, *16*, 2496–2502. [CrossRef]
9. Muller, M.; Wings, E.; Bergmann, L. Developing open source cyber-physical systems for service-oriented architectures using OPC UA. In Proceedings of the 2017 IEEE 15th International Conference on Industrial Informatics (INDIN), Emden, Germany, 24–26 July 2017; IEEE: Piscataway, NJ, USA; pp. 83–88. [CrossRef]
10. Pena, R.R.; Fernandez, R.D.; Lorenc, M.; Cadiboni, A. Gateway OPC UA/Modbus applied to an energy recovery system identification. In Proceedings of the 2019 XVIII Workshop on Information Processing and Control (RPIC), Bahia Blanca, Argentina, 18–20 September 2019; IEEE: Piscataway, NJ, USA; pp. 235–240. [CrossRef]
11. Kambezidis, H.D. The Solar Resource. In *Comprehensive Renewable Energy*; Elsevier: Amsterdam, The Netherlands, 2012; pp. 27–84. [CrossRef]

engineering proceedings

MDPI

Proceeding Paper

Advancing the Industrial Sector Energy Transition with Hybrid Solar Systems: Evaluation of Small Winemaking in Ecuador †

Andrés Villarruel-Jaramillo [1,*], Josué F. Rosales-Pérez [1], Manuel Pérez-García [2], José M. Cardemil [1,3] and Rodrigo Escobar [1,3]

1 Departamento de Ingeniería Mecánica y Metalúrgica, Escuela de Ingeniería, Pontificia Universidad Católica de Chile, Vicuña Mackenna 4860, Santiago 7820436, Chile; jhrosales@uc.cl (J.F.R.-P.); jmcardem@uc.cl (J.M.C.); rescobar@ing.puc.cl (R.E.)
2 CIESOL Research Center on Solar Energy, Joint Center UAL-CIEMAT, University of Almería, Ctra. Sacramento s/n, 04120 Almería, Spain; mperez@ual.es
3 Centro del Desierto de Atacama, Pontificia Universidad Católica de Chile, Vicuña Mackenna 4860, Santiago 7820436, Chile
* Correspondence: afvillarruel@uc.cl
† Presented at the XXXI Conference on Electrical and Electronic Engineering, Quito, Ecuador, 29 November–1 December 2023.

Abstract: Ecuador may face an accelerated energy transition in the next 15 years due to the reduction of its oil reserves, affecting industries that require oil derivatives. To achieve a transition toward renewable energy without affecting the industry, this research proposed a technoeconomic evaluation of a hybrid system with solar flat plate collectors and photovoltaic modules that drive an absorption chiller and a compression chiller. The system supplies heat, cooling, and electricity to the winemaking industry in Ecuador. The best results for the hybrid systems reached a levelized cost of energy (LCOEn) of 0.171 USD/kWh and 0.157 USD/kWh in Guayaquil and Quito, respectively. For both locations, the LCOEn for the hybrid systems represents a decrease of the LCOEn of 53% and 32% concerning the individual solar photovoltaic and solar thermal systems, respectively. Therefore, the proposed hybrid system has a significant potential to integrate solar energy into the industry sector.

Keywords: hybrid solar cooling system; solar heating and cooling systems; solar photovoltaic; solar thermal; absorption chillers; sustainable industry

Citation: Villarruel-Jaramillo, A.; Rosales-Pérez, J.F.; Pérez-García, M.; Cardemil, J.M.; Escobar, R. Advancing the Industrial Sector Energy Transition with Hybrid Solar Systems: Evaluation of Small Winemaking in Ecuador. *Eng. Proc.* **2023**, *47*, 21. https://doi.org/10.3390/engproc2023047021

Academic Editor: Jaime Cepeda

Published: 7 December 2023

1. Introduction

The worldwide main goals for energetic transition are focused on achieving energy decarbonization by 2050 and, thus, on reducing the gas emissions responsible for global warming [1]. Therefore, it is necessary to attenuate the growing energy demand since developing countries will project the most significant increase in energy consumption in 2050, and currently, the industry represents 30% of energy consumption, being the most critical sector in carbon emissions [2]. According to Sikder et al. [3], in developing countries, the industrial sector tends to grow together with population and economy. In the case of Ecuador, although the industry represents 17% of the total energy consumption, it is expected that energy consumption in the industry will continue growing in the future [4]. In addition, it is relevant to consider that Ecuador is an oil-producing country in which 76% of its energy demand comes from oil derivatives, concentrating maximum demand on fuels with a high carbon footprint [4]. According to Espinoza et al. [5], it is expected that between 2035 and 2045, oil production will cease due to the end of the useful life of oil wells and commercial shortage produced by changes in the energy matrix worldwide. Consequently, Ecuador faces a potential deterioration of economic activity in sectors with high oil dependence such as the industrial and transport sectors. Therefore, establishing technological options to cover energy consumption in the industrial sector is essential to guarantee economic growth during the period of energy transition.

The exploration of nonconventional renewable energy for industrial and residential processes has been extensively detailed for heat processes, and solar energy is the most accessible technology for micro and medium plants in comparison to geothermal or wind energy due to its worldwide availability [6]. Solar energy stands out for being available anywhere in the world, with ease of installation; additionally, technology such as photovoltaics has maintained a sustained decrease in costs compared to other renewables [7,8]. However, solar energy requires detailed analysis to determine its potential application due to the high variability in terms of location, weather, and efficiency. Regarding the industrial sector, industrial processes require a supply of heat and cooling to manufacture the product; for example, the automotive industry or materials industry (bricks and blocks or woods) requires temperatures over 160 °C and the food industry requires heat and cooling processes between −15 °C to 220 °C [9]. In most cases, the main source of energy in small and medium industries tends to be electricity to drive chillers and liquefied petroleum gas (GLP) to supply heat.

Solar energy has been evaluated for heat supply in the mining sector for processes that need low and medium temperatures using flat plate solar collectors (FPCs) [10]. Applications in northern Chile with high solar availability and solar fields between 10,000 to 170,000 m^2 have achieved a levelized cost of heat (LCOH) between 62 at 103 USD/MWh [11]. Furthermore, the integration of solar energy in cooling processes has been highly detailed for the residential sector, in which projects such as IEA-Task-53 have developed guides for its application to building air conditioning processes [12–14]. In this sense, Cabrera et al. evaluated FPC solar fields with absorption chillers (ABCHs), finding that the levelized cost of energy in different European cities varies between 0.06 to 1.6 EUR/kWh [15]. However, integrating solar photovoltaic (PV) fields with solar-assisted heat pumps to provide heat, cooling, and the use of excess electricity for self-consumption (sfc) in the residential sector has reduced the levelized energy cost to 0.04 EUR/kWh [16]. The cost reduction comes mainly from eliminating the battery pack from the photovoltaic system and replacing it with thermal storage [17]. Additionally, one of the advantages of air conditioning systems in the residential sector is that the demand for cooling tends to concentrate during summer, which is a substantial difference compared to the demand profile of the industrial sector [18]. In this sense, the industry during the manufacturing process requires temperature controls throughout the year, requiring a cooling supply even during winter months [19]. In this aspect, Pino et al. [20] evaluated a PV system integrating an air-to-water chiller (AWCH) for cooling and thermal resistance in a hot thermal energy storage tank (*htes*) for beer production in Spain, achieving a levelized cost of cooling and heating (LCOCH) between 0.2 to 0.22 EUR/kWh [20]. The use of solar thermal (ST) systems with ABS for milk production in a small industry in India was also evaluated, reaching a levelized cost of energy (LCOEn) of 0.177 USD/kWh [21].

The industry requires cooling, heat, and electricity supply (trigeneration). However, the integration of trigeneration solar systems at medium and small scales has focused on the residential sector [22]. Solar hybrid (HYB) systems integrated with ST and PV to drive an ABCH and AWCH have been evaluated energetically, comprising individual ST and PV solar fields to drive an ABCH and AWCH. In this sense, the HYB systems to supply cooling, heat, and electricity have been evaluated for greenhouse crops production in Spain, achieving a solar fraction (SF) of 0.85, which represents a reduction of 27% compared to an individual ST-ABCH system and a fractional energy saving (f_{sav}) of 83% [23]. The HYB systems have an important potential to be applied in the industrial sector for heat and cooling processes with FPC and PV technologies. However, the studies mentioned above lack a technoeconomical comparison between individual ST, PV, and HYB systems to define the most appropriate technological selection based on the efficiency and economic assumptions of the systems. In the case of Ecuador, despite the energy problem that the country is facing, little exploration exists of the behavior of solar systems integrated into industrial processes that detail the feasibility of these technologies considering the economic parameters, which vary greatly depending on the location. Therefore, the objective of the research is to evaluate the technoeconomic behavior of HYB systems combined with PV–ST and AWCH–ABCH to supply heat, cooling, and electricity for industrial applications. The selected industrial sector is the wine industry located in the rural area of Quito and Guayaquil.

2. Methodology

2.1. System Description

The hybrid system comprises an FPC and PV solar field that produces heat and electricity. In the heat circuit, the FPC system is predesigned to reach a maximum of 95 °C and a minimum of 75 °C to charge the *htes*. The state of charge (SOC) of the *htes* is controlled based on the average temperature ($T_{avg,htes}$) and the outlet temperature ($T_{out,htes}$) of the htes toward the divergent valve. The main objective of the useful heat from the FPC system ($q_{us,st}$) is to supply heat to the generator (q_g) in the ABS chiller through the heat exchanger 1 (hx_1) and when the SOC of the *htes* reaches a value of zero, the heat flow of the gas boiler 1 (q_{aux1}) will be activated. In contrast, in periods where the ABS is deactivated, heat is delivered through heat exchanger 2 (hx_2). The hx_2 receives water at ambient temperature, which will be heated to a set temperature ($T_{set,hp}$) of 60 °C. If the temperature is less than 60 °C, it will be heated by gas boiler 2 to supply the heat flow required by the production processes (q_{hp}). The electricity circuit supplies electric power from the PV system ($p_{us,PV}$) for self-consumption ($p_{PV,sfc}$) and the consumption of the AWCH (p_{AWCH}), the pumps of the HYB system (p_{pumps}), the cooling tower fan (p_{cwt}), and the panel control of the ABCH system (p_{ABCH}). If the $p_{pv,us}$ exceeds the $p_{pv,sfc}$ and p_{AWCH} then the electricity excess is sold to the electrical grid (p_{grid}), which is considered as a revenue ($p_{PV,rv}$) in the calculation of the LCOCH. Furthermore, the cold circuit comprises the ABCH with a cooling tower (cwt), the AWCH chiller, and two cold storage tanks (*ctes*) that extract heat from the production process. The chillers alternate their activation depending on the SOC of $ctes_1$ and $ctes_2$. If the SOC of $ctes_1$ and $ctes_2$ is near zero, the ABCH will activate the direct flow to supply the cooling demand (q_{cp}). Figure 1 shows the scheme of the HYB system that will be compared with the individual ST–ABCH and PV–AWCH.

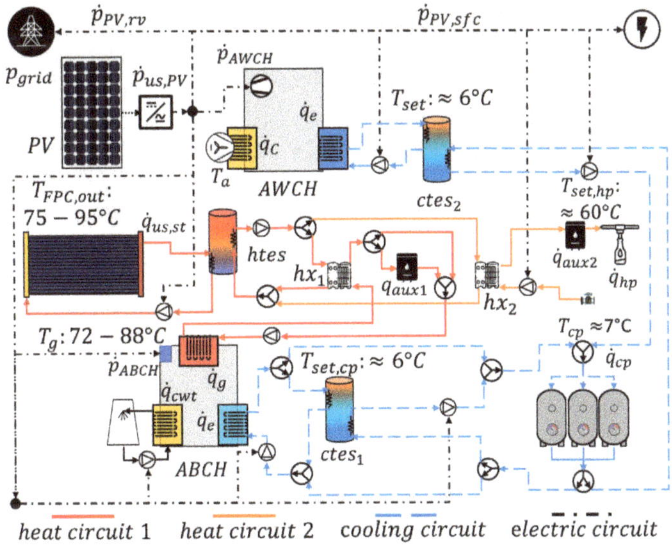

Figure 1. Hybrid solar cooling system scheme to supply heat, electricity, and cooling coverage for the manufacturing process.

The solar radiation on an inclined plane (I_T) is calculated using the HDKR method, and the effective absorbed solar radiation (S) is calculated assuming that the diffuse and ground-reflected radiation are isotropic [24]. The PV field was simulated with MATLAB R2021a using the five-parameter diode method to estimate the PV current (I_{PV}) and voltage (V_{PV}) [25], as shown in Equation (1), and the SANDIA temperature method to estimate the cell temperature (T_{ce}) of the PV module [26],

$$I_{PV} = I_L - I_o \cdot \left[e^{\left(\frac{V + I \cdot R_s}{nd \cdot \frac{N_{cells,PV} \cdot k \cdot T_{ce}}{q}} \right)} - 1 \right] - \frac{V + I \cdot R_s}{R_{sh}} \tag{1}$$

where I_L is the light current, I_o is the diode reverse saturation current, R_s is the serial resistance, nd is the diode ideality factor, k is the Boltzmann constant, $N_{cells,PV}$ is the number of PV cells in series, R_{sh} is the shunt resistance, and V is voltage. To model the inverter, the SANDIA methodology is used, which estimates the production of alternating current power ($\dot{p}_{us,PV}$) converted from the direct current power ($\dot{p}_{PV,dc}$) provided by the PV field [26]. The inverter is selected according to the maximum voltage of the PV array and the maximum power of the system for the day with the highest radiation from the available library of NREL [27]. The main characteristics in state test conditions (stc) of the selected PV module are shown in Table 1; these are efficiency (η_{PV}), module area (A_{PV}), maximum power (p_{mp}), maximum power current ($I_{mp,stc}$), and maximum power voltage ($V_{mp,stc}$). The results of the PV system simulation are loaded onto TRNSYS with TYPE-9.

Table 1. Main characteristics of the mono-c-Si PV module obtained from the SAM library [27].

Material	η_{PV} (%)	A_{PV} (m^2)	p_{mp} (W)	$I_{mp,stc}$ (A)	$V_{mp,stc}$ (V)
Mono-c-Si	23.8	1.87	445	10.7	41.6

The ST system with FPCs is simulated on TRNSYS using TYPE-1289 which is based on the EN12975-2 Dynamic Efficiency Approach (ASHRAE IAMs) to predict the energy behavior of the solar collector (sc) [28]. The method consists of resolving Equation (2) from coefficients obtained in a test performance of solar collectors,

$$\frac{\dot{q}_{us,sc}}{A_{sc}} = \eta_o \cdot K_{\theta b}(\theta) \cdot I_b + \eta_o \cdot K_{\theta b}(\theta) \cdot I_d - c_1 \cdot (T_{sc} - T_a) - c_2 \cdot (T_{sc} - T_a)^2 - c_3 \cdot \frac{dT_{sc}}{dt} \tag{2}$$

where A_{sc} is the area of the sc, η_o is the zero-loss efficiency, c_1 is the linear thermal loss coefficient, c_2 is the quadratic thermal loss coefficient, c_3 is the wind speed dependence of the heat loss coefficient, and $K_{\theta b}$ and $K_{\theta d}$ are the incidence angle modifiers for beam and diffuse radiation, respectively. The FPC selected is the HEATboost 35/10 collector whose main characteristics are detailed in Table 2 [28].

Table 2. Main characteristics of the HEATboost 35/10 collector [27].

A_{sc} (m^2)	η_o (−)	c_1 (W/m^{2}K)	c_2 (W/m^{2}K^2)	c_3 (kg/m^{2}K)	K_d
13.57	0.779	2.41	0.015	6.798	0.98

The htes and ctes are stratified storage tanks that are simulated with TYPE-158. The auxiliary gas boiler is simulated using TYPE-1222, which is a simplified model to predict the $\dot{q}_{aux}$ required to cover the heat demand and considers a constant boiler efficiency ($\eta_{gb} = 0.8$) to calculate the gas energy input needed for the boiler ($\dot{q}_{gb}$). The simulation of the chillers necessitates the application of a polynomial regression based on the data reported in the catalog of the manufacturer. In the case of the AWCH, York chillers are used, with a nominal capacity ranging from 44.7 to 111 kW [29]. The simulation is carried out with TYPE-118 to predict the $\dot{q}_{e,AWCH}$ and $\dot{p}_{c,AWCH}$ according to the T_a, the input temperature of the evaporator ($T_{in,AWCH}$), and $T_{set,AWCH}$. The single-effect water/lithium bromide ABCH requires applying the SEABC method using MATLAB to predict the high nonlinear performance of the q_{cwt}, q_g, and $q_{e,ABCH}$ that is based on input temperatures of the generator ($T_{in,g}$), cooling water tower ($T_{in,cwt}$), evaporator ($T_{e,in}$), and the $T_{set,ABCH}$. The ABCH is simulated in TRNSYS using TYPE-117, considering equipment with a nominal capacity ranging from 17.7 kW to 105 kW of the YAZAKI catalog [30]. The detailed methodology of the validation of each component in the systems is described in [23].

The balance energy to supply the demand with the HYB system is shown in Equations (3)–(5), where $\dot{q}_{loss}$ is the heat losses, $\dot{p}_{loss}$ is the electric power losses, and $\dot{p}_{el}$ is the total electric power demand of the industry.

$$\dot{q}_{cp} \cong \dot{q}_{AWCH} + \dot{q}_{ABCH} - \dot{q}_{loss,ctes1} - \dot{q}_{loss,ctes2} \tag{3}$$

$$\dot{q}_{hp} \cong \dot{q}_{us,ST} + \dot{q}_{qux1} + \dot{q}_{aux2} - \dot{q}_{loss,htes} - \dot{q}_{loss,hx1} - \dot{q}_{loss,hx2} - \dot{q}_g \tag{4}$$

$$\dot{p}_{el} \cong \dot{p}_{us,PV} + \dot{p}_{grid} - \dot{p}_{AWCH} - \dot{p}_{ABCH} - \dot{p}_{loss,inv} - \dot{p}_{sfc} - \dot{p}_{PV,rv} - \dot{p}_{pumps} \tag{5}$$

2.2. *Key Performance Indicators*

The energy evaluation is carried out based on the total annual thermal energy (Q) and electric energy (P) of the systems. Equations (6) and (7) show the SF and the fractional savings of CO_2 ($f_{sav,CO2}$):

$$SF = \frac{\sum Q_{g,ST} + \sum Q_{hp,ST} + \sum P_{pr,PV}}{\sum Q_{g,ST} + \sum Q_{g,aux1} + \sum Q_{hp,ST} + \sum Q_{hp,ST} + \sum P_{pr}} \tag{6}$$

$$f_{sav,CO2} = \left(1 - \frac{e_{rn}}{e_{co}}\right) \cdot 100 \tag{7}$$

where e is the annual emissions for the renewable system (rn) and the conventional system (co). Q_{cp} is prioritized to reduce electricity consumption. When the CTES is fully charged, the ST system uses its energy production to cover Q_{hp}, while the $P_{us,PV}$ uses energy for $P_{PV,sfc}$ and $P_{PV,pr}$. The $P_{PV,rv}$ is considered a positive revenue in the calculation of the LCOEn, which is calculated with Equation (8), where CAPEX is capital expenditures, O&M is operation and maintenance (2% of CAPEX), r is discount rate ($r = 9\%$), and n is the project duration in years (25 years).

$$LCOEn = \frac{CAPEX + \sum_{t=1}^{n} (Cost_{O\&M} + Cost_{GLP} + Cost_{el} - Cost_{PV,\,rv}) \cdot (1+r)^{-n}}{\sum_{t=1}^{n} \left(Q_{cp} + Q_{hp} + P_{PV,pr} + P_{PV,sfc}\right) \cdot (1+r)^{-n}} \tag{8}$$

2.3. *Study Case*

The wine industry was evaluated due to its characteristic heating and cooling demands. The evaluation focused on two wine industries located in Guayaquil (Morro) and Quito (Tababela). The $\dot{q}_{cp}$ for fermentation (fr), aging (ag), and storage (stg) were calculated from the energy balance of Equations (9)–(10) [31]:

$$\dot{q}_{fr} = \left[\pm U_{stl} \cdot A_{stl} \cdot \left(T_{room} - T_{fr}\right) + \frac{v_m \cdot sw \cdot \left(Cp_{fr} - Cp_{CO2} - hf_{H2O} \cdot Cp_{H2O} - hf_{et} \cdot Cp_{et}\right)}{Time_{fr}} \right] \cdot N_{stl} \tag{9}$$

$$\dot{q}_i = \left(\dot{m}_{cj,i} \cdot Cp_{rf,i} \cdot \left(T_{fr} - T_{set}\right) \cdot N_{stl,i}\right. \tag{10}$$

where U_{stl} is overall heat transfer coefficient of steel, A_{stl} is area of contact with the external medium, T_{room} is internal temperature of the industrial shed, T_{wy} is fermentation control temperature, $Time_{fr}$ is the fermentation time, i represents ag or st, $\dot{m}_{cj}$ is refrigerant mass flow in the cooling jacket, Cp_{rf} is refrigerant specific heat (glycol 20%), N_{stl} is number of steel tanks, v_m is must volume (200 hl) reported by Catrini et al. [19], sw is sugar concentration (238 g), T_{fr} is fermentation time (144 h), Cp is specific heat for fermentation ($Cp_{fer} = 0.59(\text{kJ/kg})$), water ($Cp_{H2O} = 0.031(\text{kJ/kg})$), carbon dioxide ($Cp_{CO2} = 0.011(\text{kJ/kg})$), ethanol ($Cp_{et} = 0.0135(\text{kJ/kg})$), and hf is latent heat of water evaporation ($hf_{H2O} = 2437.6\text{kJ/kg}; hf_{et} = 840.18\text{kJ/kg}$) [31]. To solve the equations, the data reported by Catrini et al. [19] were taken as input parameters. It was assumed that the production volume as a function of N_{stl} and T_{room} is constant, varying only when changing months. Therefore, $\dot{q}_{cp}$ was corrected using the radiant time series (RTMS) methodology, which is based on generating a representative load for a typical day, selecting roof and walls material coefficients for hourly load distribution [32].

The $\dot{p}_{sfc}$ was calculated considering that the electric power of the cooling process ($\dot{p}_{cp}$) represents 46% of the total electric consumption ($\dot{p}_{el}$), assuming a COP of 2, $\dot{p}_{sfc} = (q_{cp}/COP)\cdot 0.46^{-1} - (q_{cp}/COP)$, while $q_{hp} = Cp_w \cdot \dot{m}_{hp} \cdot \left(T_{set,hp} - T_a\right)$ where Cp_w is the specific heat of water and $\dot{m}_{hp}$ is the mass flow of bottles obtained from the volume of N_{stl}. The meteorological data were obtained from the National Solar Radiation Database (NSRBD) [33] (Table 3 and Figure 2): latitude (Lat), longitude (Long), direct normal irradiation (DNI), yearly cooling load (Q_{cp}), yearly heat load (Q_{hp}), and yearly electricity consumption (P_{el}.) The materials are concrete block walls (type 21) and metal deck roof (type 13) according to base data for the periodic response factor method [32].

Table 3. Summary of the annual energy demand and location of the small wine industries.

Location	Latitude (°)	Longitude (°)	DNI (kWh-y)	Q_{cp} (MWh-y)	Q_{hp} (MWh-y)	P_{el} (MWh-y)
Guayaquil (Morro)	−2.51	−80.3	1152	125	50	125
Quito (Tababela)	−0.15	−78.3	1697	45	62	64

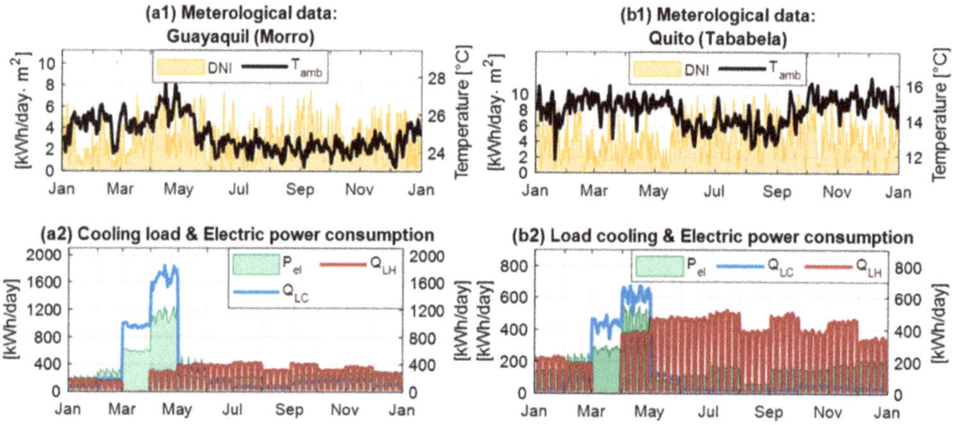

Figure 2. Meteorological data and energy demand for Guayaquil (a) and Quito (b): (**a1**) daily DNI and temperature for Guayaquil; (**a2**) daily electric consumption, daily cooling load, and heat load for Guayaquil; (**b1**) daily DNI and temperature for Quito; and (**b2**) daily electric consumption, daily cooling load, and heat load for Quito.

According to regulations issued by the Energy Ministry of Ecuador [34], the cost of electricity sold to the distributor was established with Equation (11), and the economic assumptions are shown in Table 4. Finally, the systems were evaluated based on the area fraction (FA) that compares the FPC area (A_{FPC}) and total solar field area (A_{sf}), where $FA = 100 \cdot A_{FPC}/A_{sf}$.

$$Cost_{PV,rv\ month} = \frac{\frac{CAPEX_{PV}}{12} \cdot \frac{r}{(1-r)^{-n}} + Cost_{O\&M,month}}{P_{PV,rv\ month}} \quad (11)$$

Table 4. Summary of economic assumptions and equipment cost.

Description	Guayaquil	Quito	Unit	Ref.
Electric cost	0.100	0.094	(USD/kWh)	[35]
GLP cost	0.067		(USD/kWh)	[36]
kgCO2(GLP)	0.200		(kgCO2/kWh)	[37]
kgCO2(El.)	0.270		(kgCO2/Wh)	[37]
PV cost	200		(USD/kW$_{peak}$)	[7]
Inverter cost	40		(USD/kW$_{NC}$)	[7]
Boiler cost	45		(USD/Boiler$_{NC}$)	[38]
FPC cost	$815 * (A_{FPC})^{-0.355}$		(USD/m^2)	[38]
ABCH cost	$6240 + (292 \cdot ABCH_{NC}) - \left(0.121 \cdot ABCH_{NC}^2\right)$		(USD)	[38]
cwt cost	$10000 + 20 \cdot (cwt_{NC} - 233)$		(USD)	[38]

3. Results and Discussion

This section describes the yearly results of the PV–AWCH, ST–ABCH and HYB systems with solar field areas from 14 m^2 to 1700 m^2. Figure 3 shows the SF on the y-axis and the total area on the x-axis, and the colors of the scatter points represent the percentage of the $f_{sav,CO2}$ for each system. Additionally, the A_{FPC} is fixed at 24 m^2, 41m^2, and 109 m^2 to show the performance of the HYB system for Guayaquil (Figure 3a) and Quito (Figure 3b). In the case of Guayaquil, the ST system presents a maximum SF of 0.74, the HYB system reaches an SF of 0.69, and the PV system has an SF of 0.3. Comparatively, in Quito, the ST system has the maximum SF of 0.87, while the HYB system reaches an SF of 0.77, and the PV system presents the lowest SF of 0.17. The low SF of the PV systems in Quito and Guayaquil results from the system only covering the Q_{cp} with the AWCH, while the gas boiler supplies the heat for the Q_{hp}. Quito has the maximum Q_{hp} and the lowest Q_{cp} as compared to Guayaquil; as a consequence, the efficiency of the PV–AWCH system is highly affected, reaching a low SF. However, the SF of the ST system in Quito is beneficial due to having a low demand of Q_{cp}, which, added to high solar radiation, achieves a 15% increase in SF as compared to Guayaquil. The HYB system reached the maximum SF with an area of 109 m^2 in Guayaquil and 420 m^2 in Quito. From the energy point of view, the HYB system presents an increase in the SF of 56% and 78% as compared to the PV system for Guayaquil and Quito, respectively. However, in both Guayaquil and Quito, the HYB system presents a minor SF as compared to the ST. It is important to notice that in Guayaquil the HYB system tends to slightly overcome the SF of the ST with 109 m^2 of FPC and an FA of 40%. Regarding the emissions savings, the ST systems in Guayaquil and Quito reached $f_{sav,CO2}$ close to 83%, and the PV system tends to have a $f_{sav,CO2}$ lower than 30%, while for the HYB systems, the maximum $f_{sav,CO2}$ is 59% and 73% for Guayaquil and Quito.

The energy results obtained under the weather conditions and demand profile in Quito and Guayaquil show that the ST system is the best technological solution compared to the HYB and PV systems. However, to define a selection criterion from a technoeconomic perspective, it is necessary to compare the system performance with the cost of energy production. In this sense, Figure 4 shows the area of solar systems on the x-axis, and the y-axis shows the LCOEn. At the same time, the color bar indicates the percentage of FA, where 0% is the field area of the individual PV system, and 100% is the field area of the individual ST system. In the case of Guayaquil, the minimum LCOEn for the ST system is 0.37 USD/kWh and 0.25 USD/kWh for the PV system. In contrast, the HYB system reaches a minimum LCOEn of 0.171 USD/kWh, which is a better performance concerning individual technologies. This trend is repeated in Quito, where the minimum LCOEn is 0.34 USD/kWh for the ST systems, 0.23 USD/kWh for the PV system, and 0.157 USD/kWh for the HYB systems. The high LCOEn for the ST systems results from the higher CAPEX related to the cost of the solar field, the ABCH, the increase of the Q_{gb}, and the inability to supply the P_{el} and sell electricity in relation to the PV system. In addition, to explain the economic performance, it is necessary to observe the economic assumptions related to energy cost. For both locations, electricity cost is higher than the GLP by about 30%. In that sense, the ST system only covers part of the $\dot{q}_{cp}$, and the electric grid covers the total electric consumption. Additionally, the $\dot{q}_g$ to drive the ABCH increases the GLP consumption, increasing the LCOEn. In contrast, the PV system presents a significant decrease in the LCOEn concerning the ST system, due to covering the power electric demand related to the $\dot{p}_{pumps}$, $\dot{q}_{cp}$, and $\dot{p}_{sfc}$. Despite the increase in $\dot{p}_{el}$ for the AWCH, the amount of energy is low due to the high COP of the AWCH, and together with the low cost of the GLP to supply heat to the bottling process, the PV systems reach a 32% reduction in the LCOEn compared to the ST system.

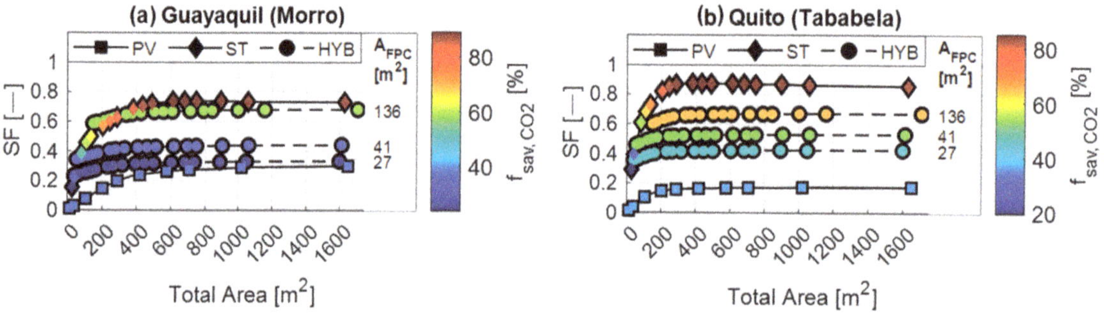

Figure 3. Annual SF of PV, ST, and HYB systems as a function of total area and $f_{sav,e(CO2)}$ evaluated in two locations of Ecuador: (**a**) Guayaquil and (**b**) Quito.

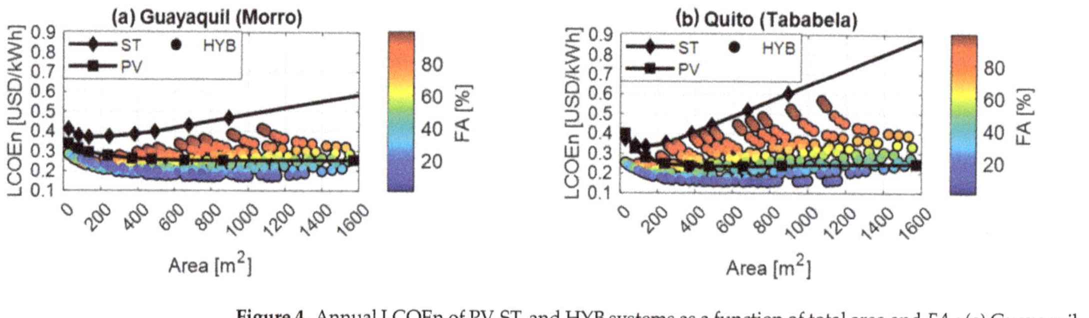

Figure 4. Annual LCOEn of PV, ST, and HYB systems as a function of total area and *FA* : (**a**) Guayaquil and (**b**) Quito.

To understand the technoeconomic performance of the HYB systems, Figure 5 shows the results of the HYB systems that have lower LCOEn than the PV systems, where the x-axis is the LCOEn, the y-axis is the SF and the color bar represents the percentage of FA for Guayaquil (Figure 5a) and Quito (Figure 5b). The results show that the best performance in Guayaquil is 0.171 USD/kWh with an SF of 0.67, while in Quito, the best result for the HYB systems is a 0.64 SF with an LCOEn of 0.57 USD/kWh. In both locations, the reduction of the LCOEn for the HYB systems with respect to the best result of the PV system is close to 30% and 54% for the ST systems. However, in both cases, the SF for the HYB systems is lower than that in the ST systems. The main limit of both individual systems is related to the capacity to cover the cooling demand with solar energy. In that sense, the HYB system has a significant advantage because using two small chillers instead of a big one to cover the cooling demand allows for the deactivation of one chiller in periods of low cooling demand, thus increasing the possibility of leveraging the solar energy for $\dot{p}_{el}$ and $\dot{q}_{hp}$. Compared with studies that integrated solar energy to supply cooling and heating for industrial processes, in the case of milk production in India, the ST-ABCH reached an LCOEn of 0.177 USD/kWh with an off-grid configuration [21]. In contrast, the PV-AWCH systems for microbreweries in Spain reached an LCOEn between 0.2 and 0.22. At the same time, the HYB system reached LCOEn values of 0.171 USD/kWh in Guayaquil and 0.157 USD/kWh in Quito. Therefore, it is shown that the hybrid system has a high application potential for small industries that need to cover cooling, heat, and electricity demands during the manufacturing process.

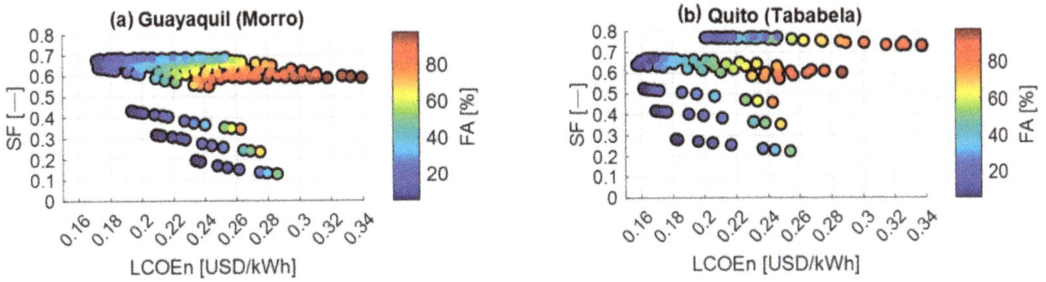

Figure 5. Annual SF of HYB system as a function of LCOEn and *FA*: (**a**) Guayaquil and (**b**) Quito.

4. Conclusions

The findings of this research show significant potential for integrating solar systems within the industry. Although the individual FPC system reached the highest SF with respect to the PV systems, the best economical solution is the PV system due to the capability of selling the energy excess produced by the system. However, the use of HYB systems focused on cooling loads, plus the coverage of heat and electricity consumption, can reduce the LCOEn by 30% when compared to the PV systems, with the advantage that they increase the SF by over 0.6 with solar field areas less than 600 m^2. Consequently, the HYB system is presented as an alternative that allows the reduction in fossil energy consumption and CO_2 emissions with a significant improvement with respect to individual solar technologies. However, it is necessary to research the performance based on the available solar radiation and the effects produced by the variability under different weather zones to determine the possible technoeconomical limits. At the same time, the evaluation of the HYB system proposed, which includes a solar-assisted heat pump, could contribute to leveraging the PV system for heat processes and clarify the potential to integrate this technology in the industry. Although the HYB system was proposed to evaluate the winemaking process, the HYB system could be applied to the dairy, beverages, soap and detergents industries, etc., or any others that require cooling for the manufacturing process. Therefore, the HYB system is a potential technology to be integrated into the industrial sector to ensure energy availability during the transition from a fossil to a renewable energy matrix.

Author Contributions: Conceptualization, A.V.-J. and R.E.; methodology, A.V.-J. and R.E.; formal analysis, A.V.-J., J.F.R.-P., M.P.-G., J.M.C. and R.E.; investigation, A.V.-J.; resources, R.E.; writing—original draft preparation, A.V.-J. and J.F.R.-P.; writing—review and editing, A.V.-J., J.F.R.-P., M.P.-G., J.M.C. and R.E.; visualization, A.V.-J.; supervision, R.E.; funding acquisition M.P.-G., J.M.C. and R.E. All authors have read and agreed to the published version of the manuscript.

Funding: This research was partially funded by National Agency for Research and Development ANID BECAS/DOCTORADO NACIONAL 21200821 and 21200844.

Institutional Review Board Statement: Not applicable.

Informed Consent Statement: Not applicable.

Data Availability Statement: Data are contained within the article.

Acknowledgments: Andrés Villarruel-Jaramillo would like to acknowledge the funding from the National Agency for Research and Development ANID BECAS/DOCTORADO NACIONAL 21200821, and Josué F. Rosales-Pérez the funding from ANID BECAS/DOCTORADO NACIONAL 21200844. The authors wish to express their gratitude to the projects ANID/FONDAP 1522A0006 "Solar Energy Research Center"—SERC-Chile.

Conflicts of Interest: The authors declare no conflict of interest.

Nomenclature

Abbreviations	
ABCH	Absorption chiller
AWCH	Air-to-water chiller
CAPEX	Capital expenditures
COP	Coefficient of performance
DNI	Direct normal irradiation (W/m^2)
FPC	Flat plate solar collectors
GLP	Liquefied petroleum gas
HYB	Solar hybrid
LCOCH	Levelized cost of cooling and heating
LCOEn	Levelized cost of energy
LCOH	Levelized cost of heat
NSRBD	National Solar Radiation Database
O&M	Operation and maintenance
PV	Photovoltaic
SF	Solar fraction
SOC	State of charge
ST	Solar thermal
Latin Symbols	
A	Area (m^2)
c_1	Linear thermal loss coefficient (W/m^2K)
c_2	Quadratic thermal loss coefficient (W/m^2K^2)
c_3	Wind speed dependence of the heat loss coefficient (kg/m^2K)
Cp	Specific heat (kJ/kg)
$ctes$	Cold thermal energy storage tank
e	Annual emissions (CO2 ton eq.)
η	Efficiency
f_{sav}	Fractional energy saving
FA	Area fraction
hf	Latent heat of evaporation (kJ/kg)
$htes$	Hot thermal energy storage tank
hx	Heat exchanger
I	Irradiation (W/m^2) ∣ current (A)
I_L	Light current (A)
I_o	Diode reverse saturation current (A)
k	Boltzmann constant (J/K)
K	Incidence angle modifiers
$\dot{m}$	Mass flow rate (kg/h)
N	Number
n	Project duration in years
nd	Diode ideality factor
$\dot{p}$	Electric power (kW)
P	Monthly or yearly electric energy (kWh)
Q	Monthly or yearly energy (kWh)

$\dot{q}$	Heat flow rate (kW)
r	Discount rate (%)
Rs	Serial resistance (Ω)
Rsh	Shunt resistance (Ω)
sw	Sugar concentration (g)
T	Temperature (°C)
T_{set}	Set temperature objective (°C)
U	Overall heat transfer coefficient $\left(W/m^2K^{-1}\right)$
V	Voltage (V)
v	Volume $\left(m^3\right)$
$tetha$	Incidence angle (°)
Subscripts	
a	Ambient
ag	Aging
aux	Auxiliar
avg	Average
b	Beam
$cells$	Photovoltaic cells
co	Conventional system
$CO2$	Carbon dioxide
cp	Cooling process
cwt	Cooling water tower
d	Diffuse
e	Evaporator
el	Electric
et	Ethanol
fr	Fermentation
g	Generator
gb	Gas boiler
$grid$	Electric grid
$H2O$	Water
hp	Heat process
in	Inlet
$loss$	Losses
mp	Maximum power
out	Outlet
pr	Process
$pumps$	Pumps of the heat and cooling circuit
rn	Renewable system
$room$	Workspace of the manufacturing plant
rv	Revenue
sc	Solar collector
sf	Solar field area
sfc	Self-consumption
st	Solar thermal
stc	State test conditions
stg	Storage
stl	Steel
us	Useful
wy	Wine

References

1. United Nations. *Paris Agreement*; United Nations: New York, NY, USA, 2015.
2. IEA. *The Future of Cooling*; IEA: Paris, France, 2018.
3. Sikder, M.; Wang, C.; Yao, X.; Huai, X.; Wu, L.; Kwame Yeboah, F.; Wood, J.; Zhao, Y.; Dou, X. The integrated impact of GDP growth, industrialization, energy use, and urbanization on CO_2 emissions in developing countries: Evidence from the panel ARDL approach. *Sci. Total Environ.* **2022**, *837*, 155795. [CrossRef] [PubMed]
4. IIGE. *Balance Energético Nacional 2021*; IIGE: Quito, Ecuador, 2022.

5. Espinoza, V.S.; Fontalvo, J.; Martí-Herrero, J.; Miguel, L.J.; Mediavilla, M. Analysis of energy future pathways for Ecuador facing the prospects of oil availability using a system dynamics model. Is degrowth inevitable? *Energy* **2022**, *259*, 124963. [CrossRef]
6. Settino, J.; Sant, T.; Micallef, C.; Farrugia, M.; Staines, C.S.; Licari, J.; Micallef, A. Overview of solar technologies for electricity, heating and cooling production. *Renew. Sustain. Energy Rev.* **2018**, *90*, 892–909. [CrossRef]
7. Solar Energy Industries Association. Solar Market Insight Report 2020 Q3 2020. Available online: https://www.seia.org/research-resources/solar-market-insight-report-2020-q3 (accessed on 2 April 2023).
8. IRENA. *Renewable Generation Costs in 2022*; International Renewable Energy Agency: Abu Dhabi, United Arab Emirates, 2022.
9. Farjana, S.H.; Huda, N.; Parvez Mahmud, M.A.; Saidur, R. Solar process heat in industrial systems-A global review. *Renew. Sustain. Energy Rev.* **2018**, *82*, 2270–2286. [CrossRef]
10. Rosales-Pérez, J.F.; Villarruel-Jaramillo, A.; Romero-Ramos, J.A.; Pérez-García, M.; Cardemil, J.M.; Escobar, R. Hybrid System of Photovoltaic and Solar Thermal Technologies for Industrial Process Heat. *Energies* **2023**, *16*, 2220. [CrossRef]
11. Quiñones, G.; Felbol, C.; Valenzuela, C.; Cardemil, J.M.; Escobar, R.A. Analyzing the potential for solar thermal energy utilization in the Chilean copper mining industry. *Sol. Energy* **2020**, *197*, 292–310. [CrossRef]
12. Neyer, D.; Ostheimer, M.; Dipasquale, C.; Köll, R. Technical and economic assessment of solar heating and cooling—Methodology and examples of IEA SHC Task 53. *Sol. Energy* **2018**, *172*, 90–101. [CrossRef]
13. Neyer, D.; Ostheimer, M.; Mugnier, D.; White, S. 10 key principles for successful solar air conditioning design—A compendium of IEA SHC Task 48 experiences. *Sol. Energy* **2018**, *172*, 78–89. [CrossRef]
14. Neyer, D.; Ostheimer, M.; Hauer, N.; Halmdienst, C.; Pink, W. Application of an adapted single-/half- effect NH_3/H_2O absorption chiller in tri-generation and solar cooling systems. *Sol. Energy* **2018**, *173*, 715–727. [CrossRef]
15. Cabrera, F.; Fernández-García, A.; Silva, R.; Pérez-García, M. Use of parabolic trough solar collectors for solar refrigeration and air-conditioning applications. *Renew. Sustain. Energy Rev.* **2013**, *20*, 103–118. [CrossRef]
16. Ayadi, O.; Al-dahidi, S. Comparison of solar thermal and solar electric space heating and cooling systems for buildings in different climatic regions. *Sol. Energy* **2019**, *188*, 545–560. [CrossRef]
17. Luerssen, C.; Gandhi, O.; Reindl, T.; Sekhar, C.; Cheong, D. Levelised Cost of Storage (LCOS) for solar-PV-powered cooling in the tropics. *Appl. Energy* **2019**, *242*, 640–654. [CrossRef]
18. Villarruel-Jaramillo, A.; Pérez-García, M.; Cardemil, J.M.; Escobar, R.A. Review of Polygeneration Schemes with Solar Cooling Technologies and Potential Industrial Applications. *Energies* **2021**, *14*, 6450. [CrossRef]
19. Catrini, P.; Panno, D.; Cardona, F.; Piacentino, A. Characterization of cooling loads in the wine industry and novel seasonal indicator for reliable assessment of energy saving through retrofit of chillers. *Appl. Energy* **2020**, *266*, 114856. [CrossRef]
20. Pino, A.; Pino, F.J.; Guerra, J. Integration of solar energy in Small-scale Industries: Application to microbreweries. *Sustain. Energy Technol. Assess.* **2023**, *57*, 103276. [CrossRef]
21. Kumar, D.; Sharma, D.; Hamza, A.; Ali, H. Optimization and thermo-economic performance of a solar-powered vapor absorption cooling system integrated with sensible thermal energy storage. *Energy Convers. Manag. X* **2023**, *20*, 100440. [CrossRef]
22. Buonomano, A.; Calise, F.; Palombo, A. Solar heating and cooling systems by absorption and adsorption chillers driven by stationary and concentrating photovoltaic/thermal solar collectors: Modelling and simulation. *Renew. Sustain. Energy Rev.* **2018**, *82*, 1874–1908. [CrossRef]
23. Villarruel-Jaramillo, A.; Rosales-Pérez, J.F.; Pérez-García, M.; Cardemil, J.M.; Escobar, R. Modeling and Performance Evaluation of Hybrid Solar Cooling Systems Driven by Photovoltaic and Solar Thermal Collectors—Case Study: Greenhouses of Andalusia. *Energies* **2023**, *16*, 4888. [CrossRef]
24. Duffie, J.A.; Beckman, W.A. *Solar Engineering of Thermal Processes*, 4th ed.; John Wiley & Sons, Ltd.: Madison, WI, USA, 2013. [CrossRef]
25. De Soto, W.; Klein, S.A.; Beckman, W.A. Improvement and validation of a model for photovoltaic array performance. *Sol. Energy* **2006**, *80*, 78–88. [CrossRef]
26. King, D.L.; Boyson, W.E.; Kratochvi, J.A. *Photovoltaic Array Performance Model*; Sandia National Laboratories: Albuquerque, NM, USA, 2004.
27. NREL. *System Advisor Model Version (SAM 2020.11.29)*; NREL: Lakewood, CO, USA, 2020.
28. Tian, Z.; Perers, B.; Furbo, S.; Fan, J. Analysis and validation of a quasi-dynamic model for a solar collector field with flat plate collectors and parabolic trough collectors in series for district heating. *Energy* **2018**, *142*, 130–138. [CrossRef]
29. YORK. Model YCAL Style C Air-Cooled Scroll Chiller Engineering Guide. n.d. Available online: https://goo.su/fBxG (accessed on 7 July 2022).
30. YAZAKI Energy Systems. Water-Fired Chiller/Chiller-Heater. Available online: http://jmp.sh/FacJtkG (accessed on 5 November 2021).
31. Álvarez, A. *Diseño y Calculo de las Instalaciones Necesarias Para Una Bodega de Elaboración de Vinos Blancos, Tintos Jóvenes y Crianza Ecológicos Con D.O Ribera del Duero Para Una Capacidad de 200.000 Kg de Uva al año en la Localidad de Adrada de Haza*; Universidad Politécnica de Madrid: Burgos, Spain, 2021.
32. Spitler, J.D.; Michaels, C.S.; Boyle, S.; Ramsdell, L.; Becker, T. *Load Calculation Applications Manual*, 2nd ed.; SI Edition; ASHRAE: Atlanta, Georgia, 2014.
33. NSRDB National Solar Radiation Database. NREL n.d. Available online: https://nsrdb.nrel.gov/ (accessed on 3 April 2023).

34. Agencia de Regulación y Control de Energía y Recursos Naturales No Renovables. *Resolución Nro. ARCERNNR-XX/2022*; Agencia de Regulación y Control de Energía y Recursos Naturales No Renovables: Quito, Ecuador, 2022.
35. Agencia de Regulación y Control de Electricidad. *Pliego Tarifario Para Las Empresas Eléctricas de Distribución—Servicio Público de Energía Eléctrica*; Periodo: Enero-Diciembre 2020; Arconel; Agencia de Regulación y Control de Electricidad: Quito, Ecuador, 2019; Volume 19.
36. LOJAGAS. Precio Glp 2023. Available online: https://lojagas.com/precio-glp/ (accessed on 1 July 2023).
37. Brander, A.M.; Sood, A.; Wylie, C.; Haughton, A.; Lovell, J. Electricity-Specific Emission Factors for Grid Electricity. Ecometrica 2011. Available online: http://ecometrica.com/assets//Electricity-specific-emission-factors-for-grid-electricity.pdf (accessed on 10 July 2023).
38. Chahartaghi, M.; Baghaee, A. Technical and Economic Analyses of a Combined Cooling, Heating and Power System Based on a Hybrid Microturbine (Solar-Gas) for a Residential Building. *Energy Build.* **2020**, *217*, 110005. [CrossRef]

engineering proceedings

MDPI

Proceeding Paper

A Spectrum Analyzer in the 470 to 698 MHz Band Using Software Defined Radio for the Analysis of Digital Terrestrial Television Signals (DTTs) †

Karla Andrade [1,*], Fausto Patiño [1,*] and Tarquino Sánchez-Almeida [2,*]

1 Department of Electronics, Telecommunication and Information Networks, Escuela Politécnica Nacional, Ladrón de Guevara, Quito E11-253, Ecuador
2 Educational Research Group INEDU-EPN, Escuela Politécnica Nacional, Ladrón de Guevara, Quito E11-253, Ecuador
* Correspondence: karla.andrade@epn.edu.ec (K.A.); fausto.patino@epn.edu.ec (F.P.); tarquino.sanchez@epn.edu.ec (T.S.-A.)

† Presented at the XXXI Conference on Electrical and Electronic Engineering, Quito, Ecuador, 29 November–1 December 2023.

Abstract: This article presents the development of a prototype software for a terrestrial digital TV signal spectrum analyzer using software-defined radio (SDR). This study involves four phases: signal reception from Quito's TV channels, configuring the software and hardware to work within the 470–698 MHz frequency range using the One-seg service of the ISDB-Tb standard, setting up the spectrum analyzer with a low-pass filter to reduce noise, and analyzing the results by visualizing digital TV signals from various areas of Quito with the SMPlayer software. While capturing TDT signals in Ecuador, it was discovered that some channels' signals lacked playable audiovisual content on video players during testing.

Keywords: spectrum analyzer; SDR; Digital Terrestrial Television; One-seg

Citation: Andrade, K.; Patiño, F.; Sánchez-Almeida, T. A Spectrum Analyzer in the 470 to 698 MHz Band Using Software Defined Radio for the Analysis of Digital Terrestrial Television Signals (DTTs). *Eng. Proc.* **2023**, *47*, 22. https://doi.org/10.3390/engproc2023047022

Academic Editor: Soraya Lucia Sinche Maita

Published: 6 December 2023

1. Introduction

Most radio signal analysis systems rely on hardware and operate within specific frequency ranges, making modifications costly and inefficient. The pandemic has increased the demand for virtual labs, especially in educational institutions, particularly with engineering majors such as the Electronics and Telecommunications program of the Escuela Politécnica Nacional, where the lack of virtual labs for analog and digital communications has hindered students' academic progress [1]. SDR technology is ideal for space applications because space communications increasingly require more flexible, adaptable, and reconfigurable radios. This has led to software-defined radios beginning to be used in some missions. For example, NASA's Curiosity rover uses a software-defined radio to establish communications in the UHF band with the Mars Reconnaissance Orbiter (MRO). The incorporation of SDR technology in the Curiosity–MRO link has managed to increase the data transmission rate between the two systems from 256 kbps to 2048 kbps by implementing algorithms for adapting the data rate depending on the conditions of the communication channel [2].

This work aims to develop a system for spectrum analysis using software-defined radio (SDR) technology to observe the behavior of systems such as Digital Terrestrial Television (DTT), which, in Ecuador, is in the process of implementation, with the purpose of analyzing different points and detecting possible problems. The system developed with SDR is low cost and constitutes an excellent teaching tool for learning this new technology [3]. The SDR-based spectrum analyzer will enable the analysis of signals within the 470–698 MHz frequency range, like a physical spectrum analyzer, without the need for in-person communication labs. The goal is to facilitate the study of the transmission and reception parameters of authorized TV channels in Quito, complying with the regulation of the Agency for the Regulation and Control of Telecommunications of Ecuador (ARCOTEL) and improving the capacities of education and research in telecommunications [4].

2. Objectives and Contribution

In this work, the following objectives are established:

- Implement a spectrum analyzer in the 470–698 MHz band using an SDR system for the analysis of terrestrial digital television signal parameters in sectors of the city of Quito. The specific objectives are the following:
- Study the basic concepts and operation of an SDR system. Design the spectral analysis of the 470–698 MHz frequency band using an SDR system.
- Implement a system to visualize radio spectrum parameters for receiving terrestrial digital television (TDT) signals.
- Perform measurements at different strategic points in the city of Quito to verify the functionality of the implemented spectrum analyzer.

The significant contribution is supported by the inclusion of the application of the ISDB-Tb standard for the real-time processing of television channel signals, the setup and operation of an SDR system using GNU Radio, and compliance with ARCOTEL regulations for signal modeling and processing [5]. Furthermore, the establishment of a valuable, cost-effective, and efficient virtual communication laboratory enables the in-depth analysis of transmission and reception parameters for authorized channels in Quito, supporting research and academic training in this critical field.

3. Other Investigations

While it is true that the research area in terms of systems and applications with SDR has grown and evolved, there are works in this area.

One work involves developing an algorithm to analyze and display data collected from the NOAA satellite. The interface is designed and programmed in NI LabVIEW, where satellite signals are taken from the antenna using a USRP N2920 device, which allows for signal conversion and transport to the interface [6], or the reception of DTT signals using an SDR device through Matlab to finally visualize the received multiplex audiovisual programs [7]. There is even a work on the creation of a prototype spectrum analyzer using software-defined radio techniques, a single-board computer, and an RTL-SDR blog V3 tuner device for the reception of electromagnetic signals [8].

It is worth noting that these are important applications, but in the current research, efforts have been made to keep implementation costs low by using an SDR-RTL device, which has basic electronic components, rather than a USRP, which can cost up to 20 times more and may have an onboard programmable FPGA card. The goal is to minimize hardware usage so that everything can be completed through software and to use a free programming platform. This involves applying the knowledge of signal reception and processing, and although it is primarily used for DTT due to its current popularity, it is also applicable for receiving any type of RF signal.

4. Project Development

4.1. Methodology

This article focused on reviewing concepts related to terrestrial digital transmission signals, as well as the standard used in Ecuador and the functioning of One-seg. Subsequently, a problem that could be addressed using software-defined radio (SDR) 2832U and signal analysis equipment currently accessible for enhancing learning and research was investigated. The article proposes a prototype that aids in signal analysis, enabling the visualization of the signal spectrum and conducting measurements to achieve these objectives. The necessary steps for project development were established, and its implementation was carried out.

4.2. Project Overview

The block diagram in Figure 1 outlines the necessary steps to implement and program a prototype that fulfills the established objectives. The prototype has the following functions.

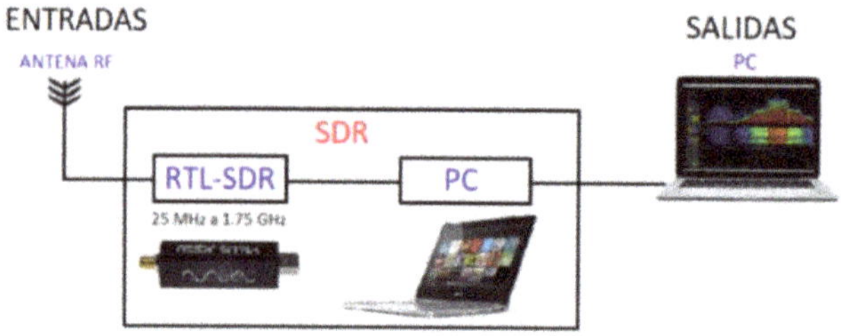

Figure 1. General scheme of the spectral analysis process using an SDR system.

4.3. Implementation of an ISDBT-TB Receiver/Decoder in GNU Radio for Terrestrial Digital Television Transmission Analysis Using NESDR SMART

After analyzing some options and due to the nature of the free software under which the GNU Radio software was released, the adaptability, direct compatibility with NESDR or any RTL2832U based device, and the feasibility of modifying its modules for the generation of specific solutions were considered. The GNU Radio and NESDR link were the options with the best expectations and is compatible with both Windows-based operating systems as well as Linux. However, due to the closed-source nature of Windows, the direct modification of modules and dependencies that the solution may require becomes challenging. Given this circumstance, the version for GNU-Linux was selected over an intuitive and user-friendly operating system like UBUNTU 18. This decision was made because, starting from version 20, there are changes in the operating system kernel, including Python. Taking as a reference the implementation by the IIE (Institute of Electrical Engineering), Faculty of Engineering, University of the Republic, Montevideo, Uruguay [9], it was necessary to create a development environment compatible with ISDB-Tb with the aim of achieving a software-level tool to reach the objective. Consequently, GNU Radio version 3.7 and the necessary dependencies were compiled to achieve the desired compatibility [9].

The implementation process, shown in Figure 2, involves the acquisition and processing of radiofrequency (RF) signals for the reception and decoding of terrestrial digital television (TDT) transmissions in the 1-seg standard. Several GNU Radio blocks are used to carry out the different stages of the process.

Blocks That Make up the Receiver.

- RTL-SDR SOURCE: This allows for the reception of radiofrequency signals using an SDR device based on the RTL2832U chip. Parameters such as tuning frequency and sampling rate are configured. It provides a complex sample stream output for further processing.
- Rational Resampler: This performs the resampling of the obtained RF signal to obtain segments of the primary signal without overloading the reception system. It uses a polyphase interpolation and decimation algorithm to change the signal's sampling rate.
- Low Pass Filter: This filters the sampled signal to eliminate unwanted frequencies and improve signal quality. It enables the implementation of digital low-pass filters, and parameters such as cutoff frequency and gain are configured.
- Sync Dem OFDM 1seg: This performs the synchronization of the OFDM symbols from the captured signal. It extracts the OFDM symbols, synchronizes the frequency and timing of the input signal, and decodes the transmitted data.
- OFDM Demod 1seg: This decodes the OFDM data transmitted in the received signal. It uses signal processing techniques for time and frequency synchronization, channel error removal, and data recovery.

- TMCC Decoder 1seg: This decodes the Transmission and Multiplexing Configuration Channel (TMCC) in the 1-seg TDT standard. It allows for the extraction and decoding of TMCC information for transmission management and control.
- Frequency Deinterleaver: This corrects the effects of dispersion in the received signal. It rearranges the received data into a sequence that minimizes decoding errors using the Fourier transform.
- Time Deinterleaver: This corrects the effects of selective fading in the received signal. It rearranges the received data into a sequence that minimizes decoding errors in the time domain.
- MER One-seg: This calculates the Modulation Error Ratio (MER) of the digital television signal in the 1-seg standard. It provides a metric to evaluate the signal quality.
- Decoder 1seg: This decodes and recovers the audio and video signal from the transmission channel. It performs the reverse process of signal compression, corrects errors, and synchronizes the signal for data playback.
- Vector to Stream: This converts an input signal in vector format to an output signal in stream format. It allows for the real-time transmission and processing of vector signals.
- File Sink: This writes signal data to a file on the hard drive. It enables the storing data for further processing or analysis.
- QT GUI Frequency Sink, QT GUI Number Sink, QT GUI Time Sink, and QT GUI Constellation Sink: These output blocks for the real-time visualization of the signal in different domains, such as frequency spectrum, numerical values, time domain, and constellation diagram.

These blocks are used in a processing chain that spans from the acquisition of the RF signal to the decoding and playback of the audio and video signal. Graphical tools such as QT GUI Chooser, Entry, or Range are used to allow the user to adjust settings in real-time.

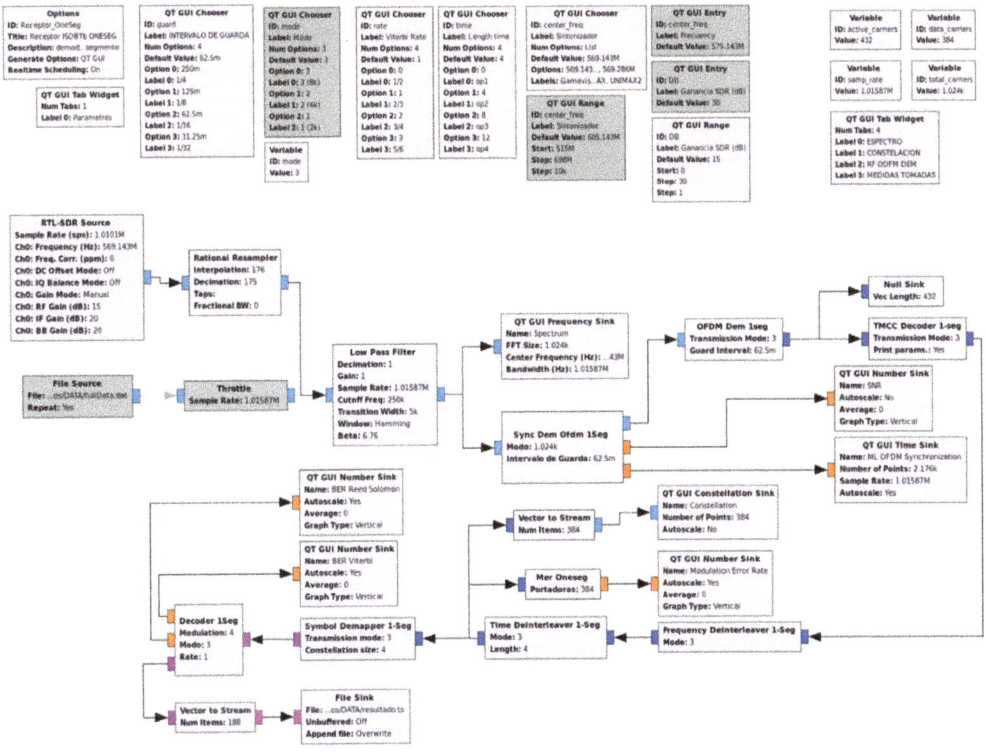

Figure 2. Block diagram for signal reception.

The gray-colored blocks are currently disabled with minor modifications for signal power measurements. Meanwhile, the purple-colored blocks are enabled and operational. You can toggle these blocks on and off using the E (Enable) and D (Disable) keys as needed.

5. Results

Initial tests were conducted to assess the quality of digital television signals on different channels. The obtained results indicate the feasibility of recovering audio and video from signals of good and medium quality. However, channels with poor-quality signals were found where it was not possible to recover the audio and video signal.

5.1. ISDB-TB Spectrum, Channel with Good-Quality Signal

Through the implemented spectrum analyzer, the signal transmitted by the national television channel Teleamazonas was collected. The power intensity was measured using the spectrogram available in the GNU Radio library, and a measurement of −55.16 dB was observed at a frequency of 580.897 MHz, which falls within the regulatory ranges, indicating a signal of good quality suitable for recovering and analyzing its audio and video content. The amplitude vs. time peaks in the OFDM synchronization diagram showed clear definition, suggesting good audio and video quality.

The measurements taken, including the MER (Modulation Error Ratio) of 24.61 dB and the SNR (Signal-to-Noise Ratio) of 20.62 dB, indicate low dispersion and a high-quality signal (Figure 3), even after error correction, which is highly favorable.

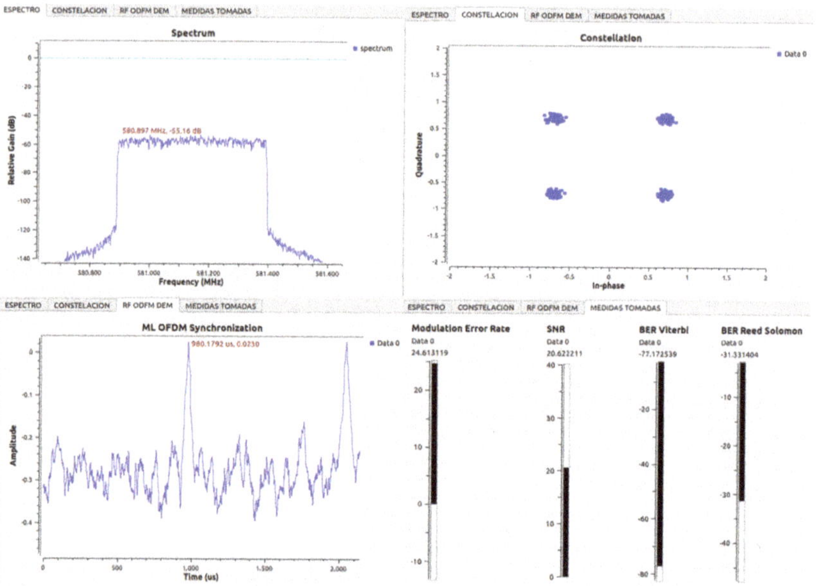

Figure 3. Spectrum, synchronization start symbols, constellation diagram, and measurements of ISDB-Tb channel Teleamazonas in Barrio Miraflores.

5.2. ISDB-TB Spectrum Channel with Medium Quality Signal

A measurement of −74.49 dB was observed at a frequency of 670.985 MHz, within the regulatory ranges, indicating an acceptable but medium-quality signal (Figure 4). The bandwidth used falls within the One-seg parameters. The constellation diagram showed an asymmetrical and scattered constellation pattern, indicating a reception of signal, but of low quality. Despite this, acceptable results were achieved, enabling synchronization. The OFDM synchronization displayed well-defined peaks in the amplitude vs. time diagram, indicating good audio and video quality.

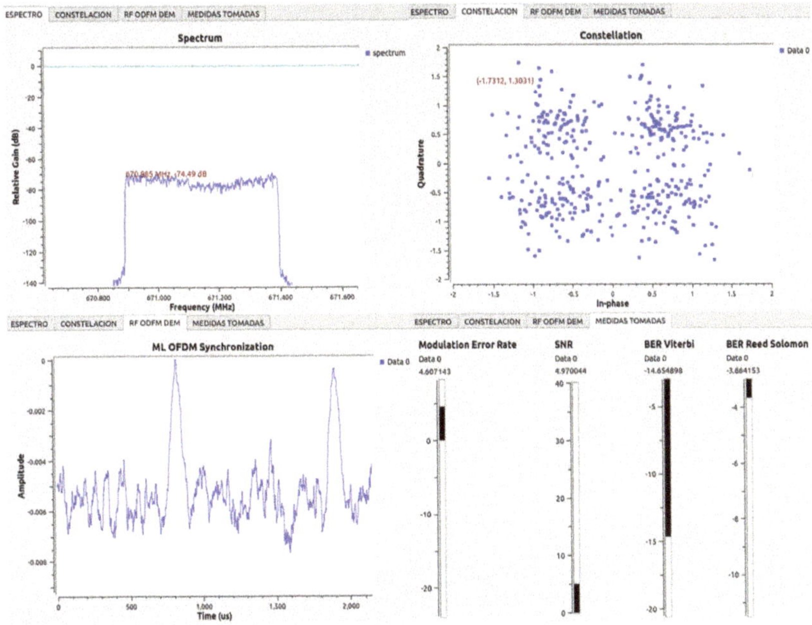

Figure 4. Spectrum, synchronization start symbols, constellation diagram, and measurements of ISDB-Tb channel TC Televisión, Barrio El Conde.

In the measurement diagram, the obtained MER was 4.6 dB, signifying significant dispersion and a low-quality signal. The measured SNR was 4.97 dB, confirming the low image quality. Even after error correction, these measurements still showed low values, indicating poor signal quality.

5.3. ISDB-TB Spectrum, Channel with Bad Quality Signal

A measurement of −81.29 dB was observed at a frequency of 580.905 MHz, which is outside the regulatory ranges. No signal was obtained for recovery (Figure 5), indicating that audio and video cannot be retrieved. The constellation diagram showed a non-symmetric and scattered constellation pattern, indicating a lack of signal reception. The results for this TV channel are unacceptable, meaning that audio and video cannot be obtained. The OFDM synchronization displayed poorly defined peaks in the amplitude vs. time diagram, indicating poor audio and video quality.

In the measurement diagram, the obtained MER was −2.0 dB, indicating significant dispersion and a low-quality signal. The measured SNR was 7.24 dB, confirming the poor image quality. These measurements, even after error correction, also showed low values, implying that the signal cannot be recovered.

Audio and video reception was successfully achieved for the Ecuavisa channel once the frequency was detected and the spectrum was analyzed. The audiovisual material was obtained using the appropriate bandwidth and meeting the power, MER, and SNR parameters. However, for channels such as TVC or Canal Uno that are not transmitting information, no audio and video signal could be obtained.

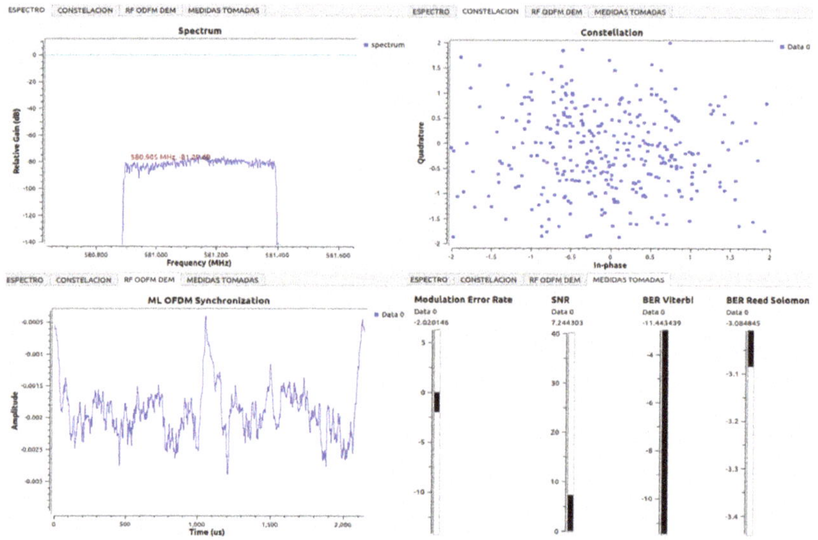

Figure 5. Spectrum, synchronization start symbols, constellation diagram, and measurements of ISDB-Tb channel Teleamazonas, Barrio El Conde.

6. Conclusions

This study established the different stages of the spectral analysis process using an SDR system involving a fixed antenna, SDR, and a computer to obtain audio and video as the output. Each phase of the process, from signal acquisition to data conversion into audio and video, was identified and deemed crucial to ensure signal quality and system efficiency. The knowledge gained from this study can be applied in various fields, from amateur radio to the telecommunications industry, and can serve as a foundation for future research in the field of spectral analysis using SDR systems.

Signal power analysis, MER, SNR, spectrum, and constellation diagrams are essential elements for evaluating the performance and quality of a communication system. Signal power measurement allows us to determine the amount of energy transmitted and received, which is crucial for ensuring system robustness and coverage. On the other hand, MER and SNR serve as indicators of signal quality, and their measurements are fundamental to ensure reliable and error-free communication.

Author Contributions: Conceptualization, K.A. and F.P.; methodology, K.A. and F.P.; software, K.A. and F.P.; validation, T.S.-A.; formal analysis, K.A. and F.P.; investigation, K.A. and F.P.; resources, K.A. and F.P. and T.S.-A.; data curation, K.A. and F.P.; writing—original draft preparation, K.A. and F.P. and T.S.-A.; writing—review and editing, K.A. and F.P. and T.S.-A.; visualization, K.A. and F.P. and T.S.-A.; supervision, T.S.-A.; project administration, T.S.-A.; funding acquisition, K.A. and F.P. and T.S.-A. All authors have read and agreed to the published version of the manuscript.

Funding: Escuela Politécnica Nacional.

Institutional Review Board Statement: Not applicable.

Informed Consent Statement: Not applicable.

Data Availability Statement: The datasets generated for this study are available on request to the corresponding author.

Acknowledgments: The authors thank the Escuela Politécnica Nacional for their support in the use of DTT testing laboratories.

Conflicts of Interest: The authors declare that the research was conducted in the absence of any commercial or financial relationships that could be construed as potential conflicts of interest.

References

1. Universidad Técnica Particular de Loja (UTPL). Virtual Laboratory Practices Ensure a Successful Career. UTPL. 10 March 2021. Available online: https://noticias.utpl.edu.ec/las-practicas-en-laboratorios-virtuales-aseguran-una-carrera-exitosa (accessed on 20 July 2022).
2. Muñoz, C. Software Defined Radio, a New Option in Space Communications. Gobierno de México. 2017. Available online: https://haciaelespacio.aem.gob.mx/revistadigital/articul.php?interior=498 (accessed on 19 September 2023).
3. Ministry of Telecommunications and Information Society. Master Plan of Transition to Digital Terrestrial Television. Quito. 2018. Available online: https://www.telecomunicaciones.gob.ec/wp-content/uploads/2018/10/PLANMAESTRO-DE-TRANSICI%25C3%2593N-A-LA-TELEVISI%25C3%2593N-DIGITAL-TERRESTRE-2018-2021.pdf (accessed on 20 July 2022).
4. Enríquez Chicaiza, P.E.; Rivadeneira Obregón, E.D. *Comparative Analysis for Digital Terrestrial Television Service in Quito*; Salesian Polytechnic University: Quito, Ecuador, 2018. Available online: http://dspace.ups.edu.ec/handle/123456789/15508 (accessed on 8 March 2023).
5. Balseca, V. *Communications System Using SDR for Practical Multidisciplinary Courses in the Electronics Engineering and Communications of Technical University of Ambato*; Technical University of Ambato: Ambato, Ecuador, 2021. Available online: https://repositorio.uta.edu.ec/bitstream/123456789/32718/1/t1807ec.pdf (accessed on 19 September 2023).
6. Martínez, M. *Design and Implementation of an Algorithm for Satellite Signal Processing Using the Software-Defined Radio Plataform USRP*; University of the Americas UDLA: Quito, Ecuador, 2017. Available online: http://dspace.udla.edu.ec/handle/33000/7934 (accessed on 19 September 2023).
7. Terán, M. *DVB-T Signal Reception with Software Defined Radio*; University of Cantabria: Santander, Spain, 2016. Available online: http://hdl.handle.net/10902/9045 (accessed on 19 September 2023).
8. Rivera, S.; Beltrán, V.; Pérez, G. Spectrum Analyzer Using Software Defined Radio. Higher School of Mechanical and Electrical Engineering Zacateco Unit. 2019. Available online: http://tesis.ipn.mx/handle/123456789/30136 (accessed on 19 September 2023).
9. Belzarena, P.; Flores Guridi, P.; Gómez Sena, G.; Gonzalez Barbone, V. Implementation of an Open ISDB-T Transceiver for Metrology under the Software Defined Radio Paradigm. ANII. Available online: https://iie.fing.edu.uy/investigacion/grupos/artes/en/projects/gr-isdbt/ (accessed on 28 March 2023).

engineering proceedings

Proceeding Paper

Hourly Load Curves Disaggregated by Type of Consumer Using A Density-Based Spatial Clustering Technique †

Carlos Andrés Peñaloza and Patricia Elizabeth Otero-Valladares *

Department of Electrical Energy, Escuela Politécnica Nacional, Quito 170525, Ecuador; carlos.penaloza@epn.edu.ec
* Correspondence: patricia.otero@epn.edu.ec

† Presented at the XXXI Conference on Electrical and Electronic Engineering, Quito, Ecuador, 29 November–1 December 2023.

Abstract: This paper discusses the innovative application of a new methodology for acquiring load curves in energy systems such as Azogues Electric Company (EEA). The proposed approach imports the database from Excel and, through an iterative clustering algorithm based on density with noise, generates daily load curves with a breakdown of weekdays and weekends, identifying one curve by type of consumer. Moreover, the groups obtained are validated by means of a Silhouette verification index (IS) identifying bad groupings, which are discarded to obtain results. Per unit value responses are presented through tables with hourly values on weekdays and weekends. The graphical comparison with the previous methodology of real measurements in Excel is also added.

Keywords: load curves; type of consumer; clustering algorithm; verification index

Citation: Peñaloza, C.A.; Otero-Valladares, P.E. Hourly Load Curves Disaggregated by Type of Consumer Using A Density-Based Spatial Clustering Technique. *Eng. Proc.* **2023**, *47*, 23. https://doi.org/10.3390/engproc2023047023

Academic Editor: Jaime Cepeda

Published: 7 December 2023

1. Introduction

Electric companies carry out studies in the distribution system periodically to identify and execute the necessary investments in infrastructure, with the aim of reducing technical losses and maintaining quality standards [1]. These studies involve handling various data from low voltage measurements and measurements at the feeder's head to develop electrical studies as load characterization and support the system operator. Certain measured data may present great similarity or homogeneity, so they are analyzed in daily load curves 0–23 h.

Currently, the Excel tool is used with average or average calculations to obtain load curves by consumption group [1], but there are negative and zero value measurements that modify the final shape of the group of curves, requiring previous work to manually cluster the load curves, eliminating atypical measurement. For this reason, it is worth having an algorithm based on data mining techniques to obtain one load curve broken down by each type of consumer and use them as the starting point for expansion planning and loss calculations at medium and low voltage levels [2].

Nowadays, clustering algorithms are used in data mining tools to identify important patterns and similar distributions in large information databases. There are some methods for obtaining groups or clusters that use averages, dendrograms, and grids, but the one chosen in this work is the method based on densities that identify groups with arbitrary shapes in the presence of atypical data or noise [3].

2. Materials and Methods

2.1. Meter Measurement Data

The 236 m with measurement data are obtained from the main feeders of the EEA distribution subsystem, as well as from users connected to the secondary circuits in the years 2018 and 2019. These meters are classified by type of consumer in the cadaster of users for the EEA. Therefore, Table 1 shows the description of meters and their voltage level. Low voltage meters measure in a 220 V grid, while de medium voltage meters are located in a 13.8 kV grid. The total values for each type of consumer are shown in the right column.

Table 1. Meter information by type of consumer.

Type of Consumer	Low Voltage 220 V	Medium Voltage 13.8 kV	Total
Commercial	25	20	45
Industrial	18	14	32
Others	9	9	18
Residential	131	0	131
No Identified	5	5	10
Total	188	48	236

2.2. Per unit System of Data

The real daily values of average power obtained from the meter are normalized, taking as the base power the maximum on each day and in the upper part the measured data in a per unit system (*pu*) of 24 h (0–23 h). This is represented in Equation (1):

$$P_{pu} = \frac{Promedio\ P}{P_{maxPorDia}} \tag{1}$$

P_{pu}: Dimensionless value in per unit system.
Promedio P: Average value of active power in kW.
$P_{maxPorDia}$: Maximum value of active power in kW per day.

2.3. DBSCAN in MATLAB

Two types of characteristic load curves are grouped according to the day of the week, a group from Monday to Friday and another from Saturdays to Sundays. Also, the types of consumers are identified as residential, commercial, industrial, and others.

In this process, the KNNsearch function is used to find a k-distances graphic of the curves; then the knee_pt function is used to find the knee of this graphic, that is, the value of Eps and its typical zone, which is a necessary argument to establish the clusters in the DBSCAN function. The Mpts argument is set to a relatively low value so that all points belonging to the same group are included. This algorithm requires three input parameters and returns an idx vector with the resulting grouping or cluster. Next, Equation (2) in MATLAB R2021b:

$$idx = DBSCAN\ (X,\ Eps,\ Mpts) \tag{2}$$

X: Size of the neighbor list or data matrix.
Eps: Radius that delimits the neighborhood area of a point (neighborhood-Eps).
$Mpts$: Minimum number of data or objects around neighborhood-Eps.

2.4. KNNSearch

This function finds the K nearest neighbors, according to Euclidean distances, and returns their indices in a column vector d and their respective distances kD. It uses input data or a database. Equation (3) shows its structure in MATLAB R2021b for its correct use.

$$[d,\ kD] = knnsearch\ (X,\ Y, Name,\ Value) \tag{3}$$

X: Values in pu of data 0–23 h for the measurement days.
Y: Values in pu 0–23 h.
Name: Write "K" and calculate the nearest neighbor distances.
Value: Matrix size of how many nearest neighbors in the distance metric.

2.5. Kneepoint

The kneepoint function advances along the K-dist plot of distances, one bisection point at a time, fitting two lines, these being the first derivative and the second derivative. The knee is at a bisection or threshold point that minimizes the sum of errors for the two adjustments. Any value less than this threshold density Eps can efficiently cluster patterns because these would lie in typical k-dist plot territory [4]. Equation (4) in MATLAB:

$$x = knee\_pt(distances) \quad (4)$$

$distances$: Euclidean distances of the K-dist graph.
x: Value in x of the elbow of the K-dist type graph.

2.6. Validation Index Silhouette (IS)

Each group can be represented by a silhouette, which is based on the comparison of their closeness and separation. This silhouette shows which objects are well classified within their group and which are simply infiltrating between the groups. The average width of the silhouette provides an assessment of clustering validity and could be used to select an "appropriate" number of clusters [5]. Equation (5) in MATLAB:

$$s = silhouette(X, idx) \quad (5)$$

X: Data between objects.
idx: is the partition obtained (by applying some grouping or cluster technique).
s: Value between −1 and 1, denoting 1 as belonging to the cluster and −1 not belonging.

3. Discussion

3.1. Clustering on Weekdays and Weekends

Figure 1a is a K-dist type graph made to find Epsilon (Eps) with the function Kneepoint. The respective knee of the function of distances is denoted with a red circle and any point up to that curve will represent a correct value to use it as an entry parameter in DBSCAN. Eps is established with the kneepoint function, and Mpts is set to a value greater than 1. Groupings such as the one in Figure 1b begin to be obtained, which contain descriptions on the bottom, top, and lateral sides.

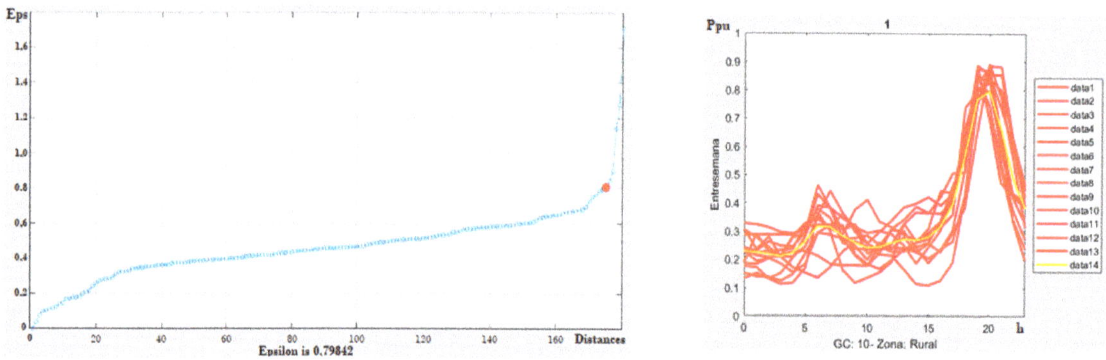

(a) (b)

Figure 1. Complementary algorithms and clustering (**a**) Kneepoint and Knnsearch on finding Eps (**b**) Partial result of a residential group in the clustering algorithm.

Then, Figure 2a shows the grouping of meters considered noise. Given the amount of data in this figure, a similar behavior is not distinguished in the characteristic curves, and a resulting yellow curve is also observed, which is not entirely true for this grouping. Then, the algorithm validates the grouping performance. For this task, the index IS, shown in

Figure 2b, is the verification in MATLAB and detects which cluster is incorrectly grouped, having index values of −1.

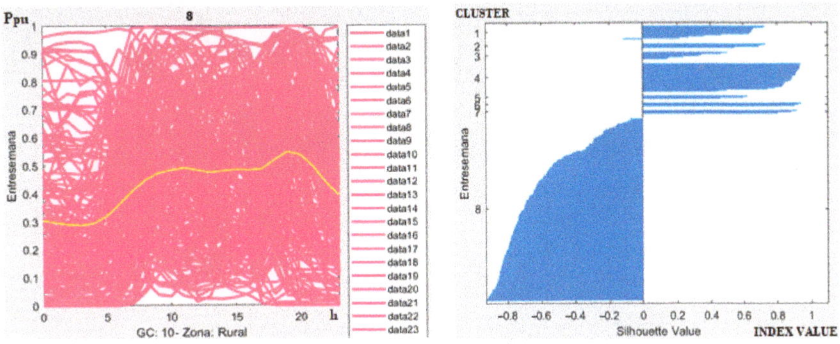

(**a**) (**b**)

Figure 2. Noise and validation (**a**) Cluster 8 on weekdays NOISE; (**b**) SILHOUETTE validation index, identifying cluster 8 as noise.

The same procedure is applied for the weekends, obtaining different values on the kneepoint, the grouping, and the validation. It is important to mention that noise is treated again with filters that take the eliminated curves and apply the clustering technique to obtain even more results from the system.

3.2. System Results

After submitting the database to the algorithm proposed in MATLAB R2021b, load curves for weekdays and weekends were found. The objective is to establish only one curve per type of consumer and a system curve. For this, reference [6] is used. Figure 3 shows the characteristic curves of the different consumption groups and a total curve of the system, having in the x-axis the hour of the day 0–23 h and in the y-axis the per unit value of the measurement.

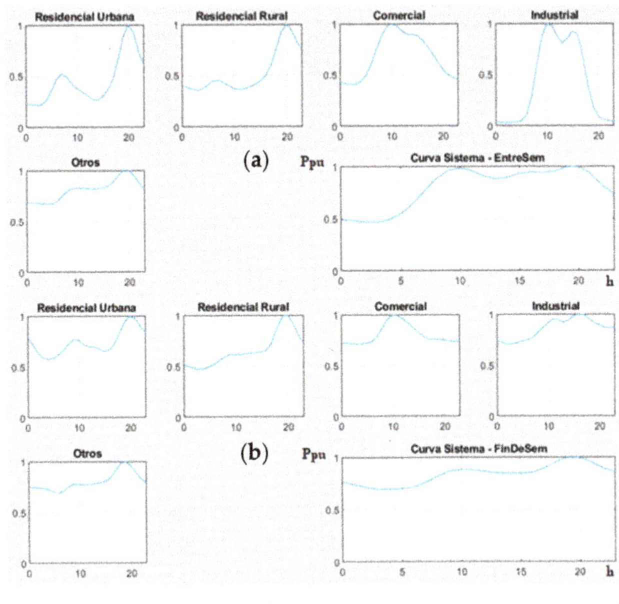

Figure 3. Graphical results in P_{pu} of each consumer and system curve. (**a**) Weekdays, (**b**) Weekends.

Table 2 presents the numerical results of the curves shown in Figure 3, they are separated in weekdays to the left side and the clustering results for weekends on the right.

Table 2. Load curves numerical results imported from MATLAB (**a**) weekdays; (**b**) weekends.

(a) weekdays						
Daily Hour	**Residential U**	**Residential R**	**Commercial**	**Industrial**	**Others**	**System Curve-Weekdays**
0	0.2295	0.4077	0.4293	0.0369	0.6848	0.4910
1	0.2234	0.3907	0.4179	0.0363	0.6836	0.4810
2	0.2166	0.3690	0.4070	0.0360	0.6790	0.4689
3	0.2203	0.3577	0.4083	0.0372	0.6730	0.4658
4	0.2592	0.3706	0.4330	0.0411	0.6695	0.4869
5	0.3544	0.4089	0.4916	0.0524	0.6723	0.5435
6	0.4687	0.4478	0.5940	0.1105	0.6973	0.6365
7	0.5211	0.4553	0.7355	0.2970	0.7510	0.7578
8	0.4886	0.4338	0.8741	0.6037	0.8016	0.8791
9	0.4213	0.4020	0.9673	0.8748	0.8257	0.9586
10	0.3743	0.3749	1.0000	1.0000	0.8305	0.9829
11	0.3463	0.3649	0.9785	0.9806	0.8287	0.9607
12	0.3098	0.3726	0.9308	0.8761	0.8245	0.9099
13	0.2718	0.3909	0.8990	0.8071	0.8219	0.8761
14	0.2661	0.4143	0.8957	0.8488	0.8245	0.8922
15	0.3017	0.4465	0.8853	0.9244	0.8312	0.9305
16	0.3647	0.4969	0.8404	0.8927	0.8480	0.9453
17	0.4668	0.5931	0.7768	0.6857	0.8909	0.9372
18	0.6531	0.7578	0.7039	0.3999	0.9550	0.9526
19	0.8800	0.9303	0.6224	0.1884	1.0000	0.9942
20	1.0000	1.0000	0.5499	0.0934	0.9988	1.0000
21	0.9326	0.9422	0.4998	0.0625	0.9487	0.9296
22	0.7421	0.8288	0.4698	0.0517	0.8692	0.8131
23	0.5992	0.7542	0.4571	0.0476	0.8148	0.7339
(b) weekends						
Daily Hour	**Residential U**	**Residential R**	**Commercial**	**Industrial**	**Others**	**System Curve-Weekends**
0	0.7927	0.5184	0.7238	0.7468	0.7484	0.7648
1	0.7219	0.5000	0.7206	0.7263	0.7454	0.7397
2	0.6370	0.4811	0.7157	0.7064	0.7401	0.7107
3	0.5861	0.4721	0.7115	0.7083	0.7351	0.6961
4	0.5732	0.4762	0.7121	0.7213	0.7233	0.6945
5	0.5842	0.4936	0.7195	0.7379	0.6984	0.7005
6	0.6141	0.5212	0.7385	0.7494	0.6853	0.7168
7	0.6598	0.5554	0.7867	0.7726	0.7129	0.7555
8	0.7203	0.5941	0.8722	0.8246	0.7563	0.8162
9	0.7689	0.6157	0.9593	0.8731	0.7767	0.8652
10	0.7579	0.6118	1.0000	0.9177	0.7750	0.8801

Table 2. *Cont.*

(b) weekends						
Daily Hour	**Residential U**	**Residential R**	**Commercial**	**Industrial**	**Others**	**System Curve-Weekends**
11	0.7140	0.6155	0.9901	0.9538	0.7702	0.8760
12	0.6959	0.6265	0.9531	0.9438	0.7722	0.8647
13	0.6944	0.6290	0.9091	0.9199	0.7798	0.8519
14	0.6704	0.6347	0.8653	0.9420	0.7888	0.8452
15	0.6497	0.6425	0.8223	0.9834	0.8039	0.8453
16	0.6661	0.6558	0.7814	1.0000	0.8397	0.8542
17	0.7262	0.7342	0.7578	0.9952	0.9103	0.8934
18	0.8355	0.8788	0.7555	0.9549	0.9839	0.9551
19	0.9500	0.9916	0.7555	0.9188	1.0000	1.0000
20	1.0000	1.0000	0.7485	0.8959	0.9585	0.9972
21	0.9745	0.9164	0.7398	0.8728	0.8937	0.9526
22	0.8981	0.7987	0.7341	0.8659	0.8272	0.8935
23	0.8380	0.7223	0.7322	0.8689	0.7893	0.8559

3.3. *Comparison with Previous Method*

The method of obtaining load curves used by the EEA focuses on using the real measurements, exporting them to an Excel file, and then, with a series of filters on georeferencing, voltage level, or feeders, obtaining similar curves in real values of Active Power kWh. Then, empirically, eliminates curves that do not represent a characteristic behavior of the type of consumer to finally group and obtain the average of the similar curves. These curves are later used in distribution network simulation programs and must be in pu to obtain electrical analysis results, power flows, and losses.

Weekdays Comparison

Figures 4 and 5 show the final curve during the weekdays where the left side is the result with the previous method of the electric company in Excel, while the right side shows the resulting curve obtained with the DBSCAN algorithm in Matlab. In addition, the new method changes the y-axis from real value in kW to pu values while it maintains the x-axis in daily hours.

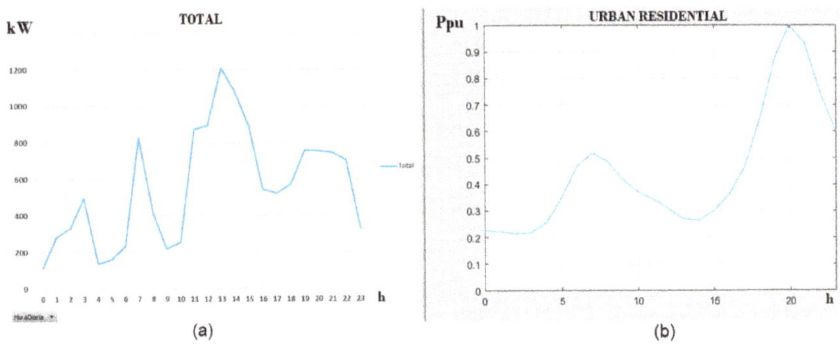

Figure 4. Final load curve on weekdays Urban residential. (**a**) Previous method; (**b**) DBSCAN.

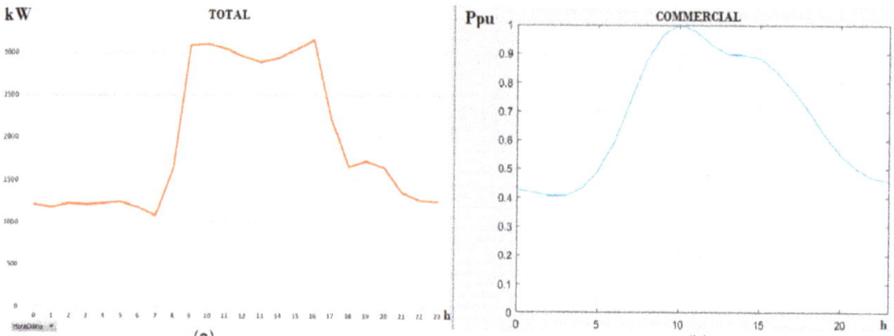

(a) (b)

Figure 5. Final load curve on weekdays Commercial. (**a**) Previous method; (**b**) DBSCAN.

The previous curves indicate different behavior for the same type of Urban Residential consumers, and the peaks are given in other hourly values. This is because in the old method, there is a lack of data to carry out filters in Excel, while the new one provides the direct grouping of similar curves with the density algorithm.

Figure 5 exhibits similar results at consumption peaks. This time, the database in Excel included many similar curves for the commercial consumer, and the grouping step was successful. In the same way, the values in pu are convenient for better identification of the resulting curve and its maximum demand.

4. Conclusions

The advantage of the method presented in this written work is the use of a clustering algorithm to obtain load curves in pu and the validation tool to discard atypical measurements. These two advantages outperform the previous EEA methodology by obtaining only one curve by type of consumer using a clustering technique instead of considering the analyst experiences.

During the clustering procedure, load curves that have the behavior of public lighting or industries that only operate in the early morning hours or at night were identified by the noise application of the algorithm. These types of consumers did not enter the obtaining of final curves due to their little intervention in the distribution electrical system.

Author Contributions: Conceptualization, P.E.O.-V. and C.A.P.; methodology, C.A.P.; software, C.A.P.; validation, C.A.P., P.E.O.-V.; formal analysis, P.E.O.-V.; investigation, C.A.P.; data curation, C.A.P.; writing—original draft preparation, C.A.P.; writing—review and editing, P.E.O.-V.; visualization, P.E.O.-V.; supervision, P.E.O.-V.; project administration, P.E.O.-V.; All authors have read and agreed to the published version of the manuscript.

Funding: This research received no external funding.

Institutional Review Board Statement: Not applicable.

Informed Consent Statement: Informed consent was obtained from all subjects involved in the study.

Data Availability Statement: Data is unavailable due to privacy restrictions.

Conflicts of Interest: The authors declare no conflict of interest.

References

1. Ministerio de Electricidad y Energías Naturales No Renovables, "Plan Maestro de Electricidad 2018–2027", Chapter 3, Estudio de la Demanda. Available online: https://www.recursosyenergia.gob.ec/plan-maestro-de-electricidad/ (accessed on 13 March 2023).
2. ARCONEL. Plan Maestro de Electrificación 2013–2022, Perspectiva y Expansión del Sistema Eléctrico Ecuatoriano. Available online: https://www.regulacionelectrica.gob.ec/plan-maestro-de-electrificacion-2013-2022/ (accessed on 15 April 2023).
3. Fong, S.; Rehman, S.U. DBSCAN: Past, Present and future. In Proceedings of the Fifth International Conference on the Applications of Digital Information and Web Technologies (ICADIWT 2014), Chennai, India, 17–19 February 2014.
4. Irwin, D.; Albrecht, J.; Satopa, V. Finding a "Kneedle" in a Haystack: Detecting Knee Points in System Behavior. In Proceedings of the 31st International Conference on Distributed Computing Systems Workshops, Minneapolis, MN, USA, 20–24 June 2011.

5. Rousseeuw, P.J. Silhouettes: A graphical aid to the interpretation and validation of cluster analysis. *J. Comput. Appl. Math.* **1999**, *20*, 53–65. [CrossRef]
6. Gönen, T. *Electric Power Distribution Engineering*; CRC Press: Boca Raton, FL, USA, 2014.

engineering proceedings

MDPI

Proceeding Paper

5G NR Radio Network Planning at 3.5 GHz and 28 GHz in a Business/Dense Urban Area from the North Zone in Quito City †

Valdemar Farré [1], José David Vega Sánchez [1] and Henry Carvajal Mora [2,*]

1 Departamento de Electrónica, Telecomunicaciones y Redes de la Información, Escuela Politécnica Nacional, Quito 170525, Ecuador; valdemar.farre@epn.edu.ec (V.F.); jose.vega01@epn.edu.ec (J.D.V.S.)
2 Faculty of Engineering and Applied Sciences, Telecommunications Engineering, Universidad de Las Américas (UDLA), Quito 170503, Ecuador
* Correspondence: henry.carvajal@udla.edu.ec
† Presented at the XXXI Conference on Electrical and Electronic Engineering, Quito, Ecuador, 27 August 2023.

Abstract: This work presents an overview of fifth-generation (5G) technology, its key functionalities, and its applications. Then, we introduce the topic of 5G radio network planning, making analysis and design methods for coverage and capacity determinations, considering 3GPP standards. With these results, we apply the standards and procedures for a business/dense urban area in Quito city, considering the relevant parameters for calculations and simulations using the Atoll tool. Specifically, we consider the following: the key propagation models Macro/Micro, bandwidths, use cases, data rates, frequency bands (3.5/28 GHz), and modulation orders, etc. Finally, we discuss the proposed network planning and provide some useful insights regarding the 5G mobile network in Quito.

Keywords: 5G NR; radio network planning; coverage; 3GPP; capacity; link budget; simulation; atoll software

Citation: Farré, V.; Vega Sánchez, J.D.; Carvajal Mora, H. 5G NR Radio Network Planning at 3.5 GHz and 28 GHz in a Business/Dense Urban Area from the North Zone in Quito City. *Eng. Proc.* **2023**, *47*, 24. https://doi.org/10.3390/engproc2023047024

Academic Editor: Soraya Lucia Sinche Maita

Published: 5 December 2023

1. Introduction

In recent years, the global demand for data traffic from mobile networks and higher and faster data rates and very low latency have increased due to the high demand for various applications and the ability to connect thousands of devices simultaneously. In this fact, the fifth generation (5G) of the cellular communication system, i.e., New Radio (NR), offers these features and requirements through specific and different scenarios, such as ultra-high mobility, ultra-high connection, and ultra-high traffic [1–5], compared to legacy networks such as LTE [6]. Generally, radio network planning (RNP) is a critical topic and is considered to be the primary step in implementing a wireless network such as a 5G network. The services and applications provided by 5G NR networks are grouped into three use case scenarios, namely enhanced mobile broadband (eMBB), ultra-reliable and low-latency communications (uRLLC), and massive machine-type communication (mMTC), and all these use cases are applied to the Internet of Things (IoT) [7–9]. The level of importance of each capability in the different use cases is shown in Figure 1.

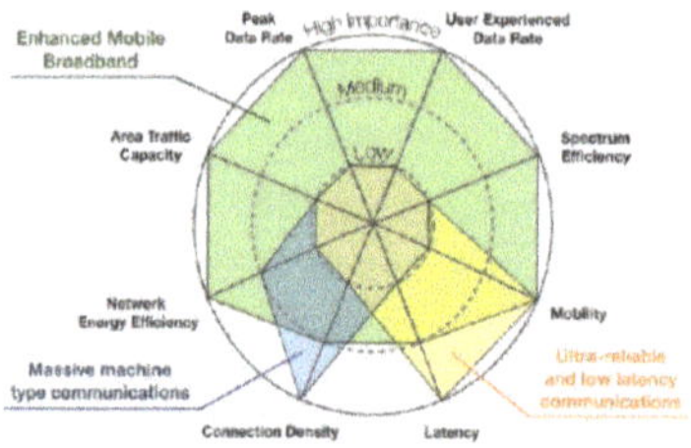

Figure 1. Key capabilities in different use cases and scenarios [7].

The frequency bands allocated to the 5G NR network are classified into three main categories according to Table 1.

Table 1. 5G spectrum band for each scenario and use cases [7].

Category	Frequency and Bandwidth	Usage Scenario Typical	Frequency Center	Range
Low-Frequency Bands Coverage Layer	Below 1 GHz bandwidth up to 20 MHz	mMTC	Band of 0.7–1.9 GHz	0.7–1.990 GHz
Medium-Frequency Bands Coverage Layer	1–6 GHz bandwidth up to 100 MHz	mMTC, eMBB	Band of 3.5–15 GHz	3.3–4.2 GHz
High-Frequency Bands Coverage Layer	Above 6 GHz bandwidth up to 800 MHz	eMBB, uRLLC	Band of 26–28 GHz	24.25–27.5 GHz 26.5–29.5 GHz

The main contribution of this paper is that, until now, 5G NR network planning with a C-band and millimeter wave (mm-wave) band was yet to be performed in Ecuador for the metropolitan and business areas in the north zone in Quito city [10–13]. Our goal is to show the resumed planning, designing, and evaluation of a 5G NR network in the dense urban and urban area of Quito city in Ecuador by using a C-band/mm-wave band applying a densification process using macro/micro cells in the network planning [14]. For this purpose, we use the Link budget and propagation model calculations to determine the path loss values, propagation values, and cell radius in order to calculate the number of sites required in the target zones. All these activities are applied by considering the coverage and capacity planning, dense urban/urban areas, and frequency of operation. The data calculated is tested using a planning tool called Atoll ver. 3.4.0, and our results are reflected in the simulation results from the synchronization signal—RSRP (SS-RSRP), synchronization signal to noise ratio (SS-SNR), and downlink/uplink throughput data rate plots for each scenario.

This paper is organized as follows: Section 2 defines the architecture of the 5G network, methodology, planning concepts, and employed system parameters. Section 2 also introduces 5G network planning into zones of Quito city; here, the results of the link budget and each parameter for the simulations are also presented. Section 3 presents the simulation results to verify the effectiveness of the suggested planning in Quito city, and Section 4 shows the conclusions.

2. Methodology of Research, Radio Planning, and Design of 5G Network

2.1. Methodology and General Process Flowchart for Radio Network Planning

The scheme of the research for the radio planning and design of 5G NR at the 3.5 GHz and 28 GHz frequencies for metropolitan/business areas in the north zone of Quito city is shown in Figure 2 [15]. Specifically, it presents the flowchart of the simplified basic radio planning and dimensioning from our research. This is focused on 5G NR coverage and capacity planning from the point of view that one site can serve with its resources. Coverage planning includes calculating the link budget, path losses, and throughout, as well as the attenuations and losses between the user equipment (UE) and next-generation nodeB (gnodeB) [16]. These calculations reveal the weakening signal in the wireless channel and show the maximum distances for a full-coverage radio or range from one gnodeB.

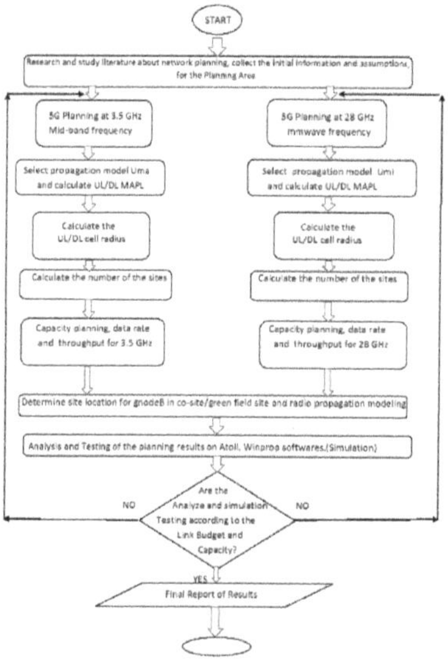

Figure 2. Flowchart of the research for 5G NR network planning [15].

Next, we identify the path loss values using Uma/Umi propagation models recommended at 3GPP 38.901; the cell radius value is estimated as the maximum distance between gnodeB and UE. The identified cell radius value is used to calculate the site's coverage area and determine the number of sites needed in the target zone as the final result of this coverage planning calculation [13]. With the link budget to hand, the next step is to analyze the planning results in each parameter obtained. In these terms, the initial stage of this work is to determine the locations of each site for the 5G NR network planning that can be deployed in the business/dense urban area of the north zone of Quito city. The data required in this study are about the target area, geographic location, and population in this zone of Quito. Furthermore, to calculate the services classification, data on the population density of the zone are additional requirements [16]. The next part of the planning is to calculate the data throughput and the capacity of the number of users for each site and use case. The final phase of this research is to analyze and determine the respective simulations about coverage, quality, and capacity based on the data obtained in the Link budget and design previews around the results from the calculation. The data input and necessary parameter are processed using Atoll version 3.4.0. These results are the essential part of this research, because the plots showed the coverage, quality, and capacity situations, and their parameters are compared with the coverage, quality, and data rates calculated previously throughout the Link budget [12]. The final result from the perspective of this work's coverage is to find the number of sites needed to cover the target area location.

2.2. Link Budget Dimensioning and Planning

The maximum allowable path loss (MAPL) is the maximum attenuation received between the mobile antenna from UE and the antennas from gnodeB for uplink and downlink links throughout the air interface. The calculation of the coverage takes into account the losses between UE and gnodeB, and the required data for the calculation of the link budget, displaying the parameters used by the UE terminal and gnodeB entities, are shown in Tables 2 and 3. First, the value of the thermal noise is calculated using (1), and the number of subcarriers is computed using (2), so [12]

$$N_{thermal} = 10 \times \log(K \times T \times B), \quad (1)$$

where

K = Boltzmann Constant (1.38×10^{-23} J/K)

Table 2. Link budget simplified From 5G New Radio planning.

Parameters for Calculate (MAPL)					
Parameter	**Notation**	**Downlink**	**Uplink**	**Downlink**	**Uplink**
		C-Band: 3.5 GHz		**mmwave Band: 28 GHz**	
gnodeB transmitter power (dBm)	a	43	-	36.5	-
SCS (KHz)/B (MHz)	-	30/100	30/100	60/200	60/200
Resource Block	RB	273	273	264	264
Subcarrier quantity	b	3276	3276	3168	3168
gnodeB antenna gain(dBi)	c	25	25	33.5	33.5
gnodeB cable loss (dBi)	d	0	0	0	0
Penetration loss (dB)	e	28	28	19	19
Folliage loss (dB)	f	10	10	10	10
Body block loss (dB)	g	3	3	7	7
Interference margin (dB)	h	6	2	1	1
Rain/Ice margin (dB)	i	0	0	3	3
Slow fading margin (dB)	j	7	7	7	7
UE antenna gain (dB)	k	0	0	0	0
Noise Bandwidth (SCS) (MHz)	B	0.03	0.03	0.06	0.06
Boltzmann Constant	K	1.38×10^{-23}	1.38×10^{-23}	1.38×10^{-23}	1.38×10^{-23}
Temperature (°K)	T	293	293	293	293
Thermal Noise Power (dBm)	l	−129.1592799	−129.1592799	−126.14898	−126.14898
UE noise figure (dB)	m	9	9	9	9
Demodulation threshold SNR (dBm)	n	−1	−1	−1	−1
Planning Area (Km^2)	A	6.72	6.72	6.72	6.72

Table 3. Resource Block For medium/high-freq. band (3GPP_TS 38.101-1,38.101-2 v17.6).

Bandwidth	**SCS (KHz)—3.5 GHz**			**Bandwidth**	**SCS (KHz)—28 GHz**	
	15 (KHz)	**30 (KHz)**	**60 (KHz)**		**60 (KHz)**	**120 (KHz)**
20 (MHz)	106	51	24	50 (MHz)	66	32
50 (MHz)	270	133	65	100 (MHz)	132	66
100 (MHz)	N/A	273	135	200 (MHz)	264	132

T = Temperature (at 293 °K),

and,

$$N_{thermal} = \text{Thermal Noise},$$

$$S_{CQ} = R_B \times S_{RB} \quad (2)$$

where

S_{CQ} = Subcarrier quantity
R_B = Resource Block
S_{RB} = Subcarrier per Resource Block

The value of the path loss is significantly affected by the parameters used in the link budget for the planning of the 5G NR network, so the path loss is calculated using the following expression [12]:

$$\text{Pathloss (dBm)} = a - 10\log(b) + c - d - e - f - g - h - i - j + k - l - m - n, \quad (3)$$

where the required values in (3) are given in Table 2. Next, in order to obtain the value of d_{3D} using (3), the following values need to be calculated from the 3GPP standards [14]

$$h'_{BS} = h_{BS} - h_E \quad (4)$$

$$h'_{UT} = h_{UT} - h_E \quad (5)$$

$$d'_{BP} = (4 \times h'_{BS} \times h'_{UT} \times f_c)/c, \quad (6)$$

where h_{BS} and h_{UT} are the heights of the gNodeB (m) and UE (m), f_c denotes the frequency of operation (GHz), c is the speed of the light, and h_E = 1 m, according to (3GPP_TS_38.901).

Based on [16], the propagation model in the scenario with the Line of sight (LOS) of the Urban Macro model (Uma) and Urban Micro model (Umi), we use the following formulations, considering that 10 m < d_{2D} < d'_{BP}:

$$Lp1 = 28 + 22\log(d_{3D}) + 20\log(f_c)\ \text{(Uma with LOS)} \quad (7)$$

$$Lp2 = 32.4 + 21\log(d_{3D}) + 20\log(f_c)\ \text{(Umi with LOS)} \quad (8)$$

On the other hand, for the propagation model for NLOS, we use [16]:

$$Lp3 = 13.54 + 39.08\log(d_{3D}) + 20\log(f_c) - 0.6\,(h_{UT} - 1.5)\ \text{(Uma with NLOS)} \quad (9)$$

$$Lp4 = 22.34 + 35.3\log(d_{3D}) + 21.3\log(f_c) - 0.3\,(h_{UT} - 1.5)\ \text{(Umi with NLOS)}, \quad (10)$$

where:

Lp1, Lp2 = Value of pathloss (dBm) for scenario LOS for Uma/Umi models
Lp3, Lp4 = Value of pathloss (dBm) for scenario NLOS for Uma/Umi models
d_{3D} = result of the distance between h_{BS} and h_{UT} (m)
d'_{BP} = break point distance (m).

After the value of d_{3D} (outdoor) is obtained using the Pathloss calculation of the propagation model, the value of d_{2D} (outdoor) is the cell radius obtained from the following:

$$\text{Radius cell}(d_{2D}) = \sqrt{\left({d_{3D}}^2 - \left(h_{BS} - h_{UT}\right)^2\right)}. \quad (11)$$

From the cell radius, the coverage area of one gnodeB site can be determined using the following expression for a three-sector model [13]:

$$C_A = \text{Coverage Area from gnodeB } (m^2) = 2.6 \times (d_{2D})^2 \quad (12)$$

To determine the number of sites required in the target area of planning for coverage, this must be known for comparison with the coverage area of gnodeB. The calculation of the number of gnodeB in the case of LOS is the following [12]:

Number of site = total area surface area (Km^2)/coverage area of 1 site (Km^2)

$$N_{\_gNodeB} = \frac{A_T}{C_A} \quad (13)$$

where

A_T = Total surface area (m^2)

Cellular network planning generally covers two perspectives: coverage area planning and capacity planning [12].

2.3. Coverage Area Planning

This type of planning is radio network planning, in which calculations are based on the target zone to be covered by the gnodeB dimensioned. In fact, any parameters can influence this planning, such as the reception sensitivity from mobile devices, path loss, transmit power, link budget calculation, and cell radius. In this work, the analysis from the Link budget is used to determine the MAPL; meanwhile, the cell radius is determined using propagation models selected adequately, namely Uma and Umi. The network planning aims to determine the ideal number of sites for the business/metropolitan and dense urban areas of the north zone of Quito by calculating the coverage area for each site, considering the propagation model to be used in the simulation and the RSRP levels according to Table 4. Table 5 presents the d_{3D} and cell radius (d_{2D}) values using the Uma and Umi models to determine the quantity of the gnodeB needed.

Table 4. RSRP VALUE CATEGORY [12].

Category	Range of Signal Level
Good	$-90 \leq$ RSRP ≤ -70 dBm
Normal	$-110 \leq$ RSRP ≤ -90 dBm
Bad	$-130 \leq$ RSRP ≤ -110 dBm

Table 5. Results of the calculations.

Parameter	Scenario 1: 3.5 GHz (FR1 Band)		Scenario 2: 28 GHz (FR2 Band)	
	LOS	NLOS	LOS	NLOS
Thermal Noise	−129.1592799		−126.14898	
Subcarrier quantity	3276		3168	
Pathloss	100.005841		106.1411282	
h'BS	24		24	
h'UE	0.5		0.5	
d'BP	560		4480	
d_{3D}	600.3236595	85.92038528	135.90535	31.55408462
d2D/Cell Radius	600.3234512	85.91893044	135.90443	31.55012291
Coverage Area (Km^2)	0.93700944	0.019193363	0.04802204	0.002588067
Target Area (Km^2)	6.72	6.72	6.72	6.72
Number of gnodesB	7.171752721	350.1210328	139.935755	2596.532806
Data Rate Downlink (Mbps)	4.456, 60	1.114, 15	8.619, 37	2.154, 84
Data Rate Uplink (Mbps)	2.445, 70	611, 42	4.730, 14	1.182, 54
DL/UL Throughput service per user (Mbps)	50	50	100	100
Number of user per use case DL	89.13	22.28	86.19	21.55
Number of user per use case UL	48.91	12.23	47.30	11.83

Here, we chose an area of about 6.72 Km^2 in a dense urban/urban zone, as shown in Figure 3.

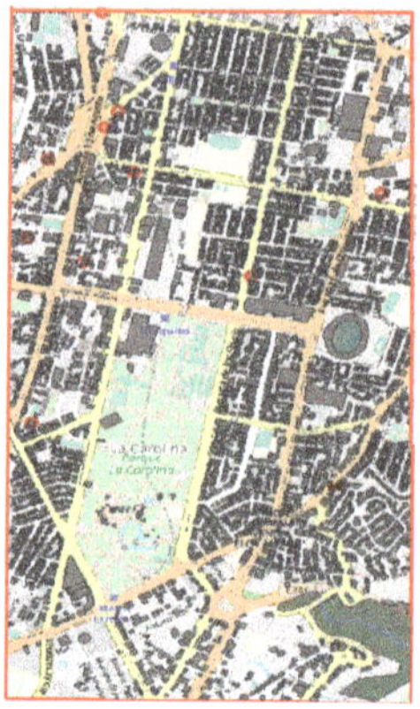

Figure 3. Target area of business/dense area of north zone in Quito city, 5G planning.

2.3.1. 5G NR Reference Signal-Received Power (SS-RSRP)-Parameter Secondary Synchronization

SS-RSRP is the average level of the received secondary synchronization signal. This parameter compares the signal strengths of the individual cells in a 5G network; SS-RSRP is equivalent to the RSRP parameter in 4G LTE networks [9]. In gnodeB, there is a processing circuit configured to code several Synchronization Signal blocks (SSBs); each SSB is related to a different beam of a number of multiple-in-multiple-outs (MIMOs) from antennas [11]. Thus, each one of the SSBs is structured with a Primary Synchronization Signal (PSS), Secondary Synchronization signal (SSS), and Physical Broadcast Channel (PBCH).

2.3.2. 5G NR Signal-to-Noise and Interference Ratio (SS-SINR)

This parameter is the mean value of the power contribution in Watts (W) or dB of the resource elements carrying the SSS divided by the linear average of the noise and noise power contributions in W or dB, respectively. The metropolitan/business zone of the north zone in the Quito city area covered a total area of 6.72 Km^2, with a percentage of 60% penetration. In the 5G network, UE measures the signal strength accepted in the SSS for the closest cell transmitter.

2.4. Capacity Planning

In this paper, capacity planning is focused on calculating the data rate and throughput. These parameters are measures that show the bits in data in the transmission process that can be sent per unit time, measured in bits per second. The maximum downlink and uplink data rates supported by the UE are calculated according to the user device's mean category. For the design of a 5G NR network, the approximate data rate and throughput for a determined number of carriers in a bandwidth or bandwidth combination are calculated using the expressions in (14) and (15) [15], with assumptions of throughput services per use case (eMBB) and modulation order being max/min for LOS/NLOS scenarios:

$$\text{Data Rate(Mbps)} = 10^{-6}\sum_{j=1}^{j}\left(V_{Layers}^{(P)}\, Q_m^{(j)} \times f^{(j)} \times R_{max} \times \frac{N_{PRB}^{BW(j),\mu}\, x\, 12}{T_S^{\mu}}\left(1 - OH^{(j)}\right)\right), \quad (14)$$

where

P = Number of carrier components

Rmax = 948/1024

$V_{Layers}^{(P)}$ = Maximum number of MIMO layers

$Q_m^{(j)}$ = Modulation order

$f^{(j)}$ = Scaling factor,
μ = Numerology
T_S^{μ} = OFDM Symbol
$N_{PRB}^{BW(j),\mu}$ = Maximum number of resource blocks PRB (according to 3GPP 38.104) for selected bandwidth, frequency range, and numerology per selected carrier.
$OH^{(j)}$= Overhead
BW = Bandwidth.

Number of user per use case DL/UL = Data Rate (Mbps)/DL/UL Throughput service per user (Mbps). (15)

3. Results, Analysis, and Simulations

3.1. Model Propagation Uma and Umi for 5G NR Radio Network Planning

The number of sites needed in the area planning for the business/dense urban zone of the north zone of Quito city is first calculated by selecting adequate propagation models. As mentioned in the previous section, the selected propagation models are Uma and Umi, based on the link budget formulas for 5G planning from 3GPP TS 38.901, considering a transmission antenna height of gnodeB of 25 m, a reception height from UE of 1.5 m, and an inter-site distance of 500 m [14] for a middle band and 30–100 m for a high band.

3.2. Link Budget, Coverage, and Capacity Calculation Results

In our setup, we use MIMO antennas of 64 × 64 Tx-Rx for coverage and capacity and a modulation order of 4 for the NLOS scenario and 8 for the LOS scenario. In addition, we use one carrier component, with numerology of 1 and 2 for 3.5 GHz/28 GHz, respectively. For more details about these configurations, the reader can refer to Table 5. In Table 6, we show the results of the Link Budget, coverage, and capacity planning for 3.5 GHz/28 GHz in LOS/NLOS scenarios.

Table 6. The main system parameter [12].

Key Parameter	Spectrum Details			
Technology/Duplex	5G NR/TDD			
Carrier Frequency/Bandwidth	3.5 GHz	100 MHz	28 GHz	200 MHz
Start Frequency–End Frequency	3300 MHz	4200 MHz	26,500 MHz	29,500 MHz

3.3. Simulation Planning Results Based on Link Budget, Coverage, and Capacity Calculations

Two scenarios are used for the simulations, i.e., outdoor to outdoor (O2O) with LOS and NLOS for 3.5 GHz and 28 GHz. The new sites are allocated according to the cell radius of the coverage distance. This radio planning uses existing sites and non-existent sites. Therefore, a recommendation for new site placement per the coverage area is calculated. The site allocation is set automatically using the Automatic Coverage Planning (ACP) tool available in the Atoll software. The results of the Atoll RNP tool simulation show the number of gnodeB required to cover the target area, the values of the SS-RSRP for coverage and the SS-SINR parameters for quality, and the average throughput data rate for capacity obtained in this research, and these are shown in the following plots.

3.3.1. Scenario 1–2: O2O Downlink—SS-RSRP

In Figure 4, the average SS-RSRP obtained from the prediction using eight sites of gnodeB is −85 dBm. This means the signal strength is allocated into the mean values; meanwhile, the minimum is −110 dBm and the maximum value is −70 dBm.

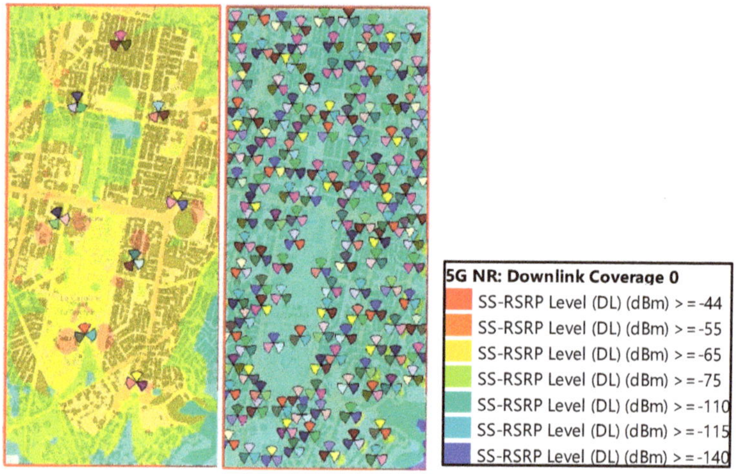

Figure 4. SS-RSRP coverage plot with ACP: 3.5 GHz (**left**) and 28 GHz (**right**).

3.3.2. Scenario 1–2: O2O Downlink—SS-SNR

In Figure 5, the average SS-SINR obtained from the prediction using X sites of gnodeB is 17 dB. In other words, the SINR ratio is allocated into the mean values; meanwhile, this paper revealed an SS-SINR with a minimum value of −5 dB and a maximum value of 30 dB.

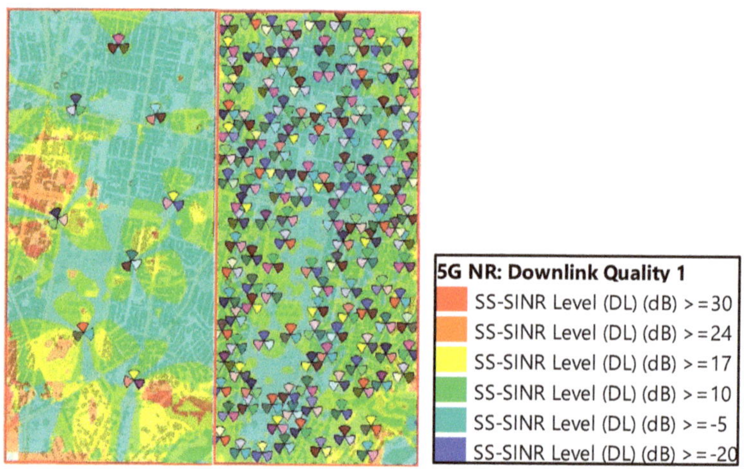

Figure 5. SS-SINR quality plot with ACP: 3.5 GHz (**left**) and 28 GHz (**right**).

3.3.3. Scenario 1–2: O2O Downlink—Data Rate Throughput

In Figure 6, the average throughput data rate obtained from the prediction using X sites of gnodeB is 50 Mbps. Meanwhile, this research shows a throughput data rate with a minimum value of 30 Mbps and a maximum value of 100 Mbps. By comparing the above findings with those results between scenario one and scenario two, the simulation outcomes are very similar to the calculation results.

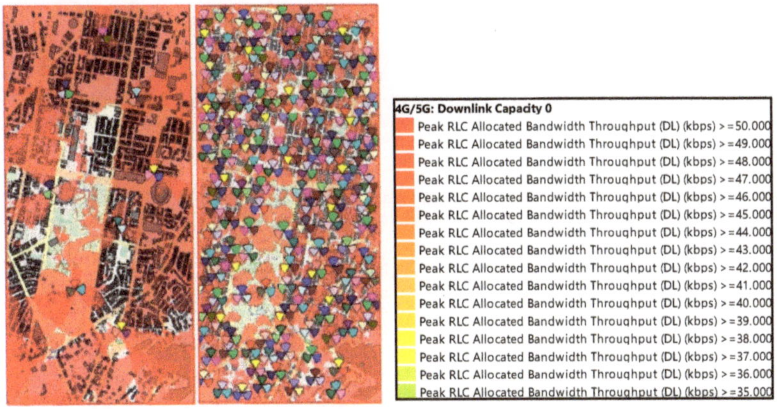

Figure 6. DL Throughput data rate plot with ACP: 3.5 GHz (**left**) and 28 GHz (**right**).

4. Conclusions

1. We have significantly contributed with a speedy and efficient way of performing the designs, calculations, and planning of 5G NR networks for business/dense urban areas in compliance with coverage, quality, and capacity objectives. In our case, this was accomplished by applying the existing 3GPP standards for 5G through the use and selection of the correct criteria, adequately applying the propagation models, urban density scenarios type, topography, LOS/NLOS criteria, service use cases, bandwidths for the user, link budget parameters, and frequency spectrum, and following best practices from commercial 5G network deployments.
2. The link budget calculations and coverage, quality, and capacity simulations obtained with the Atoll tool were highly accurate and consistent with the theoretical calculations obtained. This can be verified and confirmed by the average levels achieved for the coverage SS-RSRP (>90 dBm) and quality SS-SNR (>10 dB), which ensures a good data experience, good coverage for the target area, and outstanding data throughput levels (>50 Mbps per user), with all these being observed in both planned scenarios (3.5/28 GHz).
3. The results obtained from our contribution can make it easier for local mobile network operators to plan their 5G networks. This allows them to approach their designs more effectively, conducting their commercial deployments with successful commercial frequency bands in 5G networks.

Author Contributions: Conceptualization, V.F. and J.D.V.S.; methodology, J.D.V.S. and H.C.M.; software, V.F.; validation, J.D.V.S.; formal analysis, V.F. and H.C.M.; investigation, V.F. and J.D.V.S.; resources, V.F.; data curation, V.F. and J.D.V.S.; writing—original draft preparation, V.F. and J.D.V.S.; writing—review and editing, J.D.V.S. and H.C.M.; visualization, V.F; supervision, J.D.V.S.; project administration, H.C.M.; funding acquisition, V.F, J.D.V.S. and H.C.M. All authors have read and agreed to the published version of the manuscript.

Funding: This research received no external funding.

Institutional Review Board Statement: Not applicable.

Informed Consent Statement: Not applicable.

Data Availability Statement: Data are contained within the article.

Conflicts of Interest: The authors declare no conflict of interest.

References

1. Shaddad, R.Q.; Al-Kmali, F.S.; Noman, M.A.; Ahmed, N.K.; Marish, E.M.; Al-Duais, A.M.; Al-Yafrsi, A.A.; Al-Sabri, F.A. Planning of 5G Millimeterwave Wireless Access Network for Dense Urban Area. In Proceedings of the 2019 First International Conference of Intelligent Computing and Engineering (ICOICE), Hadhramout, Yemen, 15–16 December 2019; pp. 1–4. [CrossRef]
2. Shah SA, W.; Khattab, T.; Shakir, M.Z.; Hasna, M.O. A distributed approach for networked flying platform association with small cells in 5G+ networks. In Proceedings of the GLOBECOM 2017—2017 IEEE Global Communications Conference, Singapore, 4–8 December 2017.
3. Chen, Z.; Lee, J.; Quek, T.Q.; Kountouris, M. Cooperative caching and transmission design in cluster-centric small cell networks. *IEEE Trans. Wirel. Commun.* **2017**, *16*, 3401–3415. [CrossRef]
4. Fallgren, M.; Timus, B. Scenarios Requirements and KPIs for 5G Mobile and Wireless System. Available online: https://cordis.europa.eu/docs/projects/cnect/9/317669/080/deliverables/001-METISD11v1pdf.pdf (accessed on 7 August 2023).
5. Jamalzadeh, M.; Ong, L.-D.; Mohd, M.N.B. 5GTechnologies: A New Network Architectures and Design. *J. Internet Technol.* **2018**, *19*, 1983–1991.
6. Sukarno, A.; Hikmaturokhman, A.; Rachmawaty, D. Comparison of 5G NR Planning in Mid-Band and High-Band in Jababeka Industrial Estate. In Proceedings of the 2020 IEEE International Conference on Communication, Networks and Satellite (Comnetsat), Batam, Indonesia, 17–18 December 2020; pp. 12–17. [CrossRef]
7. Situmorang, A.C.; Gunawan, D.; Anggraini, V.G. 5G Trials on 28 GHz Band in Indonesia. In Proceedings of the 28th Wireless and Optical Communication Conference (WOCC 2019), Beijing, China, 9–10 May 2019.
8. Al-Mekhlafi, A.A.; Al-Kamali, O.M.; Shaddad, R.Q.; Abduljalil, M.A. Dense 5G Network Planning for Coastal Area at Aden City, Yemen. In Proceedings of the 2019 First International Conference of Intelligent Computing and Engineering (ICOICE), Hadhramout, Yemen, 15–16 December 2019; pp. 1–7. [CrossRef]
9. Rianti, D.; Hikmaturokhman, A.; Rachmawaty, D. Techno-Economic 5G New Radio Planning Using 26 GHz Frequency at Pulogadung Industrial Area. In Proceedings of the 2020 3rd International Seminar on Research of Information Technology and Intelligent Systems (ISRITI), Yogyakarta, Indonesia, 10 December 2020; pp. 272–277. [CrossRef]
10. Ekawibowo, S.A.; Pamungkas, M.P.; Hakimi, R. Analysis of 5G Band Candidates for Initial Deployment in Indonesia. In Proceedings of the 2018 4th International Conference on Wireless and Telematics (ICWT), Nusa Dua, Bali, 12–13 July 2018; pp. 1–6.
11. Fahira, G.; Hikmaturokhman, A.; Danisya, A.R. 5G NR Planning at mmWave Frequency: Study Case in Indonesia Industrial Area. In Proceedings of the 2020 2nd International Conference on Industrial Electrical and Electronics (ICIEE), Lombok, Indonesia, 10–11 June 2020; pp. 205–210. [CrossRef]
12. Esa, R.N.; Hikmaturokhman, A.; Danisya, A.R. 5G NR Planning at Frequency 3.5 GHz: Study Case in Indonesia Industrial Area. In Proceedings of the 2020 2nd International Conference on Industrial Electrical and Electronics (ICIEE), Lombok, Indonesia, 10–11 June 2020; pp. 187–193. [CrossRef]
13. Karo, F.K.; Hikmaturokhman, A.; Amanaf, M.A. 5G New Radio (NR) Network Planning at Frequency of 2.6 GHz in Golden Triangle of Jakarta. In Proceedings of the 2020 3rd International Seminar on Research of Information Technology and Intelligent Systems (ISRITI), Yogyakarta, Indonesia, 10 December 2020; pp. 278–283. [CrossRef]
14. Huawei. 5G Link Budget 'Best Partner for Innovation. Available online: https://telecomfiles.com/5g-link-budget (accessed on 1 August 2023).
15. 3GPP, 5G. *Study on Channel Model for Frequencies from 0.5 to 100 GHz (3GPP TR 38.901 version 14.0.0 Release 14). ETSI TR138.901*; 3GPP: Valbonne, France, 2017.
16. 3GPP, 3rd Generation Partnership Project. In *Technical Specification Group Radio Access Network. NR. User Equipment (UE) Radio Access Capabilities*; 3GPP: Valbonne, France, 2019.

MDPI
St. Alban-Anlage 66
4052 Basel
Switzerland
www.mdpi.com

Engineering Proceedings Editorial Office
E-mail: engproc@mdpi.com
www.mdpi.com/journal/engproc

www.ingramcontent.com/pod-product-compliance
Lightning Source LLC
Chambersburg PA
CBHW040838250926
51850CB00007BA/1349
9783725803712